THIRD EDITION

Introduction to High Energy Physics

Now the smallest Particles of Matter may cohere by the strongest
Attractions, and compose bigger Particles of weaker Virtue.... There are
therefore Agents in Nature able to make the Particles of Bodies stick
together by very strong Attractions. And it is the Business of experimental
Philosophy to find them out.

Newton, *Optics* (1680)

Les Philosophes qui font des systèmes sur la secrète construction de
l'univers, sont comme nos voyageurs qui vont a Constantinople, et qui
parlent du Sérail: Ils n'en ont vu que les dehors, et ils prétendent savoir ce
que fait le Sultan avec ses Favorites.

Voltaire, *Pensées Philosophiques* (1766)

THIRD EDITION

Introduction to
High Energy Physics

DONALD H. PERKINS
University of Oxford

Addison-Wesley Publishing Company, Inc.
The Advanced Book Program
Menlo Park, California • Reading, Massachusetts
Don Mills, Ontario • Wokingham, U.K. • Amsterdam • Sydney
Singapore • Tokyo • Madrid • Bogota • Santiago • San Juan

Sponsoring Editors: Allan M. Wylde, Richard W. Mixter
Production Coordinator: Kristina M. Montague
Copy Editor: Linda Thompson
Cover Designer: Victoria Ann Philp

The cover image shows the interaction of a 200-GeV proton in a xenon gas target placed in a streamer chamber. See page 54 for further description. *Courtesy of V. Eckhardt, MPI, Munich.*

8 9 10 MA 95949392

Library of Congress Cataloging-in-Publication Data
Perkins, Donald H.
    Introduction to high energy physics.
    Includes bibliographies and index.
    1. Particles (Nuclear Physics) I. Title.
QC793.2.P47   1986   539.7'21   86-10916
ISBN 0-201-12105-0

To My Family

For their patience, forbearance, and encouragement

PREFACE TO THE THIRD EDITION

The main intention behind this book has been to present the more important aspects of the field of high-energy physics, or particle physics, at an elementary level. The content is based on courses of lectures given to undergraduates in Oxford specializing in nuclear physics, but the book would also serve as an introductory text for first-year graduate students in experimental high-energy physics. I have tried to make the coverage as broad as possible while keeping the text to a reasonable length.

Since the first edition was written sixteen years ago, high-energy physics has undergone many revolutionary developments, and both the volume and range of the subject has increased many times. This has meant a substantial rewriting of the text and a modest expansion in length. The interrelation between different aspects of the subject is now so strong that the division of the material under the various chapter headings has perforce been rather arbitrary.

The first chapter presents basic introductory ideas, the historical development, and a brief overview of the subject; the second and third chapters deal with experimental methods, conservation laws, and invariance principles—just as in the first edition. The following chapters deal in turn with the main features of the interactions between hadrons; the description of the hadrons in terms of quark constituents, and discussion of the basic interactions–electromagnetic, weak and strong–between the lepton and quark constituents. The final chapter discusses unification of the various interactions. During the last few years, the astrophysical and cosmological implications of results and ideas from high-energy physics have become important and indeed vital to our understanding of the development of the universe. I have tried to convey some of the flavor of this connection, since it will clearly help to shape the trends in high-energy physics in the foreseeable future.

As in the first edition, the interplay between experiment and theory has been emphasized, and some discussion given of key experiments in the field. Long theoretical treatments have been avoided, and for much of the mathematical detail the student is referred to Appendices or other texts. Some knowledge of elementary quantum mechanics is assumed, but generally the material has been presented from the empirical viewpoint, with a minimum of formalism and using an intuitive approach. Physics is about numbers, and I have taken the view that it was more important that a student should know how to calculate a cross-section or a decay rate, in order of magnitude, than how to derive a complicated formula (usually based on assumptions of questionable validity) without any real idea on how to confront it with experiment. In the same spirit, I have included a list of (mostly numerical) problems for each chapter, together with worked solutions at the end of the book.

ACKNOWLEDGMENTS

For permission to reproduce various photographs, figures and diagrams, I am indebted to the authors cited in the text and to the following laboratories and publishers: Brookhaven National Laboratory, Long Island, New York; CERN Information Services, Geneva; Rutherford and Appleton Laboratories, Chilton, Didcot, England; Stanford Linear Accelerator Laboratory, Stanford, Calif.; DESY Laboratory, Hamburg; Max Planck Institute, Munich; Annual Reviews Inc., Palo Alto, Calif.; the American Institute of Physics, New York, publishers of *The Physical Review, Physical Review Letters*, and *Reviews of Modern Physics*; The Italian Physical Society, Bologna, publishers of *Il Nuovo Cimento*; North Holland Publishing Co., Amsterdam, publishers of *Physics Letters, Nuclear Physics*, and *Physics Reports*; the Institute of Physics, Bristol, England, publishers of *Reports on Progress in Physics*; Pergamon Press Ltd., Oxford.

Many people helped me with suggestions and advice during the preparation of the text, and were kind enough to point out errors in the first edition. I owe a great debt to Chris Llewellyn-Smith (Oxford) and Chris Quigg (Fermilab) for their careful reading of the manuscript, detailed criticisms and comments, and their many suggestions for improvements. Several people were very kind in supplying me with original photographs and diagrams, and I should like especially to thank the staff of the CERN Information Services; Fred Combley, of the University of Sheffield; and Bärbel Lücke, Sabine Platz and Rolf Felst, of the DESY laboratory, Hamburg.

Finally, I wish to record my special thanks to Irmegarde Smith for her preparation of the line drawings, Cyril Band and his colleagues for the photographic work, and Suzanne Motyka for her careful typing of the manuscript.

DONALD H. PERKINS

CONTENTS

QC793 PER

DATE DUE FOR RETURN

- 5 MAY 2004

CHAPTER 1

Introduction and Overview

1.1. INTRODUCTION

High-energy physics deals basically with the study of the ultimate constituents of matter and the nature of the interactions between them. Experimental research in this field of science is carried out with giant particle accelerators and their associated detection equipment. High energies are necesssary for two reasons: First, in order to localize the investigations to the very small scales of distance associated with the elementary constituents, one requires radiation of the smallest possible wavelength and highest possible energy; second, many of the fundamental constituents have large masses and require correspondingly high energies for their creation and study.

Fifty years ago, only a few "elementary" particles—the proton and neutron, the electron and neutrino, together with the electromagnetic field quantum (the photon)—were known. The universe as we know it today appears indeed to be composed almost entirely of these particles. However, attempts to understand the details of the nuclear force between protons and neutrons, as well as to follow up the pioneering discoveries of new, unstable particles observed in the cosmic rays, led to the construction of ever larger accelerators and to the observation of many hundreds of new unstable particle states, collectively called *hadrons* (strongly interacting particles).

Out of this seeming chaos has emerged a rather simple picture:

(i)   All matter is composed of fundamental spin-$\frac{1}{2}$ fermion consti-
      tuents—the *quarks*, with fractional electric charges ($+\frac{2}{3}e$ and
      $-\frac{1}{3}e$), and the *leptons*, like the electron and neutrino, carrying
      integral electric charges. Neutrons and protons are built from
      quarks, three at a time.

(ii) These constituents can interact by exchange of various funda-
      mental *bosons* (integral spin particles) which are the carriers or
      quanta of four distinct types of fundamental interaction or field.
      Gravity is familiar to everyone, yet on the scales of mass and
      distance involved in particle physics, it is by far the least
      important of the four. Apart from gravity, *electromagnetic*
      interactions account for most extranuclear phenomena in phys-
      ics (because electromagnetic forces have the longest range) and
      lead to the bound states of atoms and molecules. *Weak* inter-
      actions are exemplified by the extremely slow process of radio-
      active β-decay of nuclei. *Strong* interactions are postulated to
      hold together the quarks in a proton, and their residual effects
      apparently account for the interactions between neutrons and
      protons, that is, for the nuclear binding force. Both weak and
      strong interactions are of short range (less than or of order one
      fermi or *femtometer*, 1 fm $= 10^{-15}$ m).

There are many unusual, even bizarre, aspects of this picture. The
fractionally charged quarks have not been observed as free particles, and
seem to be permanently confined in hadrons. Quarks come in a variety of
types or *flavors* (six are known) as do the leptons (three types of charged and
of neutral leptons). We neither understand the mechanism of confinement,
nor the real reason for the "Xerox copies" of quark and lepton flavors, when
the universe, on the basis of what we see today, seems to be constructed
predominantly from just two types of quark and one neutral and one charged
lepton.

The multiplicity of quark and lepton flavors is paralleled by the
existence of the four types of fundamental interaction. Here, some real
progress has been made. There are good grounds for supposing that some,
perhaps all, the interactions are *unified*, that is, different aspects of one single
interaction. The weak and electromagnetic interactions appear to have the
same intrinsic coupling of fermion constituents to the respective mediating
bosons—they are different aspects of a single *electroweak* interaction. Compared with electromagnetism (mediated by the massless photon field with
infinite range), the weakness of the weak interactions is ascribed to their
short-range nature (they are mediated by massive bosons $W^{\pm}$, Z^0, whose
mass is found to be of order 100 proton masses). At high enough energies and

momentum transfers, well above such a mass scale, electromagnetic and weak interactions should have the same actual strength.

Why the high-energy symmetry is badly broken at low energy, and the respective bosons have such widely differing masses, is still an unsolved problem. The important point however is that the strengths of the different interactions are not fixed once and for all; they depend on energy scales. At high energies, strong interactions appear to grow weaker, and the strong and electroweak interactions may also merge at the colossal energy of 10^{15} GeV.

The study of particle physics is considered to be intimately connected with evolution of the universe. We believe the universe originated in a "big bang" expansion of an energy bubble, from which all types of particles —quarks, leptons and quanta—were created. Today, we are left with the expanded, cooled remnant. So, our search toward higher energies is also a look backward in time to the very earliest stages of creation, which determined the characteristics of the universe we find today.

1.2. FERMIONS AND BOSONS

One of the most fundamental concepts underlying our analysis of the interactions of particles and fields is the spin-statistics theorem (Pauli 1940), connecting the statistics obeyed by a particle with its spin angular momentum. Particles with half-integral spin ($\frac{1}{2}\hbar$, $\frac{3}{2}\hbar$, ...) obey Fermi-Dirac statistics and are thus called fermions, while those with integral spin (0, $\hbar$, $2\hbar$, ...) obey Bose-Einstein statistics and are called bosons.

The statistics obeyed by a particle determines the symmetry of the wavefunction ψ describing a pair of identical particles, say 1 and 2, under interchange. If the particles are identical, then the square of the wavefunction, $|\psi|^2$, giving the probability of particle 1 at one coordinate and particle 2 at another, will be unaltered by the interchange $1 \leftrightarrow 2$. Thus,

$$\psi \overset{1 \leftrightarrow 2}{\to} \pm \psi.$$

The following rule holds:

$$\begin{aligned}
\text{Identical bosons:} \quad & \psi \overset{1 \leftrightarrow 2}{\to} +\psi \quad && \text{symmetric} \\
& && \qquad\qquad (1.1)\\
\text{Identical fermions:} \quad & \psi \overset{1 \leftrightarrow 2}{\to} -\psi \quad && \text{antisymmetric}
\end{aligned}$$

In order to make use of this rule, the total wavefunction of the pair can be expressed as a product of functions depending on spatial coordinates and spin orientation:

$$\psi = \alpha(\text{space})\, \beta(\text{spin}). \qquad (1.2)$$

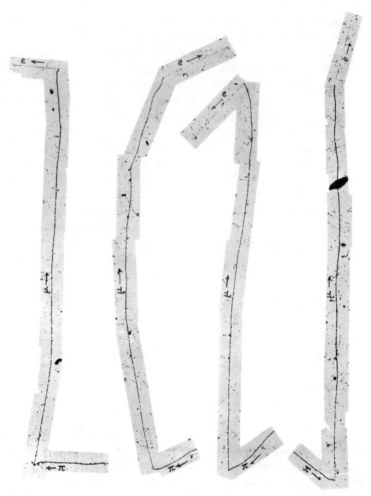

Figure 1.1 Examples of the decay sequence $\pi^+ \to \mu^+ \to e^+$ in G5 emulsion exposed at Pic du Midi. The constancy of range ($\simeq 600 \ \mu$m) of the muon implies two-body decay at rest of the pion: $\pi^+ \to \mu^+ + \nu_\mu$. The first examples of pion decay were observed by Lattes, Muirhead, Occhialini, and Powell in 1947. Note the very dense ionization of both pion and muon tracks near the end of the range, compared with the thin track of the relativistic electron, as well as the lateral deflections (Coulomb scattering) of the muon as it traverses the emulsion.

The spatial part, α, could describe orbital motion of one particle about the other, and can then be represented by a spherical harmonic function $Y_l^m(\theta, \phi)$, as described in Chapter 3. Interchange of the space coordinates of particles 1 and 2 (leaving spin alone) is equivalent to the replacement $\theta \to \pi - \theta$, $\phi \to \phi + \pi$, and introduces a factor $(-1)^l$ multiplying α, where l is the orbital quantum number. Thus, if l is even (odd), the function α is symmetric (antisymmetric) under interchange. As also indicated in

Chapter 3, the spin function β may be symmetric (spins parallel) or antisymmetric (spins antiparallel) under interchange. Equation (1.2) implies that, for identical bosons, α and β must be both symmetric or both antisymmetric; while for fermions, a symmetric α implies an antisymmetric β and vice versa.

As an example, consider the decay of the neutral ρ-meson of spin $J = 1$ into two neutral pions: $\rho^0 \to 2\pi^0$. Both pions are uncharged and spinless. Since β is necessarily symmetric, the rule for identical bosons means α must be symmetric, and thus the two pions can exist only in a state with even total angular momentum J. Hence the decay $\rho^0 \to 2\pi^0$ is forbidden by angular-momentum conservation and Bose symmetry. Decay into charged (nonidentical) pions does take place, however: $\rho^0 \to \pi^+\pi^-$, $\rho^\pm \to \pi^\pm\pi^0$.

The Pauli principle is a well-known application of the antisymmetry of the wavefunction of two identical fermions under interchange. Suppose two identical particles are in the same quantum state, so that ψ is necessarily symmetric. This violates the rule that two identical fermions must have ψ antisymmetric. Hence two identical fermions cannot exist in the same quantum state—the Pauli principle. On the other hand, there is no restriction on the number of bosons (photons, for example) which may exist in the same quantum state. An example of this is the laser.

1.3. PARTICLES AND ANTIPARTICLES

The relativistic wave equation proposed in 1928 by Dirac was able to account for the intrinsic angular momentum, or spin quantum number, of the electron, which had previously been postulated by Uhlenbeck and Goudsmit in order to account for the Zeeman effect in atomic physics. In the Dirac theory free electrons are described by four-component wavefunctions, corresponding to two spin substates, $J_z = \pm\frac{1}{2}\hbar$, each of positive or of negative energy. The negative-energy states are interpreted in terms of an antiparticle, the positron (see Appendix D). The existence of the positron was first demonstrated by Anderson in 1933 in a cloud chamber experiment with cosmic rays. The existence of antiparticles is a general property of both fermions and bosons, the antiparticle having the same mass as the particle, but opposite charge and magnetic moment.

Fermions and antifermions can only be created or destroyed in pairs. For example, a γ-ray, in the presence of a nucleus (to conserve momentum), can "materialize" into an electron-positron pair (see Fig. 1.2), and an e^+e^- bound state, called positronium, annihilates into two or three γ-rays. Theoretically, particle and antiparticle states are connected by the process of particle-antiparticle conjugation. Fermion number is conserved if each fermion is assigned a fermion number $+1$ and each antifermion -1. So the process of particle-antiparticle conjugation for fermions gives an antifermion with opposite charge, magnetic moment, and fermion number, but identical

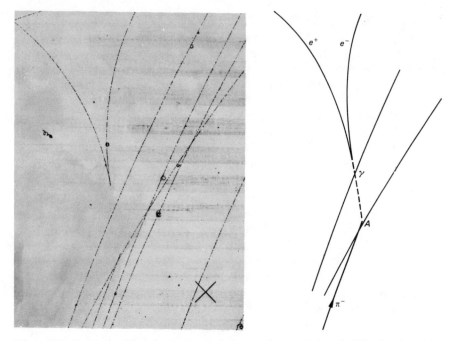

Figure 1.2 Conversion of a photon to an electron-positron pair in a bubble chamber. An incoming negative pion undergoes charge-exchange at point A: $\pi^- + p \to n + \pi^\circ$, followed by decay of the neutral pion, $\pi^\circ \to 2\gamma$. Since the π° lifetime is only 10^{-16} s, the pair appears to point straight to the interaction vertex.

mass and spin angular momentum. There is no law of number conservation for bosons, and particle-antiparticle conjugation has the same effect as charge conjugation for them.

1.4. BASIC FERMION CONSTITUENTS: QUARKS AND LEPTONS

1.4.1. Quarks

Present evidence indicates that matter is built from two types of fundamental fermion, called quarks and leptons, which are structureless and pointlike on a scale of 10^{-17} m. *Quarks* carry fractional electric charges, of $+\frac{2}{3}|e|$ and $-\frac{1}{3}|e|$. They occur in several varieties or *flavors*, distinguished by the assignment of internal quantum numbers, and are labelled u, d, s, c, b, t (Table 1.1). The u- and d- quarks are the lightest and have approximately the

TABLE 1.1 Quarks

	Quarks		Antiquarks	
$Q/\lvert e\rvert = +\frac{2}{3}$	$u, c, \ldots$		$Q/\lvert e\rvert = -\frac{2}{3}$	$\bar{u}, \bar{c}, \ldots$
$Q/\lvert e\rvert = -\frac{1}{3}$	d, s, b		$Q/\lvert e\rvert = +\frac{1}{3}$	$\bar{d}, \bar{s}, \bar{b}$

$u =$ "up" quark $\left.\right\}$ $d =$ "down" quark $I = \frac{1}{2}$ doublet	$m_u \simeq m_d \simeq 350\ \mathrm{MeV}/c^2$
$s =$ "strange" ($S = -1$),	$m_s \simeq\ \ 550\ \mathrm{MeV}/c^2$
$c =$ "charmed" ($C = +1$)	$m_c \simeq 1800\ \mathrm{MeV}/c^2$
$b =$ "bottom" ($B = -1$)	$m_b \simeq 4500\ \mathrm{MeV}/c^2$
$t =$ "top" ($T = +1$)	$m_t > 20{,}000\ \mathrm{MeV}/c^2$

same mass (within 1 MeV or so). As indicated below, protons and neutrons are considered to be built from u- and d-quarks, and consequently the near-equality of proton and neutron masses implies the same equality for u- and d-quarks. They have similar strong interactions with other quarks, and differ only in their electric charge and hence electromagnetic interactions. Historically, the equality of strong interactions of the u- and d-quark constituents appeared as the hypothesis of isospin invariance of interactions between hadrons composed of u- and d-quarks. For this reason, u- and d-quarks are sometimes grouped as an isospin doublet ($I = \frac{1}{2}$, with the third component $I_3 = +\frac{1}{2}$ for u and $-\frac{1}{2}$ for d). The s-quark is assigned an internal quantum number called strangeness, with value $S = -1$ (it is a constituent of the so-called strange particles first observed in cosmic rays in the '50s), the c-quark a charm quantum number $C = +1$, the b-quark the bottom quantum number $B = -1$, and the top quark, a top quantum number $T = +1$. c- and b-quarks were postulated as constituents of massive, short-lived hadron states observed in 1974 and 1977, respectively. These quantum numbers are discussed more fully in Chapter 5. The masses given in Table 1.1 should be taken as indicative only. Corresponding to each quark is an antiquark with opposite charge, strangeness, etc.

Hadrons are strongly interacting particles built from two types of quark combination:

$$\text{Baryon} = QQQ \quad \text{(three quarks)}$$
$$\text{Meson} = Q\bar{Q} \quad \text{(quark-antiquark pair)} \tag{1.3}$$

The fact that two and only two types of quark combination occur is successfully accounted for in the theory of the interquark forces (quantum chromodynamics). Since quarks have half-integral spin, it follows that the baryons are characterized by half-integral, and the mesons by integral, spin. Example are:

Baryons   *Mesons*

$uud = p$ (proton) $u\bar{d} = \pi^+$ (pion)

$udd = n$ (neutron) $\bar{s}d = K^0$ (kaon)

$uds = \Lambda$ (lambda hyperon) $c\bar{c} = \psi$-meson

As expected from our discussion of bosons and fermions, the conservation rule for quarks is reflected in the conservation of baryon number, while there is no conservation rule for mesons. Formally, each baryon (antibaryon) is assigned a baryon number $+1(-1)$, and the total baryon number is conserved.

The mesons are important because, although they are all unstable and do not occur in ordinary matter, the discovery of two of them, the pion and the kaon, in cosmic rays in 1947 really marks the birth of the subject of

Figure 1.3 An example of associated production, due to the interaction at A of a 4-GeV/c negative pion in a hydrogen bubble chamber: $\pi^- + p \rightarrow \Lambda + K^0$. The Λ-hyperon decays at B according to $\Lambda \rightarrow p + \pi^-$, and the K^0-meson at C according to $K^0 \rightarrow \pi^+ + \pi^-$. (Courtesy CERN.)

particle physics. Figure 1.1 shows examples of the discovery of charged pions via the weak decay, $\pi^+ \to \mu^+ + \nu_\mu$, while Fig. 1.3 shows an example of associated production of a pair of strange particles, Λ and K^0;

	π^-	+	p	$\to$	K^0	+	Λ
Quark description	$\bar{u}d$		uud		$\bar{s}d$		usd
Strangeness S	0		0		$+1$		-1

1.4.2. Leptons

The *leptons* carry integral electric charges, 0 or $\pm |e|$, and three types of each are known (see Table 1.2). The neutral leptons are called neutrinos, and have very small or zero rest mass. The electron is familiar to everyone, the muon is an unstable "heavy electron" observed in cosmic rays in the '30s and formed as a product of decay of pions produced in the atmosphere (see Fig. 1.1). The τ-lepton was first observed in accelerator experiments in 1974. The leptons appear in doublets, the neutrinos being assigned a subscript corresponding to the charged member. Charged leptons are distinguished from antileptons by the sign of charge. Neutrinos are longitudinally spin-polarized with $J_z = -\frac{1}{2}$ ("left-handed"), where z is the direction of the velocity vector, while antineutrinos have $J_z = +\frac{1}{2}$ ("right-handed").

The charged leptons have electromagnetic and weak interactions, while the neutrinos are distinguished by having only weak interactions with other particles. Quarks, in addition to weak and electromagnetic interactions, are subject to the strong (specifically quark-quark) interactions. While the strong interactions lead to quark composites (hadrons), only loosely bound and unstable combinations of charged leptons occur (for example positronium e^+e^-, bound by the Coulomb interactions).

The conservation rules for fermions apply of course to quarks and leptons. In particular, a lepton number L_e, L_μ, L_τ of $+1$ is given to each type

TABLE 1.2 Leptons

Leptons				Antileptons			
$Q/|e| = -1$	e^-	μ^-	τ^-	$Q/|e| = +1$	e^+	μ^+	τ^+
$Q/|e| = 0$	ν_e	ν_μ	ν_τ	$Q/|e| = 0$	$\bar{\nu}_e$	$\bar{\nu}_\mu$	$\bar{\nu}_\tau$

$$m_e = \quad 0.511 \text{ MeV}/c^2$$
$$m_\mu = \quad 105.6 \quad \text{MeV}/c^2$$
$$m_\tau = 1870 \quad \text{MeV}/c^2$$

of lepton, and -1 to each antilepton. Examples of lepton conservation are electron pair production by a photon (Fig. 1.2),

$$\gamma \rightarrow e^+ + e^-,$$
$$L_e \quad 0 \quad -1 \quad +1$$

pion decay:
$$\pi^+ \rightarrow \mu^+ + \nu_\mu,$$
$$L_\mu \quad 0 \quad -1 \quad +1$$

muon decay:
$$\mu^+ \rightarrow e^+ + \nu_e + \bar{\nu}_\mu, \qquad (1.4)$$
$$L_\mu \quad -1 \quad 0 \quad 0 \quad -1$$
$$L_e \quad 0 \quad -1 \quad +1 \quad 0$$

while the decay

$$\mu^+ \rightarrow e^+ + \gamma$$
$$L_\mu \quad -1 \quad 0 \quad 0$$
$$L_e \quad 0 \quad -1 \quad 0$$

is forbidden by lepton-number conservation. The limit to the branching ratio for this muon decay mode is $<10^{-9}$.

While the total quark number is conserved in all interactions, the number of quarks of a given flavor is absolutely conserved only in strong and electomagnetic interactions (equivalent to conservation of I_3, strangeness, and similar quantum numbers). In weak decay processes, the quark flavor may change ($\Delta S = 1$, $\Delta C = 1$, etc.).

1.5. INTERACTIONS AND FIELDS IN PARTICLE PHYSICS

1.5.1. Classical and Quantum Pictures

Classically, interaction at a distance is described in terms of a potential or field due to one particle acting on another. In quantum theory, it is viewed in terms of the exchange of specific quanta (bosons) associated with the particular type of interaction. Since the quantum carries energy and momentum, the conservation laws can be satisfied only if the process takes place over a time limited by the Uncertainty Principle, that is, $\Delta E \cdot \Delta t \leq \hbar$. Such transient quanta are said to be *virtual*.

That these two descriptions are equivalent on a macroscopic scale may be illustrated by considering the electrostatic field between two point charges, Q_1 and Q_2. In the classical case the force **F** on Q_2 in the diagram is ascribed to the field $E(r)$ due to Q_1; $\mathbf{F} = \mathbf{E}(r)Q_2 = \hat{\mathbf{r}}Q_1Q_2/r^2$. Quantum mechanically, the force between the charges is ascribed to exchange of virtual photons of momentum q, the change of momentum of the charge as it emits or absorbs a photon producing the force. The Uncertainty Principle links the linear dimension of the system, that is the uncertainty in position of the photon, with its momentum:

$$qr \simeq \hbar.$$

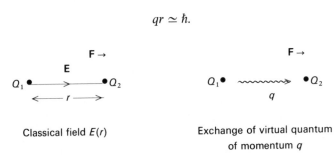

Classical field $E(r)$ Exchange of virtual quantum of momentum q

Each photon exchanged involves a momentum transfer q over a period $t = r/c$, or a force $dq/dt = \hbar c/r^2$. The number of photons emitted and absorbed by either charge is assumed to be proportional to the product of the charges, so that one obtains the Coulomb law $F = Q_1 Q_2/r^2$ as in the classical case. The quantum concept of continual emission and absorption of virtual photons by the charge source is no more or less fictitious than the classical concept of a field surrounding the source. Neither field nor virtual quanta are directly observable; it is the force that is the measured quantity. Since, however, it is observed that propagating electromagnetic fields are actually quantized in the form of free photons, the quantum description of virtual photon exchange in the static case is appropriate for discussion of interactions on a microscopic scale.

1.5.2. The Yukawa Theory

The range of the static interaction depends on the mass of the free field quantum, a connection established in 1935 by Yukawa in seeking to describe the short-range force between neutrons and protons in the atomic nucleus. Suppose that the quantum to be exchanged has mass m, then the Uncertainty Principle restricts its existence within a time $\Delta t \leq h/mc^2$, during which it could cover at most a distance $R \simeq c\Delta t \leq h/mc$. Thus, the range of the field is given by the Compton wavelength of the associated quantum. We can make this argument more quantitative with the help of the relativistic relation between the total energy E and momentum p of a free particle of mass m,

$$E^2 = p^2 c^2 + m^2 c^4. \tag{1.5}$$

The differential equation describing the wave amplitude ψ of such a free particle is obtained by substituting in (1.5) the quantum-mechanical operators

$$E_{op} = i\hbar \frac{\partial}{\partial t}, \qquad p_{op} = -i\hbar\nabla = -i\hbar \frac{\partial}{\partial \mathbf{r}}$$

which give the Klein-Gordon wave equation

$$\nabla^2\psi - \frac{m^2c^2}{\hbar^2}\psi - \frac{1}{c^2}\frac{\partial^2\psi}{\partial t^2} = 0 \qquad (1.6)$$

describing the propagation in free space of spinless particles of mass m. If we set $m = 0$, (1.6) becomes the familiar wave equation describing the propagation of an electromagnetic wave, with ψ interpreted either as the potential at a point in space and time, or as the wave amplitude of the associated free, massless photons. We are interested here not so much in the propagation of particle waves as in static potentials. If we drop the time-dependent term in (1.6), therefore, the resulting equation for the static potential U has the spherically symmetric form

$$\nabla^2 U(r) = \frac{1}{r^2}\frac{\partial}{\partial r}\left(r^2\frac{\partial U}{\partial r}\right) = \frac{m^2c^2}{\hbar^2}U(r)$$

for values of $r > 0$ from a point source at the origin, $r = 0$. Integration gives

$$U(r) = \frac{g}{4\pi r}e^{-r/R}, \qquad (1.7)$$

where

$$R = \frac{\hbar}{mc}. \qquad (1.8)$$

Here the quantity g is a constant of integration identified with the strength of the point source. The analogous equation in electromagnetism is $\nabla^2 U(r) = 0$ for $r > 0$, with solution $U = Q/4\pi r$, where Q is the charge at the origin. Thus, g in the Yukawa theory plays the same role as charge in electrostatics and measures the "strong nuclear charge".

In the historical context of nuclear forces with range $R \simeq 10^{-15}$ m, the Yukawa hypothesis predicted a spinless quantum of mass $mc^2 = \hbar c/R \simeq$ 100 MeV. The pion observed in 1947 had mass $\simeq 140$ MeV, spin 0, and strong interactions with nuclei and was assigned—as it was to turn out, somewhat too simplistically—as the nuclear-force quantum.

1.5.3. The Boson Propagator

Let us consider a particle being scattered by a potential, the effect of which is observed through the angular deflection of the particle or, equivalently, the momentum transfer $\mathbf{q}$. The potential $U(\mathbf{r})$ in coordinate space will have an associated amplitude, $f(\mathbf{q})$, for scattering of the particle, which is simply the Fourier transform of the potential (just as the angular distribution of light diffracted from an obstacle in classical optics is the Fourier transform of the obstacle). That is,*

$$f(\mathbf{q}) = g_0 \int U(\mathbf{r})e^{i\mathbf{q}\cdot\mathbf{r}}\, dV, \qquad (1.9)$$

where g_0 is the intrinsic coupling strength of the particle to the potential. For a central potential, $U(\mathbf{r}) = U(r)$, the integration over volume is easily performed, setting

$$\mathbf{q}\cdot\mathbf{r} = qr\cos\theta$$

$$dV = r^2\, d\phi \sin\theta\, d\theta\, dr,$$

where θ and ϕ are polar and azimuthal angles. Using (1.7) and (1.8) and for brevity using units $\hbar = c = 1$,†

$$f(\mathbf{q}) = 4\pi g_0 \int_0^\infty U(r)\frac{\sin qr}{qr} r^2\, dr$$

$$= g_0 g \int_0^\infty e^{-mr}\frac{(e^{iqr} - e^{-iqr})}{2iq}\, dr$$

which gives

$$f(\mathbf{q}) = \frac{g_0 g}{(|\mathbf{q}|^2 + m^2)} \qquad (1.10)$$

This equation describes, in momentum space, the law of force (1.7) in coordinate space; the two descriptions are equivalent.

The discussion here has been of the scattering of a particle of coupling g_0 by a static potential U provided by a massive source of strength g. The incident particle has been scattered but lost no energy, but in an actual collision between two particles, energy E as well as 3-momentum q will be transferred. It turns out that the result (1.10) still holds provided we interpret q as the 4-*momentum transfer* $q = (\mathbf{q}, iE)$, where $q^2 = \mathbf{q}^2 - E^2$, in units $\hbar = c = 1$ (see Appendix A for discussion of 4-vectors).

In summary, the scattering amplitude due to single boson exchange is the product of two *vertex factors* g_0, g describing the coupling of the boson to the scattered and scattering particles, and a *propagator* term $(q^2 + m^2)^{-1}$:

$$f(q^2) = \frac{g g_0}{(q^2 + m^2)} \qquad (1.11)$$

---

* See §6.2 for a derivation of this formula.
† See §1.11 for discussion of units.

The scattering cross-section itself is the product of $|f|^2$ times a phase-space factor, divided by the incident flux, as described in Chapter 4 and Appendix E. Extra factors associated with particle spin will also have to be introduced later as necessary. Nevertheless (1.11) is the basic formula describing the interaction of two particles via single boson exchange, which will be used time and again throughout the text.

As an example, for $m = 0$, the photon propagator introduces a $1/q^4$ dependence of the cross-section for scattering two charged particles, and is the basis of the Rutherford scattering formula. The representation of interactions of particles with quantized fields is frequently visualized with the aid of *Feynman diagrams*, which are associated with formal rules for assigning vertex coupling, propagator terms, etc., in order to compute matrix elements. However, in this text such diagrams will be used mainly as a pictorial representation of interactions via quantum exchange.

1.6. ELECTROMAGNETIC INTERACTIONS

The coupling constant specifying the strength of the interaction between charged particles and photons is the dimensionless fine-structure constant

$$\alpha = \frac{e^2}{4\pi\hbar c} = \frac{1}{137.0360},\qquad(1.12)$$

so called because it determines the magnitude of the fine structure (spin-orbit splitting) in atomic spectra. The quantity α enters in the matrix element for the process under consideration, which after squaring gives the decay probability or cross section.

Figure 1.4 shows diagrams depicting various electromagnetic processes. Time t flows horizontally, and space s vertically. The arrows indicate the direction of motion of particles entering or leaving vertices. Incoming electrons (momentum $\mathbf{p}$) can always be replaced by outgoing antiparticles (positrons) of momentum $-\mathbf{p}$, without changing the matrix element.

Figure 1.4(a) shows the simplest process of absorption (or emission) of a photon by an electron. This can only take place for an electron bound in an atom, to ensure momentum conservation. The photon couples to the electron with amplitude $\sqrt{\alpha}$ (or e, in units $\hbar = c = 1$), so the photoelectric cross-section, or matrix element squared, is proportional to α (or e^2). Since α occurs to the first power, this is called a first-order process.

Figure 1.4(b) indicates the second-order process of Coulomb scattering between two electrons, via exchange of a single virtual photon of momentum q, coupling at two vertices. The virtual photon introduces a so-called propagator term $1/q^2$ in the matrix element (see Eq. (1.11), which is therefore proportional to $\sqrt{\alpha}\sqrt{\alpha}/q^2$. The scattering cross-section is $d\sigma/dq^2 \propto \alpha^2/q^4$ (the Rutherford scattering formula).

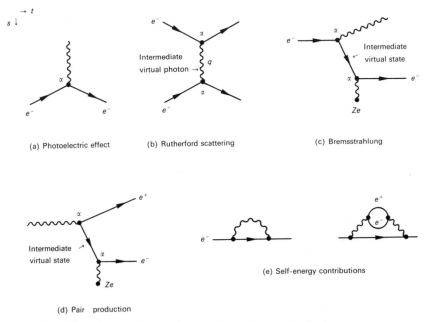

Figure 1.4 Feynman diagrams for electromagnetic processes.

Figure 1.4(c) show the emission of a real photon by an electron which has undergone acceleration in the electric field of a nucleus, charge Ze (the process of bremsstrahlung). A virtual photon must be exchanged with the nucleus, to conserve momentum, and the cross-section is of order α^3 and proportional to $Z^2\alpha^3$. Note that an intermediate virtual electron state is involved, since an electron cannot emit a real photon and conserve energy and momentum without reducing its rest mass (it is said to go "off mass shell").

Finally, Fig. 1.4(d) shows the process of e^+e^- pair creation by a photon in the field of a nucleus, also of magnitude α^3. Diagrams (c) and (d) are closely related, and the latter can be obtained from the former by replacing an incoming electron line with an outgoing positron line.

The field theory employed to compute the cross-sections for such electromagnetic processes is called quantum electrodynamics (QED). One very important property of QED is that of *renormalizability*. As shown in Fig. 1.4(e), a single electron can emit and reabsorb virtual photons (or pairs), and such "self-energy" terms contribute to the mass (and charge) of the electron: indeed, they give divergent integrals, and the theoretically calculated "bare" mass m_0 or charge e_0 become infinite. Divergent terms of this type are present in all QED calculations, for example of the processes in Fig. 1.4(a)–(d). However, it is found possible to dump all the divergences into m_0 or e_0, and then redefine the mass and charge, replacing them by their physical values e,

m (which are determined by experiment). This process is called renormalization. The result is that QED calculations, if expressed in terms of the physical quantities e and m, always give finite (and incredibly exact) values for cross-sections, decay rates, and so forth.

A second vital property of electromagnetic interactions is that of gauge invariance. In electrostatics, for example, the interaction energy which can be measured experimentally depends only on changes in the static potential and not its absolute magnitude, and is therefore invariant under arbitrary changes in the potential scale or gauge. In quantum mechanics, the phase of a fermion field (e.g., the wavefunction of an electron) is likewise arbitrary, and one could require the freedom to choose the phases of all fermion fields at all points in space-time in any way one pleases, without changing the physics. This local gauge invariance leads to conserved currents and to conservation of electric charge.

Mention has been made of renormalizability and gauge invariance because they are closely related: many particle theories in the past have possessed neither property and have run into insurmountable difficulties. The astounding success of QED, allowing exact calculations of electromagnetic processes to all orders in α, has been such that it is nowadays generally believed that all theories of fundamental fields should be renormalizable gauge theories.

1.7. WEAK INTERACTIONS

The weak interactions take place between all the quark and lepton constituents; each of them has, so to speak, a "weak charge." This interaction is so feeble however that it is usually swamped by the much stronger electromagnetic and strong interactions, unless these are forbidden by conservation rules. The observable weak interactions therefore either involve neutrinos (which have no electric or strong charges) or quarks with a flavor change ($\Delta S = 1$, $\Delta C = 1$, etc., forbidden for strong or electromagnetic interactions).

$$(a) \quad n \rightarrow p + e^- + \bar{\nu}_e \quad \text{neutron } \beta\text{-decay}$$
$$(b) \quad \bar{\nu}_e + p \rightarrow n + e^+ \quad \text{antineutrino absorption} \tag{1.13}$$

are examples of weak interactions involving neutrinos. Note that any other decay mode of a neutron is forbidden by baryon conservation and that lepton conservation requires both a neutral (ν_e) and charged (e) lepton to appear together. On the other hand the purely hadronic decay of the sigma-hyperon

$$\Sigma^- \rightarrow n + \pi^- \quad \tau = 10^{-10} \text{ s} \tag{1.14}$$
$$S \qquad -1 \quad\quad 0 \qquad 0$$

involves transformation of a strange quark ($S = -1$) to a nonstrange ($S = 0$).

One can compare this with the electromagnetic decay of Σ° (the neutral partner of the Σ^-):

$$\Sigma^0 \rightarrow \Lambda + \gamma \qquad \tau = 10^{-19}\,\text{s} \qquad (1.15)$$

$$S \qquad -1 \qquad -1$$

This is an electromagnetic decay, because although quark flavor is conserved, a member of a charged triplet of baryons ($\Sigma^{\circ,\pm}$) transforms to a charge singlet baryon (Λ) and this is forbidden for a strong interaction, (that is, by conservation of isospin, see Chapter 4). The ratio of lifetimes in (1.14) and (1.15) tells us that the weak coupling is only of order $(10^{-19}/10^{-10})^{1/2} \simeq 10^{-5}$ of the electromagnetic coupling in this case.

The weak interactions are mediated by massive bosons $W^\pm$ and Z° as shown in Fig. 1.5 in analogy with photon exchange in electromagnetic interactions. The masses of $W^\pm$ and Z° are 81 GeV and 94 GeV, respectively. $W^\pm$-exchange results in change of charge of the lepton and hadron taking part (as in (1.13)) and is called a "charged-current" reaction, while Z°-exchange does not, and is called a "neutral-current" reaction. Figures 1.6 and 1.7 show, respectively, the first example of neutral-current $\bar{\nu}_\mu e^-$ elastic

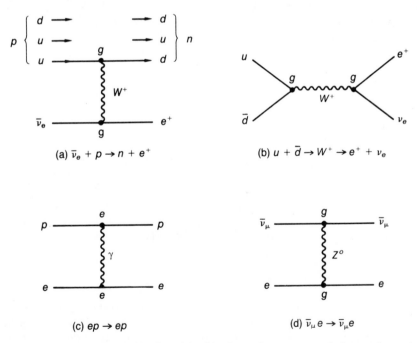

Figure 1.5 Diagrams indicating (a), (b) charged-current weak interactions, (c) electromagnetic interaction, (d) neutral-current weak interaction. (b) and (d) are the diagrams relating to the observations in Figs. 1.7 and 1.6, respectively.

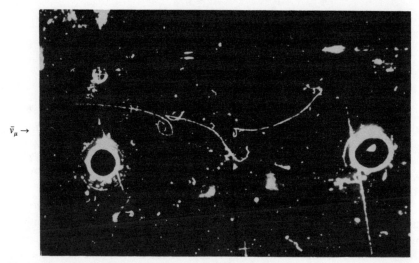

$\bar{\nu}_\mu \rightarrow$

Figure 1.6 First example of weak neutral-current process $\bar{\nu}_\mu + e \rightarrow \bar{\nu}_\mu + e$ observed in heavy-liquid bubble chamber Gargamelle at CERN irradiated with a $\bar{\nu}_\mu$ beam (Hasert *et al.*, 1973). A single electron of energy 400 MeV is projected at a small angle $(1.5 \pm 1.5°)$ to the beam, and is identified by bremsstrahlung and pair production along the track (see Chapter 2). About 10^9 $\bar{\nu}_\mu$'s traverse the chamber in each pulse and three such events were observed in 1.4 million pictures. (Courtesy CERN.)

EVENT 2958. 1279.   89578

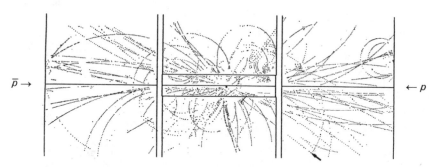

$\bar{p} \rightarrow$ $\leftarrow p$

Figure 1.7 One of the first events attributed to production and decay of a W^+-boson, $W^+ \rightarrow e^+ + \nu_e$. It originated in a collision of a 270-GeV proton (from right) with a 270-GeV antiproton (from left) in the CERN SPS collider (Arnison *et al.*, 1983). Among the 66 tracks observed outside the beam pipe is a very energetic positron (arrow) identified in a surrounding calorimeter. The positron energy is 42 GeV, its momentum transverse to the beam is 26 GeV/c. The missing transverse momentum in the whole event is 24 GeV/c, consistent with a missing neutrino from the decay $W^+ \rightarrow e^+ + \nu_e$. (Courtesy CERN.)

scattering and the first evidence of W-production and decay, $W^+ \to e^+ \nu_e$, in a high-energy $p\bar{p}$ collision. If we oversimplify the picture by denoting the W, Z couplings to quarks and leptons by a single number, g, we get from (1.11),

$$f(q^2) = \frac{g^2}{(q^2 + M_{W,z}^2)} \tag{1.16}$$

to be compared with e^2/q^2 for the electromagnetic scattering of Fig. 1.5(c). For $q^2 \ll M_{W,z}^2$ the amplitude (1.16) is independent of q^2, that is, the weak interaction is pointlike. Fermi had postulated just such a contact interaction in 1935, of strength G, between four fermions to describe β-decay. In fact, as $q^2 \to 0$,

$$G \equiv \frac{g^2}{M_W^2} \simeq 10^{-5} \text{ GeV}^{-2}, \tag{1.17}$$

where the number comes from measured β-decay rates.

The *electroweak theory* (1967/8) of Glashow, Salam and Weinberg proposed that the coupling g of $W^\pm$, Z° to leptons and quarks should be the *same as that of the photon*, i.e., $g = e$; the weak and electromagnetic interactions are unified (we omit certain numerical factors and mixing angles, to be discussed in Chapter 9). Then, from the measured value of G in (1.17) and from (1.12) it was expected that

$$M_{W,z} \sim \frac{e}{\sqrt{G}} \sim \sqrt{\frac{4\pi\alpha}{G}} \sim 90 \text{ GeV}, \tag{1.18}$$

in agreement with the values measured when the W- and Z°-particles were first oberved in 1983.

While electromagnetic and strong interactions conserve parity, that is, symmetry under spatial inversions, weak interactions do not. As an example, neutrinos are always left-handedly spin-polarized and antineutrinos, right-handedly.

1.8. STRONG INTERACTIONS BETWEEN QUARKS

The strong interactions take place between the constituent quarks which make up the hadrons. The magnitude of the coupling can be estimated from the decay probability or width Γ of unstable baryons. Consider the state Σ (1385) formed as a resonance of central mass 1385 MeV in a K^-p interaction:

$$K^- + p \to \Sigma^\circ(1385) \to \Lambda + \pi^\circ \qquad \Gamma = 36 \text{ MeV} \tag{1.19}$$

with 130 MeV Q-value in the decay and a lifetime $\tau = \hbar/\Gamma = 10^{-23}$ s estimated from the measured width Γ. If we compare this with the electromagnetic decay (1.15) with a comparable Q-value ($= 77$ MeV),

$$\Sigma^\circ(1192) \to \Lambda + \gamma \qquad \tau = 10^{-19} \text{ s} \tag{1.20}$$

we get for the strong coupling α_s, analogous to α

$$\frac{\alpha_s}{\alpha} \simeq \left(\frac{10^{-19}}{10^{-23}}\right)^{1/2} \simeq 100 \tag{1.21}$$

or

$$\alpha_s = \frac{g_s^2}{4\pi} \simeq 1$$

This is the value of the coupling of the strong charges g_s of the quarks, via an appropriate mediating boson called a *gluon*, the neutral, massless carrier of the strong force analogous to the photon in QED. In QED, there are just two types of charge, called $+$ and $-$. In the theory of interquark forces, called quantum chromodynamics (QCD) there are six types of strong charge, called "color"—which is just a name for an internal degree of freedom. A quark can carry one of three colors (say red, blue, or green) and an antiquark, the corresponding anticolor. Color symmetry is supposed to be exact, that is the quark-quark force is independent of the quark colors involved. As an example, Fig. 1.8(b) shows a red quark interacting with a blue quark via exchange of a red-antiblue gluon ($r\bar{b}$). This diagram is drawn for single gluon exchange, although if $\alpha_s \sim 1$ this cannot be the only process; multiple gluon exchange must be highly probable. It turns out that for violent collisions of very high q^2, $\alpha_s < 1$ and single gluon exchange is a good approximation, while at low q^2 (or equivalently, larger distances) the coupling α_s becomes large as in (1.21) and the theory is uncalculable. This large-distance behavior is presumably linked with confinement of quarks and gluons inside hadrons. The potential between two quarks if often taken to be of the form

$$V_s = -\frac{4}{3}\frac{\alpha_s}{r} + kr, \tag{1.22}$$

where the first term dominating at small r arises from single gluon exchange and is, apart from the numerical factor, the same as the Coulomb potential between elementary charges

$$V_{em} = \frac{-\alpha}{r}.$$

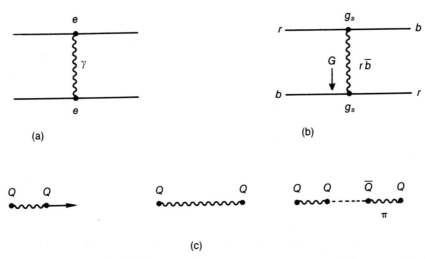

Figure 1.8 (a) Electromagnetic interaction between two charged particles, mediated by single photon exchange with coupling $|e|$. (b) Strong color force between quarks mediated by single gluon exchange with coupling g_s. A red quark interacts with a blue quark by exchange of a red-antiblue gluon. (c) Attempts to free a quark by pulling out the gluon "string" results in formation of a quark-antiquark pair (meson).

The second, linear term is associated with confinement at large r. There is direct experimental evidence for both terms in the potential (1.22).

Because of the linear term in (1.22), attempts to free a quark from a hadron simply results in production of fresh $Q\bar{Q}$ pairs (mesons). The nature of QCD is such that the lines of force of the color field are pulled together by a strong gluon-gluon interaction, so that they form a flux tube or string (see Fig. 8.23, p. 299). Pulling out this string, the stored energy kr in (1.22) eventually reaches the point where it is energetically more favorable to create a $Q\bar{Q}$ pair with two short strings rather than one long one. (Fig. 1.8(c)).

In the limit of high quark energies, the confining potential has dramatic effects. The annihilation process $e^+e^- \rightarrow$ hadrons is viewed in terms of the elementary process $e^+e^- \rightarrow Q\bar{Q}$ followed by "fragmentation" of the quark and antiquark into hadrons, by the above process with gluon strings. Since the transverse momenta involved in creating mesons is of order of a few times $m_\pi c$ only, one obtains two collimated "jets" of hadrons traveling in opposite directions and following the momentum vectors of the original quarks (see Fig. 2.23).

1.9. GRAVITATIONAL INTERACTIONS

Gravity is not an important effect in particle physics at accelerator energies but we mention it briefly for completeness. It is described in terms of the

Newtonian constant K, with the force between two equal point masses M given by KM^2/r^2, where r is their separation. By comparing with the electrostatic force between singly charged particles, e^2/r^2, the quantity $KM^2/\hbar c$ is seen to be dimensionless. For example, if M is taken as the proton mass, then

$$\frac{KM^2}{4\pi\hbar c} = 4.6 \times 10^{-40}, \tag{1.23}$$

compared with

$$\frac{e^2}{4\pi\hbar c} = \frac{1}{137}.$$

Thus, for mass scales common in high-energy physics, the gravitational coupling is negligibly small, and only approaches unity for a hypothetical elementary particle of mass equal to the Planck mass $M_p = (4\pi\hbar c/K)^{1/2} = 2.10^{19}$ GeV.

Gravity is important in the everyday world because it is cumulative. There is only one sign of gravitational charge (negative mass does not exist), so the potential on a proton is the sum of the potentials due to all nucleons and electrons in the earth. On the other hand, there are two signs of electric charge, the world is electrically neutral, and the enormously greater electrical force on a proton due to all other protons in the earth is exactly canceled by that due to the electrons.

1.10. CONSERVATION RULES IN FUNDAMENTAL INTERACTIONS

A summary of the characteristics of the fundamental interactions described above is given in Table 1.3. Included is a list of typical collision cross-sections, and of typical lifetimes for decay via the various interactions. The lifetimes and widths of hadronic states decaying by the various interactions are shown in Fig. 1.9. The width depends on the coupling involved and, to a lesser extent, on the mass of the state, that is on the phase-space available for the secondary particles produced.

Table 1.4 shows a list of some of the quantities conserved in each type of interaction. Some of the conservation rules are absolute, or very nearly so. Other properties, such as invariance under spatial inversion (parity conservation), are observed exactly in some interactions but not in others. Often the violation occurs in a regular way. For example, weak interactions may or may not conserve strangeness S. If it is violated, the rule is $\Delta S = 1$.

TABLE 1.3 Fundamental interactions (M = nucleon mass)

Interaction	Gravity	Electro-magnetic	Weak	Strong
Field quantum	Graviton	Photon	Intermediate bosons $W^\pm$, Z^0	Gluon
Spin-parity	2^+	1^-	1^-, 1^+	1^-
Mass (mc^2), GeV	0	0	80–90	0
Range, m	∞	∞	10^{-18}	$\leq 10^{-15}$
Source	Mass	Electric charge	"Weak charge"	"Color charge"
Coupling	K (Newton)	—	G (Fermi)	—
Dimensionless coupling constant	$KM^2/\hbar c = 0.53 \times 10^{-38}$	$\alpha = e^2/4\pi\hbar c = \frac{1}{137}$	$(Mc/\hbar)^2 G/\hbar c = 1.02 \times 10^{-5}$	$\alpha_s \sim 1$, large r < 1, small r
Typical cross-section, m^2 (1 GeV)	—	10^{-33}	10^{-44}	10^{-30}
Typical lifetime for decay, s	—	10^{-20}	10^{-8}	10^{-23}

TABLE 1.4 Conservation rules

Conserved quantity	Interaction		
	Strong	Electromagnetic	Weak
Energy/momentum Charge Baryon number Lepton number	Yes	Yes	Yes
I (isospin)	Yes	No	No ($\Delta I = 1$ or $\frac{1}{2}$)
S (strangeness)	Yes	Yes	No ($\Delta S = 1, 0$)
C (charm)	Yes	Yes	No ($\Delta C = 1, 0$)
P (parity)	Yes	Yes	No
C (charge-conjugation parity)	Yes	Yes	No
CP (or T)	Yes	Yes	Yes[a]
CPT	Yes	Yes	Yes

[a] But 10^{-3} violation in K^0 decay.

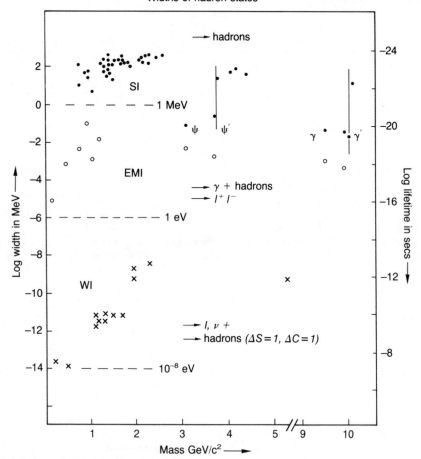

Figure 1.9 Widths and lifetimes of hadron states. The vertical scale is in powers of 10. On the left is given the width Γ in MeV, and on the right the mean lifetime in seconds, where $\tau = \hbar/\Gamma$. Since $\hbar = 6.58.10^{-22}$ MeV s, a state of width 1 MeV has $\tau = 6.6 \times 10^{-22}$ s, and a state of $\tau = 10^{-22}$ s has $\Gamma = 6.58$ MeV. ● The solid circles are for hadron states (resonances) decaying by *strong* interactions (i.e., into hadrons, and conserving I, S, C...) Note that all these states have $\Gamma > 10$ MeV, except for the anomalously narrow, lowest energy states of the Ψ- and Υ-meson resonances. For these, QCD predicts that the effective coupling (via triple gluon exchange, see Section 5.15) is weaker than for the other states. ○ The open circles denote hadrons decaying by *electromagnetic* interactions, into lepton pairs, or into a photon or photons and other hadrons. These decays do not conserve isospin, I. × The crosses refer to hadrons decaying through the *weak* interaction, either into an electron-neutrino pair, or into hadrons with a change of isospin I, strangeness S, charm C, etc. The shortest lifetime directly measurable is 10^{-16} s, for the decay $\pi^0 \rightarrow 2\gamma$. Shorter lifetimes than this are inferred from the measured width and the above formula.

1.11. UNITS IN HIGH-ENERGY PHYSICS

The fundamental units in physics are of length, mass, and time, the familiar systems (MKS) expressing these in meters, kilograms, and seconds. Such units are however not very appropriate in particle physics, where lengths are typically 10^{-15} m and masses 10^{-27} kg.

Lengths in particle physics are usually quoted in terms of the *femtometer* or *fermi* (1 fm = 10^{-15} m), and cross-sections in terms of the *barn* (1 b = 10^{-28} m^2), millibarn (1 mb = 10^{-31} m^2), or microbarn (1 μb = 10^{-34} m^2). The unit of energy is based on the electron volt (1 eV = 1.6×10^{-19} joules) with the larger units MeV (= 10^6 eV), GeV (= 10^9 eV), and TeV (= 10^{12} eV). Masses are usually measured in MeV/c^2, meaning that if the mass is M, the rest energy is Mc^2 MeV. For example, the proton has a rest energy of 938.28 MeV or 0.938 GeV. Often masses (meaning the rest-energy equivalents) are loosely quoted in MeV or GeV.

In calculations, the quantities $\hbar = h/2\pi$ and c occur frequently, and it is often advantageous to use a system of units in which $\hbar = c = 1$. We do this by choosing some standard mass m_0 (e.g., the proton mass) as the unit:

$$m_0 = 1.$$

The natural unit of length is then the Compton wavelength of the standard particle:

$$\lambdabar_0 = \frac{\hbar}{m_0 c} = 1;$$

that of time is

$$t_0 = \frac{\lambdabar_0}{c} = \frac{\hbar}{m_0 c^2} = 1,$$

and that of energy is

$$E_0 = m_0 c^2 = 1.$$

In these units, it is seen that $\hbar = c = 1$. In converting back, at the end of the calculation to the more usual units, it is useful to remember that $\hbar c = 197$ MeV fm. Thus, a particle of mass energy $m_0 c^2 = 197$ MeV has a Compton wavelength of $\hbar/m_0 c = \hbar c/m_0 c^2 = 1$ fm.

Throughout this text we shall be dealing with the coupling of charges—strong, electric and weak—to mediating bosons. In MKS units, electric charge, e, is measured in Coulombs and the fine-structure constant is then given by

$$\alpha = \frac{e^2}{4\pi\varepsilon_0 \hbar c} \simeq \frac{1}{137}.$$

For the general coupling of charges to bosons, such units are not useful and we define e in Heaviside-Lorentz units ($\varepsilon_0 = \mu_0 = 1$) so that with $\hbar = c = 1$

$$\alpha = \frac{e^2}{4\pi} \simeq \frac{1}{137},$$

as in (1.12). A similar definition is used to relate charges and coupling constants in the other interactions.

PROBLEMS

1.1. (a) Show that a negative muon captured in an S-state by a nucleus of charge Ze and mass A will spend a fraction $f \simeq 0.25A(Z/137)^3$ of its time in nuclear matter, and that in time t it will travel a total distance $fct(Z/137)$ in nuclear matter. (b) The law of radioactive decay of free muons is $dN/dt = -\lambda_d N$, where $\lambda_d = 1/\tau$ is the decay constant and the lifetime $\tau = 2.16\ \mu$s. For a negative muon captured in an atom Z, the decay constant is $\lambda = \lambda_d + \lambda_c$, where λ_c is the probability of nuclear capture per unit time. For aluminum ($Z = 13, A = 27$) the mean lifetime of negative muons is $0.88\ \mu$s. Calculate λ_c, and using the expression for f in (a), compute the interaction mean free path Λ for a muon in nuclear matter. (c) From the magnitude of Λ in (b) estimate the magnitude of the coupling constant in the reaction $\mu^- + p \rightarrow n + \nu$, assuming that a coupling constant of unity corresponds to a mean free path equal to the range of nuclear forces.

1.2. (a) Deduce an expression for the energy of a γ-ray from the decay of a neutral pion, $\pi^0 \rightarrow 2\gamma$, in terms of the mass m, energy E, and velocity βc of the pion, and of the angle of emission θ in the CMS. (b) Show that if the pion has zero spin, the distribution in θ will be isotropic and that the energy distribution of the γ-rays will be flat, extending from $E(1 + \beta)/2$ to $E(1 - \beta)/2$. (c) Find an expression for the disparity D (the ratio of energies) of the two γ-rays from π^0 decay. For the case of relativistic pions, show that $D > 3$ in half of the decays and $D > 7$ is one quarter of them.

1.3. A negative muon, when brought to rest in liquid hydrogen, can form a molecular ion H_2^+ by displacing an electron. Why? If the hydrogen contains even a tiny amount of deuterium, it is found that the negative muons eventually form molecular ions HD^+. Why? What is the typical internuclear distance in such an ion? If the two nuclei react to form ^3He, what may happen to the muon?

1.4. It has been postulated that the neutrinos ν_e, ν_μ, ν_τ, etc. may consist of linear combinations of discrete mass eigenstates $\nu_1, \nu_2, \nu_3, \ldots$. Suppose that in the decay $\pi \rightarrow \mu + \nu_\mu$, the state ν_μ consists of a combination of states of mass m_1 and m_2. Thus, in principle, the energy or momentum of the muon from pion decay at rest would consist of two discrete values. Assuming that the muon momentum can be measured with unlimited precision, show that two such discrete values could be demontrated only if $m_2^2 - m_1^2 > 1 \ (\text{eV}/c^2)^2$. Show that the irreducible error of measurement of the muon momentum places a less stringent limit on the minimum detectable value of $m_2^2 - m_1^2$.

1.5. The cross-section for the reaction $\pi^- + p \rightarrow \Lambda + K^0$ at 1-GeV/c incident momentum is approximately $1 \ \text{mb} \ (10^{-27} \ \text{cm}^2)$. Both Λ- and K^0-particles decay with a mean lifetime of about 10^{-10} s. From this information, estimate the relative magnitude of the couplings responsible for the production and decay, respectively, of the Λ- and K^0-particles.

1.6. State which of the following reactions are allowed by the conservation laws and which are forbidden, and give the reasons in either case:

$$\pi^0 \rightarrow e^+ + e^-, \tag{i}$$

$$p \rightarrow n + e^+ + \nu_e, \tag{ii}$$

$$\mu^+ \rightarrow e^+ + e^- + e^+, \tag{iii}$$

$$K^+ + n \rightarrow \Sigma^+ + \pi^0. \tag{iv}$$

1.7. It was once suggested that the numerical values of the charge of the electron and proton might differ by a small amount $|\Delta e|$, so that expansion of the universe could be attributed to an electrostatic repulsion between hydrogen atoms in space. Estimate the minimum value of $|\Delta e/e|$ required for this hypothesis. [References: Theory, H. Bondi and R. A. Lyttleton, *Nature* **184**, 974)1959); *proc. Roy. Soc.* **A252**, 313 (1959). Experimental disproof, A. M. Hillas and T. E. Cranshaw, *Nature* **184**, 892 (1959).]

CHAPTER 2

Particle Detectors and Accelerators

2.1. ACCELERATORS

All accelerators employ electric fields to accelerate stable charged particles (electrons, protons, or heavier ions) to high energy. The simplest machine would be a d.c. high-voltage source (called a Van der Graaff accelerator) which can however only achieve beam energies of about 20 MeV. To do better, one has to employ a high frequency a.c. voltage and carefully time a bunch of particles to obtain a succession of accelerating kicks. This is done in the *linear* accelerator, with a succession of accelerating elements (called drift tubes) in line, or by arranging for the particles to traverse a single voltage source repeatedly, as in the *cyclic* accelerator.

2.1.1. Linear Accelerators (linacs)

Figure 2.1 shows a sketch of a proton linac. It consists of an evacuated pipe containing a set of metal drift tubes, with alternate tubes attached to either side of a radiofrequency voltage. The proton (hydrogen ion) source is continuous, but only those protons inside a certain time bunch will be accelerated. Such protons traverse the gap between successive tubes when the field is from left to right, and are inside a tube (therefore in a field-free region) when the voltage changes sign. If the increase in length of each tube along the accelerator is correctly chosen, as the proton velocity increases

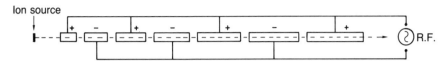

Figure 2.1 Scheme of a proton linear accelerator. Protons from the ion source traverse the line of drift tubes (all inside an evacuated pipe). The successive lengths are chosen so that as the proton velocity increases, the transit time from tube to tube remains constant.

under acceleration, the protons in a bunch receive a continuous acceleration. Typical fields are a few MeV per meter of length. Such proton linacs, reaching energies of 50 MeV or so, are used as injectors for the later stages of cyclic accelerators.

Electrons above a few MeV energy travel essentially with light velocity, so that after the first meter or so, an electron linac has tubes of uniform length. In practice, microwave frequencies are employed, the tubes being resonant cavities of a few centimeters in dimension, fed by a series of klystron oscillators, synchronized in time to provide continuous acceleration. The electrons, so to speak, ride the crest of an electromagnetic wave. The largest electron linac, at Stanford, is 3 km long and accelerates electrons to 25 GeV using 240 klystrons, which give short (2-μs) bursts of intense power 60 times per second, with a similar bunch structure for the beam.

2.1.2. Cyclic Accelerators (synchrotrons)

All modern proton accelerators and most electron machines are circular, or nearly so. The particles are constrained in a vacuum pipe bent into a torus which threads a series of electromagnets, providing a field normal to the plane of the orbit (Fig. 2.2(a)). For a proton of momentum p in GeV/c, the field must have a value of B (in Tesla), where

$$p = 0.3 \, B\rho \tag{2.1}$$

and ρ is the ring radius in meters. The particles are accelerated once or more per revolution by RF cavities. Both the field B and the RF frequency must increase and be synchronized with the particle velocity as it increases—hence the term synchrotron. Protons are injected from a linac source at low energy and at low field B, which increases to its maximum value over the accelerating cycle, typically lasting for a few seconds. Then the cycle begins again. Thus, the beam arrives in discrete pulses.

In the linac, the final beam energy depends on the voltage per cavity and the total length, while in the proton synchrotron, it is determined by the ring radius and the maximum value of B. For conventional electromagnets using copper coils, B(max) is of order 14 kgauss(1.4 T), while if superconducting coils are used, fields of 5 T or more are possible. As an example, the Fermilab synchrotron (called the Tevatron) is 1 km in radius and achieves

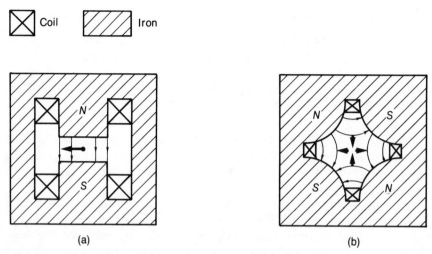

Figure 2.2 Cross-section of (a) typical bending (dipole) magnet (b) focusing (quadrupole) magnet. The light arrows indicate field directions; the heavy arrows, the force on a positive particle traveling into the paper.

400 GeV proton energy with conventional magnets and 1000 GeV = 1 TeV with superconducting magnets.

In all cyclic accelerators, protons make typically 10^5 revolutions, receiving an RF "kick" of order 0.1 MeV per turn, before achieving peak energy. In their total path of perhaps 10^6 km, the stability and focusing of the proton bunch is of paramount importance, otherwise the particles will quickly diverge and be lost. In the most recent machines, using the principle of strong focusing, the magnets are of two types: bending magnets produce a uniform vertical dipole field over the width of the beam pipe and constrain the protons in a circular path (Fig. 2.2(a)); and focusing magnets produce a quadrupole field with four poles as shown (Fig. 2.2(b)). In the sketch, the field is zero at the center and increases rapidly as one moves outward. A proton moving downward (into the paper) will be subject to magnetic forces shown by the thick arrows. The magnet shown is vertically focusing (force toward the center) and horizontally defocusing (force away from the center). Alternate quadrupoles have poles reversed so that, both horizontally and vertically, one obtains alternate focusing and defocusing effects. As anyone can demonstrate with a light beam and a succession of diverging and converging lenses of equal strength, the net effect is of focusing in both planes.

2.1.3. Focusing and Beam Stability

The particles circulating in a synchrotron do not travel in ideal circular orbits, but wander in and out from the circular path, in both horizontal and vertical planes, in what are called *betatron oscillations*. These arise from the natural divergence of the originally injected beam, from small asymmetries in fields and magnet alignments etc. The wavelength of these oscillations is related to the focal length of the quadrupoles and is short compared with the total circumference. In addition to these transverse oscillations, longitudinal oscillations, called *synchrotron oscillations*, occur as individual particles get out of step with the ideal, synchronous phase, for which the increase in momentum per turn from the RF kick exactly matches the increase in magnetic field. Thus, in Fig. 2.3, a particle *F* which lags behind an exactly synchronous particle *E* will receive a smaller RF kick, will swing into a smaller orbit and will, next time around, arrive earlier. Conversely, an early particle *D* will receive a larger impulse, move into a larger orbit and subsequently arrive later. So particles in the bunch execute synchrotron oscillations about the equilibrium position, but the bunch as a whole remains stable.

Summarizing briefly therefore: Protons are injected continuously from a linac at the beginning of the accelerator cycle when the dipole field is low. As acceleration proceeds, accelerated particles group into a number of equally spaced bunches (the spacing determined by the RF frequency). The lateral extent of the bunch initially fills the aperture of the vacuum pipe

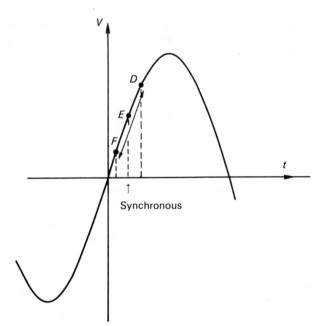

Figure 2.3 Particles arriving early (D) or late (F) receive more or less energy in RF acceleration than a synchronous particle (E). This decreases (increases) the rotation frequency, and the effect is that the particle performs oscillations (shown by arrow) with respect to the synchronous particle.

(typically 10 cm across), but becomes compacted by the focusing to 1 mm or so at the end of acceleration. Eventually the full-energy beam is extracted from the ring by a fast kicker magnet, or peeled off slowly (over a period of 1 s, say) with the aid of a thin energy-loss foil.

2.1.4. Electron Synchrotrons

We have discussed briefly the most common type of proton accelerator today, the strong-focusing synchrotron. Electron synchrotrons based on similar principles have also been built. However, they have an important limitation, absent in a proton machine. Under the circular acceleration, an electron emits synchrotron radiation, the energy radiated per particle per turn being

$$\Delta E = \frac{4\pi}{3} \frac{e^2 \beta^2 \gamma^4}{\rho}, \qquad (2.2)$$

where ρ is the bending radius, β is the particle velocity, and $\gamma = (1 - \beta^2)^{-1/2}$. Thus, for relativistic protons and electrons of the same momentum the energy loss is in the ratio $(m/M)^4$, so that it is 10^{13} times smaller for protons than electrons. For an electron of energy 10 GeV circulating in a ring of radius 1 km, this energy loss is 1 MeV per turn—rising to 16 MeV per turn at 20 GeV. Thus, even with very large rings and low guide fields, synchrotron radiation and the need to compensate this loss with large amounts of RF power become the dominant factor for an electron machine. Indeed, the large electron linac of 25 GeV was built at Stanford for this very reason.

2.2. COLLIDING-BEAM MACHINES

Much of our present experimental knowledge in the high-energy field has been obtained with proton and electron accelerators in which the beam has been extracted and directed onto an external target—the so-called fixed-target experiments. In particular, high-energy proton synchrotrons can thus provide intense secondary beams of hadrons (π, K, p, $\bar{p}$) and leptons (μ, ν), and several beam lines from one or more targets can be used simultaneously for a range of experiments, at a range of incident momenta.

During the last two decades, colliding-beam machines have become important. In these accelerators, two counterrotating beams of particles collide in several intersection regions around the ring. Their great advantage is in terms of the large center-of-mass energy available for the creation of new particles. A fixed-target proton accelerator provides particles of energy E, say, which collide with a nucleon of mass M in a target. The square of the CMS energy W is (see Appendix A)

$$s = W^2 = 2ME + 2M^2. \qquad (2.3)$$

Thus, for $E \gg M$, the kinetic energy available in the CMS for new particle creation rises only as $E^{1/2}$. *The remaining energy is not wasted*—it is converted into kinetic energy of the secondary particles in the laboratory system, and this allows the production of high-energy secondary beams.

However, if two relativistic particles (e.g., protons) of energy E_1 and E_2 and momenta p_1 and p_2 circulate in opposite directions in a storage ring, then in a head-on collision the value of W is given by

$$s = W^2 = 2(E_1 E_2 + p_1 p_2) + 2M^2 \approx 4E_1 E_2. \tag{2.4}$$

If $E_1 = E_2$, the CMS of the collision is at rest in the laboratory. Virtually all the energy is available for new-particle creation, and rises as E instead of as $E^{1/2}$ in the fixed-target case. Indeed, the value of W is the same as that in a fixed-target machine of energy $E = 2E_1 E_2/M$. As an example, the CERN ISR provided proton beams of energy 30 GeV colliding nearly head-on (at a 15° angle). Thus $W \simeq 60$ GeV, and a fixed-target proton synchrotron of energy $E = 2000$ GeV would be required to provide the same value of W.

Colliding-beam machines also possess some disadvantages. The colliding particles must be stable, limiting one to collisions of protons (or heavier nuclei), antiprotons, electrons, and positrons. All colliders so far built are of the pp, e^+e^-, or $p\bar{p}$ variety, although an ep collider is under construction (see Table 2.1). Secondly, the collision rate in the intersection region is low. The reaction rate is given by

$$R = \sigma L, \tag{2.5}$$

where σ is the interaction cross-section and L is the luminosity (in units of $cm^{-2}s^{-1}$). For two oppositely directed beams of relativistic particles the formula for L is

$$L = fn \frac{N_1 N_2}{A}, \tag{2.6}$$

where N_1 and N_2 are the numbers of particles in each bunch, n is the number of bunches in either beam around the ring, and A is the cross-sectional area of the beams, assuming them to overlap completely. f is the revolution frequency. Obviously, L is largest if the beams have small cross-sectional area A. The luminosity is however limited by the beam-beam interaction. Typical L-values are $\sim 10^{31} cm^{-2}s^{-1}$ for e^+e^- colliders and $\sim 10^{30} cm^{-2}s^{-1}$ for $p\bar{p}$ machines. These values may be compared with that of a fixed-target machine. A beam of 10^{12} protons s^{-1} from a proton synchrotron, in traversing a liquid-hydrogen target 1 m long, provides a luminosity $L \simeq 10^{37} cm^{-2}s^{-1}$.

In a pp or ep collider, two separate beam pipes and two sets of magnets are required, while e^+e^- and $p\bar{p}$ colliders have a unique feature. By the principle of charge conjugation (particle-antiparticle conjugation) invariance, it turns out that a synchrotron consisting of a set of magnets and RF cavities adjusted to accelerate, say, an electron e^-, in a clockwise

TABLE 2.1 Partial list of some present-day accelerators

Proton synchrotrons		
	Location	Energy, GeV
CERN PS	Geneva	28
BNL AGS	Brookhaven, Long Island	32
KEK	Tsukuba, Tokyo	12
Serpukhov	USSR	76
SPS	CERN, Geneva	450
Fermilab Tevatron II	Batavia, Illinois	1000

Electron accelerators		
SLAC linac	Stanford, California	25
DESY synchrotron	Hamburg	7

Colliding-beam machines				
PETRA	DESY, Hamburg	e^+e^-	22 + 22	
PEP	Stanford	e^+e^-	18 + 18	
CESR	Cornell, NY	e^+e^-	8 + 8	
TRISTAN (1986)[a]	Tsukuba	e^+e^-	30 + 30	
SLC (1987)	Stanford	e^+e^-	50 + 50	
LEP (1989)	CERN	e^+e^-	50 + 50	initially
			95 + 95	later
S$p\bar{p}$S	CERN	$p\bar{p}$	310 + 310	
Tevatron I (1986)	Fermilab	$p\bar{p}$	1000 + 1000	
HERA (1990)	Hamburg	ep	$30e + 820p$	

[a] ( ) = expected completion date.

direction, will simultaneously accelerate a positron, e^+, along the same path but anticlockwise. Thus, e^+e^- and $p\bar{p}$ colliders require only a single vacuum pipe and magnet ring.

2.2.1. Cooling in $\bar{p}p$ colliders

While it is not difficult to obtain an intense e^+(positron) source for use in e^+e^- colliders, the generation of an intense beam of $\bar{p}$(antiprotons) for a $p\bar{p}$ collider is much more difficult. The antiprotons must be created in pairs with protons in energetic proton-nucleus collisions, with a low yield and a wide spread in momentum and angle of emission from the target. In other words, viewed in the center-of-momentum frame of all the antiprotons produced, individual particles are like molecules in a very hot gas, with random motions described by a "temperature". To store enough antiprotons, the beam must be "cooled" in order to reduce its divergence and longitudinal

(a)

(b)

Figure 2.4 (a) Part of the 6-km tunnel of the CERN SPS, which accelerates protons to 450 GeV and is also used to store and collide oppositely circulating beams of 310-GeV protons and antiprotons. Magnet at lower left is a dipole (bending) magnet, that at center a quadrupole (focusing) magnet, which is followed by two more dipoles, and the pattern repeated. (b) View of antiproton accumulator ring (see also Fig. 2.5). The transfer lines carrying the signal across from pick up to kicker electrodes are visible. (Photo courtesy CERN.)

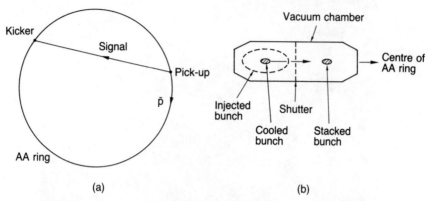

Figure 2.5 (a) Principle of stochastic cooling in an antiproton accumulator (AA) ring. A pickup coil delivers a signal depending on deviations of the antiprotons from the ideal orbit, and this activates a kicker which deflects them toward the ideal orbit as they come around the ring. (b) The vacuum chamber cross-section in the AA ring. After a mechanical shutter is opened, the cooled bunch is kicked into the inner half of the chamber and stacked together with previous bunches.

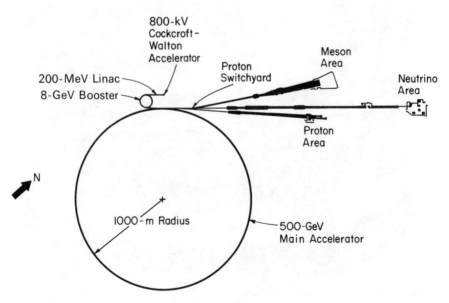

Figure 2.6 Layout of the 500-GeV proton synchrotron and beamlines at Fermilab, near Chicago. This machine has been upgraded to 1000 GeV by use of high-field superconducting magnets and will also be used as a 2-TeV proton-antiproton collider.

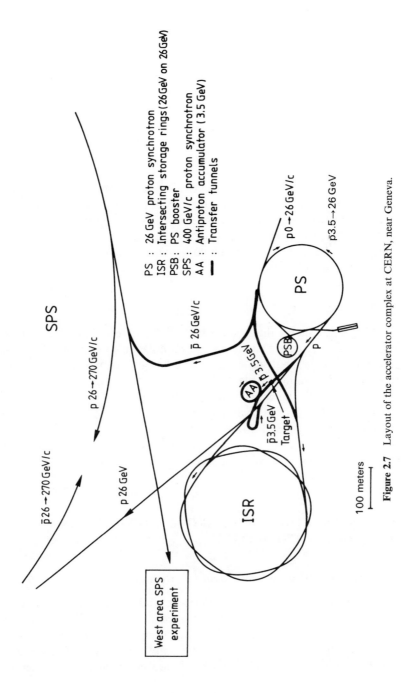

SPS

PS : 26 GeV proton synchrotron
ISR : Intersecting storage rings (26 GeV on 26 GeV)
PSB : PS booster
SPS : 400 GeV/c proton synchrotron
AA : Antiproton accumulator (3.5 GeV)
━━ : Transfer tunnels

p 26→270 GeV/c

p̄ 26→270 GeV/c

p 26 GeV

p̄ 26 GeV/c

West area SPS experiment

p0→26 GeV/c

p̄3.5→26 GeV

PS

PSB

p

p̄ 3.5 GeV

AA

p̄3.5 GeV
Target

ISR

100 meters

Figure 2.7 Layout of the accelerator complex at CERN, near Geneva.

37

momentum spread. This is a technical problem, which we mention here because its solution was crucial in the discovery of the $W^{\pm}$, Z° carriers of the weak force.

One approach, employed at the CERN $p\bar{p}$ collider, is to use a statistical method—hence it is called stochastic cooling. A bunch of $\sim 10^7$ antiprotons of momentum ~ 3.5 GeV/c emerge from a Cu target (bombarded by a pulse of 10^{13} 26-GeV protons) and are injected into the outer half of a wide-aperture toroidal vacuum chamber, divided by a mechanical shutter and placed inside a special accumulator-magnet ring (Figs. 2.4, 2.5). A pick-up coil in one section of the ring senses the average deviation of particles from the ideal orbit, and a correction signal is sent across a chord to a kicker, in time to deflect them, as they come round, toward the ideal orbit. After 2 s of circulation, lateral and longitudinal spreads have been reduced by an order of magnitude. The shutter is opened and the "cooled" bunch of antiprotons is magnetically deflected and stacked in the inside half of the vacuum chamber, where they are further cooled. The process is repeated until, after a day or so, 10^{12} antiprotons have been stacked and are then extracted for acceleration in the main SPS (collider) ring.

2.2.2. Accelerator Complexes

Figure 2.6 shows the general layout of the secondary beam lines around the 500-GeV proton accelerator at Fermilab, near Chicago. The 2.5-km-long beam to the neutrino area provides beams of muons as well as neutrinos. Figure 2.7 shows the layout of the accelerator complex at CERN, Geneva. The 26-GeV PS accelerator has been used to fill the ISR with oppositely circulating proton beams, as an injector to the 450-GeV SPS, and also to make secondary antiprotons, which are transferred, stacked, and "cooled" in a 3.5-GeV antiproton accumulator before being accelerated in the PS and injected into the SPS, used as a 310-GeV + 310-GeV $p\bar{p}$ collider.

2.3. THE INTERACTION OF CHARGED PARTICLES AND RADIATION WITH MATTER

2.3.1. Ionization Loss of Charged Particles

The detection of nuclear particles depends ultimately on the fact that, directly or indirectly, they transfer energy to the medium they are traversing via the process of ionization or excitation of the constituent atoms. This can be observed as charged ions, for example in a gas counter, or as a result of the scintillation light, Čerenkov radiation, etc., which is subsequently emitted.

The Bethe-Bloch formula for the mean rate of ionization loss of a charged particle is given by*

$$\frac{dE}{dx} = \frac{4\pi N_0 z^2 e^4}{mv^2} \frac{Z}{A} \left[\ln\left(\frac{2mv^2}{I(1 - \beta^2)} \right) - \beta^2 \right], \tag{2.7}$$

where m is the electron mass, z and v are the charge (in units of e) and velocity of the particle, $\beta = v/c$, N_0 is Avogadro's number, Z and A are the atomic number and mass number of the atoms of the medium, and x is the path length in the medium measured in $g \, cm^{-2}$ or $kg \, m^{-2}$. The quantity I is an effective ionization potential, averaged over all electrons, with approximate magnitude $I = 10Z$ eV. Equation (2.7) shows that dE/dx is independent of the mass M of the particle, varies as $1/v^2$ at nonrelativistic velocities, and, after passing through a minimum for $E \simeq 3Mc^2$, increases logarithmically with $\gamma = E/Mc^2 = (1 - \beta^2)^{-1/2}$. The dependence of dE/dx on the medium is very weak, since $Z/A \simeq 0.5$ in all but hydrogen and the heaviest elements. Numerically, $(dE/dx)_{min} \simeq 1$–$1.5 \, MeV \, cm^2 \, g^{-1}$ (or 0.1–$0.15 \, MeV \, m^2 \, kg^{-1}$).

Figure 2.8 shows the observed relativistic rise in ionization loss as a function of $p/Mc = (\gamma^2 - 1)^{1/2}$ for relativistic particles in a gas (argon-methane mixture). For $\gamma \sim 10^3$, it reaches 1.5 times the minimum value. The relativistic rise is associated with the fact that the transverse electric field of the particle is proportional to γ, so that more and more distant collisions become important as the energy increases. Eventually, when the impact parameter becomes comparable to interatomic distances, polarization effects in the medium (associated with the dielectric constant) halt any further increase. In solids, rather than gases, such effects become important at a much lower value of $\gamma \sim 10$, and this plateau value is only about 10 % larger than $(dE/dx)_{min}$. Part of the energy loss of a relativistic particle may be reemitted from excited atoms in the form of coherent radiation at a particular angle. Such Čerenkov radiation is discussed in Section 2.4.6.

The bulk of the energy loss results in the formation of ion pairs (positive ions and electrons) in the medium. One can distinguish two stages to this process. In the first stage, the incident particle produces primary ionization in atomic collisions. The electrons knocked out in this process have a distribution in energy E' roughly of the form $dE'/(E')^2$: those of higher energy (called δ-rays) can themselves produce fresh ions in traversing the medium (secondary ionization). The resultant total number of ion pairs is 3–4 times the number of primary ionizations, and is proportional to the energy loss of the incident particle in the medium. Equation (2.7) gives the *average* value of the energy loss dE in a layer dx, but there will be fluctuations about

---

* For a semiclassical derivation of this formula, see for example B. Rossi, *High Energy Particles*, Prentice-Hall Inc, Englewood Cliffs, N. J., 1961, p. 17; J. D. Jackson, *Classical Electrodynamics*, 2nd ed., John Wiley, New York, 1975, Chapter 13.

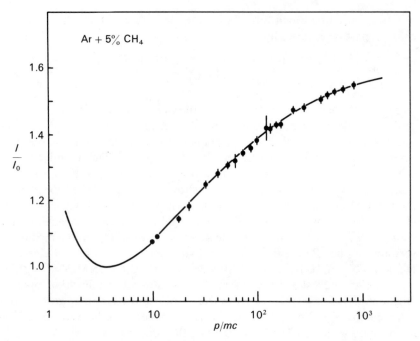

Figure 2.8 Mean ionization energy loss of charged particles in argon-methane mixture, showing relativistic rise as a function of p/mc. Measurements by multiple ionization sampling (after Lehraus *et al.* 1978).

the mean, dominated by the relatively small number of "close" primary collisions with large E'. This so-called Landau distribution about the mean value is therefore asymmetric, with a tail extending to values much greater than the average. Nevertheless, by sampling the number of ion pairs produced in many successive layers of gas and removing the "tail," the mean ionization loss can be measured within a few percent. In this way γ can be estimated from the relativistic rise, and if the momentum is known, this provides a useful method for estimating the rest mass and thus differentiating between pions, kaons, and protons.

The total number of ions produced in a medium by a high-energy particle depends on dE/dx and the energy required to liberate an ion pair. In a gas, this varies from 40 eV in helium to 26 eV in argon. In semiconductors, on the other hand, it is only about 3 eV, so the number of ion pairs is much larger. If the charged particle comes to rest in the semiconductor, the energy deposited is measured by the total number of ion pairs, and such a detector therefore not only is linear but has extremely good energy resolution (typically 10^{-4}). Because of their small size, however, such solid-state counters have not found very general applications in high-energy physics.

2.3.2. Coulomb Scattering

In traversing a medium, a charged particle suffers electromagnetic interactions with both electrons and nuclei. As Eq. (2.7) indicates, dE/dx is inversely proportional to the target mass, so that in comparison with electrons, the energy lost in Coulomb collisions with nuclei is negligible. However, because of the larger target mass, transverse *scattering* of the particle is appreciable in the Coulomb field of the nucleus, and is described by the famous Rutherford formula for the differential cross-section at scattering angle θ:

$$\frac{d\sigma(\theta)}{d\Omega} = \frac{1}{4}\left(\frac{Zze^2}{pv}\right)^2 \frac{1}{\sin^4(\theta/2)}, \tag{2.8}$$

where p, v, z are the momentum, velocity, and charge of the incident particle, and Z is the charge on the nucleus, assumed to act like a point charge. For small scattering angles the cross-section is large, so that in any given layer of material the net scattering is the result of a large number of small deviations, which are independent of one another. The resultant distribution in the net angle of *multiple* scattering follows a roughly Gaussian distribution

$$P(\phi)\,d\phi = \frac{2\phi}{\langle\phi^2\rangle}\exp\left(\frac{-\phi^2}{\langle\phi^2\rangle}\right)d\phi \tag{2.9}$$

The root-mean-square (rms) deflection in a layer t of the medium is given by

$$\phi_{\text{rms}} = \langle\phi^2\rangle^{1/2} = \frac{zE_s}{pv}\sqrt{\frac{t}{X_0}}, \tag{2.10}$$

where

$$E_s = \sqrt{4\pi \times 137}\, mc^2 = 21\text{ MeV} \tag{2.11}$$

and

$$\frac{1}{X_0} = \frac{4Z(Z+1)r_e^2 N_0}{137A}\ln\left(\frac{183}{Z^{1/3}}\right). \tag{2.12}$$

In this formula, $r_e = e^2/mc^2$ is the classical electron radius. The quantity X_0, for reasons that appear below, is called the radiation length of the medium (see Table 2.2), and incorporates all the dependence of ϕ_{rms} on the medium. Numerically therefore, a singly charged particle ($z = 1$) of momentum p and velocity v, with the product pv measured in MeV, suffers an rms deflection of $21/pv$ radians in traversing one radiation length. The rms angular deflection will be azimuthally symmetric about the trajectory: Along one axis in the plane normal to the trajectory, it will be $1/\sqrt{2}$ times the value (2.10).

Coulomb scattering is important in practice because it frequently limits the precision with which the direction of a particle can be determined.

TABLE 2.2   Radiation lengths in various elements
(after Bethe and Ashkin 1953)

Element	Z	E_c, MeV	X_0, g/cm^2
Hydrogen	1	340	58
Helium	2	220	85
Carbon	6	103	42.5
Aluminum	11	47	23.9
Iron	26	24	13.8
Lead	82	6.9	5.8

As an example, we consider the determination of the momentum of a high-energy charged particle by its deflection in the field **B** of a solid iron magnet. If there is no scattering, and the momentum does not change appreciably in traversing the magnet, the radius of curvature ρ is given by $pc = Be\rho$, so that in traversing a distance s, the deflection will be

$$\phi_{\text{mag}} = \frac{s}{\rho} = \frac{Bes}{pc} = \frac{300Bs}{pc},$$

where B is in tesla (1 T = 10 kG), s is in meters, and pc is in MeV. The rms Coulomb scattering in the plane of the trajectory will be

$$\phi_{\text{scat}} = \frac{21}{\sqrt{2}} \frac{1}{p\beta c} \sqrt{\frac{s}{X_0}},$$

so that

$$\frac{\phi_{\text{scat}}}{\phi_{\text{mag}}} = \frac{0.05}{B\beta\sqrt{X_0 s}},$$

independent of the particle momentum for a relativistic particle ($\beta = 1$). For example, in iron $X_0 = 0.02$ m, $B \simeq 1.5$ T, and $\phi_{\text{scat}}/\phi_{\text{mag}} = 0.25$ for $s = 1$ m, falling to 0.10 for $s = 6$ m.

The multiple-scattering distribution in fact is only approximately Gaussian. As indicated in Fig. 2.9, it has a long tail due to occasional single large deflections, for which the distribution varies as ϕ^{-3} (as can be deduced from (2.8)).

2.3.3. Radiation Loss of Electrons

Electrons lose energy in traversing a medium in two ways: the ionization energy loss (2.7) and the process of radiation loss or *bremsstrahlung*. The radiative collisions of electrons occur principally with the atomic nuclei of the medium. The nuclear electric field decelerates the electron, and

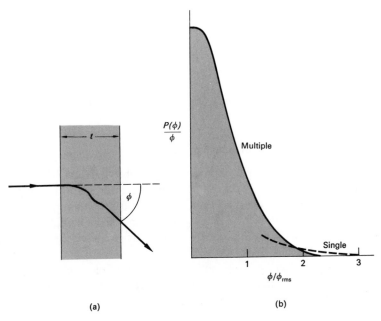

(a)   (b)

Figure 2.9 (a) Multiple scattering of a charged particle in traversing a layer of material of thickness t. (b) If the number of particles scattered through angle $\phi \to \phi + d\phi$ is $P(\phi)\,d\phi$, the distribution of $P(\phi)/\phi$ is approximately Gaussian. For very large deflections, however, the main contribution comes from single scattering (shown dashed).

the energy change appears in the form of a photon; hence the term *bremsstrahlung*, "braking radiation". The photon spectrum has the approximate form dE'/E', where E' is the photon energy. Integrated over the spectrum, the total radiation loss of an electron in traversing a thickness dx of medium is

$$\left(\frac{dE}{dx}\right)_{\text{rad}} = -\frac{E}{X_0}, \tag{2.13}$$

where X_0, the radiation length, is defined in (2.12). From (2.13) it follows that the average energy of a beam of electrons of initial energy E_0, after traversing a thickness x of medium, will be

$$\langle E \rangle = E_0 \exp\left(-\frac{x}{X_0}\right) \tag{2.14}$$

Thus, the radiation length X_0 may be simply defined as that thickness of the medium which reduces the mean energy of a beam of electrons by a factor e.

Since the rate of ionization energy loss for fast electrons, $(dE/dx)_{\text{ion}}$ is approximately constant, while the average radiation loss $(dE/dx)_{\text{rad}} \propto E$, it follows that at high energies, radiation loss dominates. The *critical energy* E_c

is defined as that at which the two are equal. From (2.7), (2.12), and (2.13) it is easy to show that, roughly,

$$E_c \simeq \frac{600}{Z} \text{ MeV.} \tag{2.15}$$

Values of X_0 and E_c in various materials are given in Table 2.2.

2.3.4. Absorption of γ-rays in Matter

There are three types of process responsible for attenuation of γ-rays in matter: photoelectric absorption, Compton scattering, and pair production. The photoelectric cross-section varies with photon energy E as $1/E^3$, and the Compton cross-section as $1/E$, so that for $E > 10$ MeV, the process of pair production, with a cross-section essentially independent of energy, is dominant (see Fig. 2.10).

The process of conversion of a high-energy photon to an electron-positron pair (in the field of a nucleus to conserve momentum) is closely related to that of electron bremsstrahlung. The attentuation of a beam of

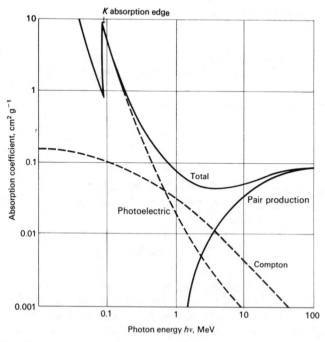

Figure 2.10   The absorption coefficient per g cm^{-2} of lead for γ-rays as a function of energy.

high-energy photons of intensity I_0 by pair production in a thickness x of absorber is described by

$$I = I_0 \exp\left(-\frac{7x}{9X_0}\right), \tag{2.16}$$

so the intensity is reduced by a factor e in a distance $9X_0/7$, sometimes called the *conversion* length. It is to be emphasized that, although the threshold for pair production is $E_{\text{th}} = 2m_e c^2 \simeq 1$ MeV, the asymptotic value (2.16) for the absorption coefficient is not attained until one reaches photon energies of almost 1 GeV.

2.4. DETECTORS OF SINGLE CHARGED PARTICLES

The detectors employed in experiments in high-energy physics are required to record the position, arrival time, and identity of charged particles. Precise evaluation of position coordinates is required to determine the particle trajectory and, in particular, its momentum (from the deflection in a magnetic field); precise timing is often required in order to associate one particle with another from the same interaction, frequently in situations where the total interaction rate per unit time may be very high. The identity of a particle may be established from simultaneous measurement of velocity (by time-of-flight or Čerenkov radiation) and momentum, and hence the rest mass; from the observation of decay modes, if the particle is unstable; and from its observed interaction with matter via strong, electromagnetic, or weak forces. Neutral particles are detected through their decay (e.g., $K^0 \rightarrow \pi^+ \pi^-$) and/or interaction with matter (e.g., $\pi^0 \rightarrow 2\gamma, \gamma \rightarrow e^+ e^-$), leading to secondary charged particles.

No single detector is in general able to meet all these requirements, and a combination of detectors of different types is required. We first discuss the principal types of detector in current use in the field, and then give some examples of how they may be combined in an integrated system.

Over the last two decades, very few radically new concepts in basic detection methods have been developed. Rather, progress has been in the exploitation and adaptation of well-established methods, a revolution in electronics and computer technology to select, record, and analyze huge amounts of data at high speed, and the development of hybrid systems involving many different types of detector, frequently on a massive scale.

2.4.1. Proportional Counters

The proportional counter is one of the oldest devices for the recording of ionization. Single counters consist of a gas-filled cylindric metal or glass tube of radius r_2 maintained at negative potential, with a fine central anode wire of radius r_1 at positive potential. The electric field in the gas for a potential difference V_0 is then

$$E(r) = \frac{V_0}{r \ln(r_2/r_1)}. \qquad (2.17)$$

An electron liberated by ionization at radius r_a will drift toward the anode, gaining an energy $T = e \int_{r_b}^{r_a} E(r)\, dr$ when it reaches radius r_b. If T exceeds the ionization energy of the gas, then fresh ions are liberated and a chain of such processes leads to an avalanche of electrons and positive ions. The gas amplification factor, equal to the total number of secondary electrons reaching the anode per initial ion pair, is typically $\sim 10^5$, independent of the number of primary ions; hence the name "proportional counter".

The most significant advance in this field was the introduction of the multiwire proportional counter (MWPC) by Charpak (1968, 1970). This device consists of many parallel anode wires stretched in a plane between two cathode planes (Fig. 2.11). The different anode wires act as independent detectors. A typical structure has wires of 20-μm diameter with 2-mm spacing, between cathode planes 12 mm apart, operating at a potential difference of 5 kV, and containing an argon-isobutane gas mixture. In general, several electrons from the primary ionization will drift toward an anode wire and create separate avalanches yielding negative pulses with very fast rise times (~ 0.1 ns). The positive ions have much lower mobility and induce pulses on both the cathode and neighboring anode wires of duration ~30 ns. The effective spatial resolution from the anode pulses is of order 0.7 mm. If the cathode is in the form of strips, the center of gravity of the cathode pulses may be used to obtain accurate spatial position of the avalanche (to within 0.05 mm).

2.4.2. Drift Chambers

As indicated above, the MWPC typically has spatial resolution of 1 mm or less, and a time resolution of 30 ns. However, to achieve this resolution over large areas, an enormous number of wires (together with amplifiers) are required. A great reduction in cost can be achieved by drifting the electrons from the primary ionization over (typically) 10 cm in a low-field region (1 kV/cm) before reaching the high-field amplification region near the anode wire; the collection time of the avalanche then gives a measure of the position of the ionization column. Figure 2.11 shows a typical arrangement of anode, cathode, and field wires required to give a uniform drift field. Space resolutions of order 0.1 mm are attained, with drift velocities of order 40 μm/ ns in argon-isobutane mixtures, almost independent of the drift field. Since, over a distance of 10 cm, the typical drift time is then 2 μs, such chambers are only useful with fairly low beam intensities or interaction rates, for example at e^+e^- colliders.

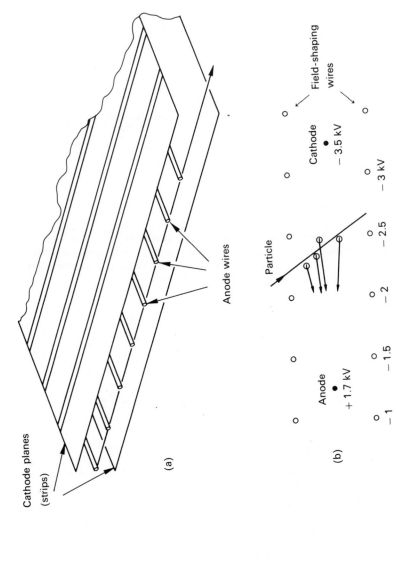

Cathode planes
(strips)

Anode wires

(a)

Particle

Field-shaping
wires

Cathode
− 3.5 kV

Anode
+ 1.7 kV

− 1   − 1.5   − 2   − 2.5   − 3 kV

(b)

Figure 2.11 (a) Schematic layout of multiwire proportional chamber. (b) Typical arrangement of electrodes for a drift-chamber cell.

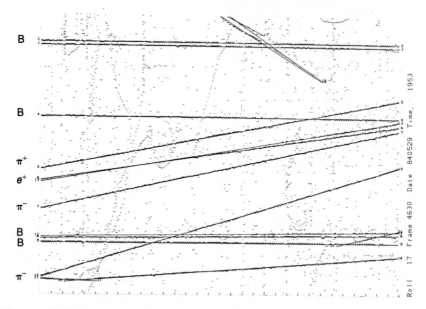

Figure 2.12 Reconstructed tracks in a large drift chamber ISIS ($4 \times 2 \times 5$ m volume of Ar + 20% CO_2) placed downstream of a small bubble chamber and spectrometer in European Hybrid Spectrometer experiment. Particles are identified from momentum and sampling of the ionization in the chamber (see Fig. 2.8). In addition to tracks fitted by lines, due to high-energy particles, several low-energy, highly scattered background tracks (electrons) are visible. In addition to beam tracks (marked B) there are secondaries from the decay of a charmed meson, $D^0 \rightarrow \pi^+\pi^-\pi^-e^+\nu$, produced in the bubble chamber upstream (to the left in the picture).

Very large volume (100 m³) drift chambers have been built. Sometimes such a device is called a *time projection chamber* (TPC). From the drift time of ions in the electric field, the track producing the ionization can be reconstructed in space, and the pulse height used to measure the ionization, and hence, from its momentum, the particle mass (see Fig. 2.8). Figure 2.12 shows an example of track reconstruction in the ISIS drift chamber detector.

2.4.3. Scintillation Counters

The scintillation counter has been a universal detector in a very wide range of high-energy physics experiments for over 30 years. The excitation of the atoms of certain media by ionizing particles results in luminescence (scintillation), which can be recorded by a photomultiplier. The scintillators in most common use are inorganic single crystals and organic liquids and plastics, although the phenomenon occurs also in liquids and gases. The decay times of the fastest (organic) scintillators are of order 1 ns.

Inorganic crystal scintillators, such as sodium iodide, are doped with activator centers (e.g., thallium). Ionizing particles traversing the crystal produce free electrons and holes, which move around until captured by an activator center. This is transformed into an excited state and decays with emission of light, over a broad spectrum in the visible region and with a decay time of order 250 ns.

In organic materials (either solid or liquid), on the other hand, the mechanism is excitation of molecular levels which decay with emission of light in the UV region. The conversion to light in the blue region is achieved via fluorescent excitation of dye molecules known as wavelength shifters, incorporated into the primary scintillator medium. Table 2.3 gives a list of a few organic and inorganic scintillators in common use.

The light from the scintillating medium is recorded by a photomultiplier tube or tubes (Fig. 2.13). These consist of a photocathode coated with alkali metals, where electrons are liberated by the photoelectric effect. The electrons travel to a chain of secondary-emission electrodes (dynodes) at successively larger potentials. Since about four secondary electrons are emitted per incident electron, amplification factors of 10^8 are achieved with 14 dynodes. The transit time from the cathode to the output dynode is typically 50 ns, with a jitter of order 1 ns. The jitter is determined mostly by the variation in transit time to the first dynode from different points on the cathode. Photocathode quantum efficiencies are typically $\leq 25\%$, peaking at $\lambda = 400$ nm.

The light from the scintillator slab travels down it by internal reflection, and the traditional way of directing this onto the photocathode is via multiple reflections down a suitably shaped plastic light guide. For very large area scintillators, the light guides consist of bent plastic rods or strips and the device can become very bulky. An alternative method is to place shifter bars along the edge of the scintillator slab. Blue light from the scintillator enters the bar, consisting of acrylic material doped with molecules (e.g., BBQ) which absorb the blue light and reemit isotropically in the green. Part of this light travels down the bar by internal reflection and is recorded by a photomultiplier glued to the end. The output is much less than that using light guides, but there is a large saving in space, number of photomultipliers, and construction of complex light-guide structures. The output pulse

TABLE 2.3 Characteristics of typical scintillators

	Pulse height (relative to anthracene)	Decay time, ns	λ_{max} Å	Density, g/cm^3
Polystyrene + p-terphenyl	0.28	3	3550	0.9
+ tetraphenylbutadiene	0.38	4.6	4800	
Sodium iodide (+ thallium)	2.1	250	4100	3.7
Anthracene	1.0	32	4100	3.7
Toluene	0.7	<3	4300	0.9

Scintillator

Light guide

Mu-metal screen

Steel outer casing

Complete scintillation counter

Photomultiplier

Associated electronics

Figure 2.13 Plastic scintillation counter, light pipe, and photomultiplier. (Photograph courtesy Rutherford Laboratory.)

from the photomultiplier is fed into suitable amplifiers, discriminators, and scalers, the essential function of which is to store the number of photomultiplier pulses of a particular magnitude.

2.4.4. Bubble Chambers

The bubble chamber has been an indispensible tool of high-energy physics for 30 years, particularly for studying complex interactions involving many secondary particles. Conceived by Glaser in 1952, it relies for its operation on the fact that in a superheated liquid, boiling will start with formation of gas bubbles at nucleation centers in the liquid, and particularly along trails of ions left by the passage of a charged particle. The liquid filling is maintained under an overpressure (typically 5–20 atmospheres), and the superheating achieved by sudden expansion of a piston, bellows, or diaphragm placed at the rear of the chamber. After expansion, bubbles along the tracks are allowed to grow over a period of order 10 ms and are photographed by an array of stereo cameras using flash illumination. The bubbles then collapse under a recompression stroke. Since the cycle time is of order 1 s, the bubble chamber is well matched to pulsed, cyclic accelerators, with repetition rates of the same order.

The most usual liquid fillings for bubble chambers are hydrogen, deuterium, and heavy liquids such as neon-hydrogen mixture, propane (C_3H_8), and Freon (CF_3Br)—see Fig. 2.14. The entire chamber is immersed in a strong magnetic field (2–3.5 T) provided by an electromagnet with conventional or superconducting coils, to permit momentum measurement from track curvature. The bubble images are recorded on photographic film from several cameras in stereo. Subsequent measurement of images on film are digitized, and a geometry program is used to reconstruct tracks and event vertices in three dimensions. The great detail which is recorded in a complex production and decay pattern is well illustrated in the example of Fig. 2.15.

The main disadvantages of a bubble chamber are, first, that it has a low repetition rate, and analysis of film is a lengthy process (typical experiments rarely involve more than 10^5 useful analyzed events); second, that many present and essentially all new accelerators are of the colliding-beam type, with an effectively d.c. interaction rate, and the very low duty cycle (10^{-2}) of bubble chambers, as well as the interaction geometry of the beams, excludes their use in this field.

2.4.5. Streamer and Flash Chambers

Proportional chambers are operated with anode-cathode potentials of order 5 kV, where the efficiency for recording charged particles reaches a plateau of practically 100%. Further increase in voltage eventually leads to electric breakdown of the gas. This takes place when the space charge inside

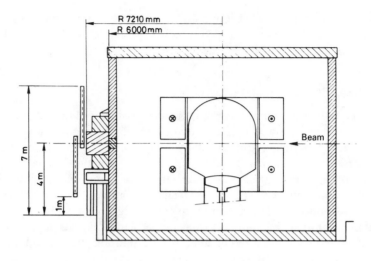

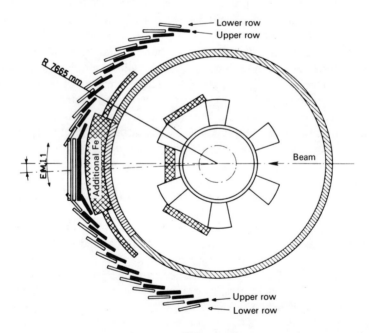

E.M.I. chambers

Figure 2.14 Elevation and plan views of the 3.7-m-diameter bubble chamber (BEBC) at CERN. The chamber is filled with liquid hydrogen, deuterium, or neon-hydrogen mixture and is equipped for neutrino experiments with an external muon identifier. This consists of 150 m² of multiwire proportional chambers placed outside the magnet yoke.

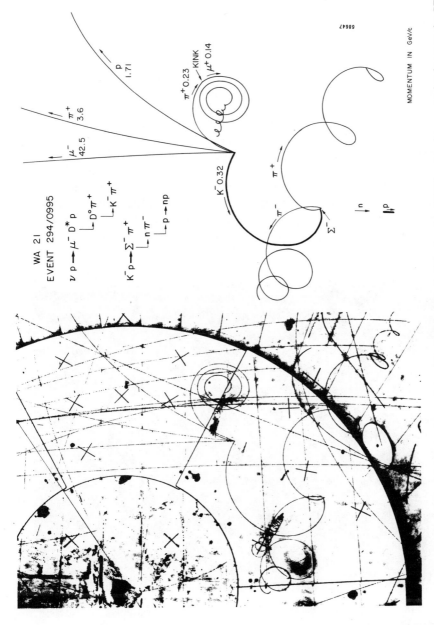

Figure 2.15 Example of charmed-particle production and decay in the hydrogen bubble chamber **BEBC** exposed to a neutrino beam at the CERN SPS. (Courtesy CERN.)

WA 21
EVENT 294/0995

$\nu p \longrightarrow \mu^- D^* p$
$\qquad\quad\mathrel{\rightharpoondown} D^0 \pi^+$
$\qquad\qquad\quad\mathrel{\rightharpoondown} K^- \pi^+$

$K^- p \longrightarrow \Sigma^- \pi^+$
$\qquad\quad\mathrel{\rightharpoondown} n \pi^-$
$\qquad\qquad\quad\mathrel{\rightharpoondown} p \longrightarrow np$

p
1.71

π^+
3.6

π^+ 0.23
KINK
μ^+ 0.14

μ^-
42.5

K^- 0.32

π^+

π^-

Σ^-

$|n$

$\|p$

686647

MOMENTUM IN GeV/c

53

the avalanche is strong enough to shield the external field; recombination of ions then occurs, resulting in photon emission and the birth of secondary ionization and new avalanches outside the initial one. This process propagates until an ion column links anode and cathode and a spark discharge occurs. The counter is then said to operate in the Geiger region.

If, however, a short (10-ns) high-voltage pulse (10–50 kV cm^{-1}) is applied between transparent parallel-plate electrodes, then only short (2–3-mm) streamer discharges develop from the ion trail of a crossing particle, and a direct track image (rather like that in a bubble chamber) can be obtained and photographed through the electrodes (Fig. 2.16). Such a *streamer chamber* possesses good multitrack efficiency and space resolution; an advantage over the bubble chamber is that the high potential is triggered (by outside scintillators) and that it has a very fast response, limited only by the speed of the film transport system.

The *flash chamber* is a simple and cheap detector, consisting of a large number of tubes or plastic channels filled with neon-helium mixture placed between planar electrodes to which a triggered high-voltage pulse is applied. A glow discharge is generated in the cells where ionization has been

Figure 2.16 Interaction of a 200-GeV proton in a xenon gas target placed in a streamer chamber. The proton enters the target at right. The streamers are normal to the plane of the picture, as is the applied magnetic field. The gas filling is 90% Ne, 10% He. (Courtesy V. Eckhardt, MPI, Munich.)

produced by a crossing charged particle. The discharge is recorded photographically or by electronic readout. Spatial resolution is determined by the tube diameter—typically several millimetres—but the method allows large-volume calorimeter-type detectors to be constructed at low cost.

2.4.6. Čerenkov Counters

When high-energy charged particles traverse dielectric media, part of the light emitted by excited atoms appears in the form of a coherent wavefront at fixed angle with respect to the trajectory—a phenomenon known as the Čerenkov effect, after its discoverer. Such radiation is produced whenever the velocity βc of the particle exceeds c/n, where n is the refractive index of the medium. From the Huyghens construction of Fig. 2.17 one sees that the wavefront forms the surface of a cone about the trajectory as axis, such that

$$\cos \theta = \frac{ct/n}{\beta ct} = \frac{1}{\beta n}, \qquad \beta > \frac{1}{n}. \tag{2.18}$$

Čerenkov radiation appears as a continuous spectrum. In a dispersive medium, both n and θ will be functions of the frequency v. The total energy content of the radiation, per unit track length, is

$$\frac{dE}{dx} = \frac{4\pi^2 z^2 e^2}{c^2} \int \left(1 - \frac{1}{\beta^2 n^2}\right) v \, dv. \tag{2.19}$$

The number of photons at a particular frequency or wavelength is proportional to dv or to $d\lambda/\lambda^2$. Thus, blue light predominates. Over a small frequency range, we can neglect the dependence of n on v and (2.19) becomes

$$\frac{dE}{dx} = \frac{z^2}{2} \left(\frac{e^2}{\hbar c}\right)^2 \left(\frac{mc^2}{e^2}\right) \left[\frac{(hv_1)^2 - (hv_2)^2}{mc^2}\right] \left(1 - \frac{1}{\beta^2 n^2}\right)_{av}. \tag{2.20}$$

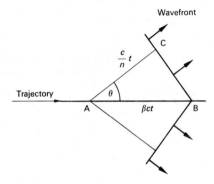

Figure 2.17

For a particle of $z = 1$ and $\beta \simeq 1$ in water ($n = 1.33$) this expression gives $dE/dx = 400\,\text{eV}\,\text{cm}^{-1}$ for visible light ($\lambda = 400$–$700\,\text{nm}$) and thus 200 photons cm^{-1}. Note that this is small compared with the total energy loss of about $2\,\text{MeV}\,\text{cm}^{-1}$.

The usefulness of the Čerenkov effect lies in the fact that measurement of the angle in (2.18) provides a direct measurement of the velocity βc. Table 2.4 gives a list of some radiating media, demonstrating that most of the range of γ-values from 1.2 to 100 can be covered by means of solids, liquids, gases, or aerogels.

Threshold Čerenkov counters can be used to discriminate between two relativistic particles of the same momentum p and different masses m_1 and m_2, if the heavier, slower particle (m_2) is just below threshold. In this case, $\beta_2^2 = 1/n^2$ and it is straightforward to show that the production rate of photons from the particle m_1 is given by the previous equation with

$$\sin^2 \theta_1 = 1 - \frac{1}{\beta_1^2 n^2} \simeq \frac{m_2^2 - m_1^2}{p^2}. \tag{2.21}$$

Thus the length of radiator to produce a given number of photoelectrons (assuming one with the right refractive index can be found) increases as the square of the momentum. For high-momentum beams, large, pressurized gas radiators several meters long may be required.

In the differential Čerenkov counter, the angle of Čerenkov emission is measured in order to identify particles. The cone of light from the radiating particle is focused by a lens or spherical mirror into a ring image, and an adjustable diaphragm at the focus transmits the light to a phototube. Differential counters with velocity resolution $\Delta\beta/\beta \simeq 10^{-7}$ have been built, and separation of charged pions, kaons, and protons up to several hundred GeV/c is achievable. As an alternative to changing the radius of the diaphragm, this may be fixed and a velocity scan carried out by varying the gas pressure. By integrating the photomultiplier output over time, such a counter may be used to measure secondary particle yields in high-intensity beams.

TABLE 2.4 Cerenkov radiators

Medium	$n - 1$	γ (threshold)
Helium (NTP)	3.3×10^{-5}	123
CO_2 (NTP)	4.3×10^{-4}	34
Pentane (NTP)	1.7×10^{-3}	17.2
Aerogel	$0.075 \rightarrow 0.025$	$2.7 \rightarrow 4.5$
H_2O	0.33	1.52
Glass	$0.75 \rightarrow 0.46$	$1.22 \rightarrow 1.37$

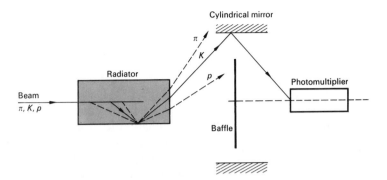

Figure 2.18 Early design of differential Čerenkov counter. The arrangement is intended to select light from one of three components of the beam (*K*-mesons, in the case shown).

It is also possible to dispense with photomultipliers for light collection. In one recent design, photons emitted in the Čerenkov cone from a liquid radiator traverse a suitable gas, and the electrons liberated by photoionization are then drifted in an electric field onto a plane multiwire proportional chamber, so that a ring image (with diameter determined by particle velocity) of electronic signals is found on the anode-cathode plane.

2.5. SHOWER DETECTORS AND CALORIMETERS

The energy and position coordinates of secondaries from high-energy interactions can also, under suitable conditions, be measured by total-absorption methods. In the absorption process, the incident particle interacts in a large detector mass, generating secondary particles which in turn generate tertiary particles, and so on, so that all (or most) of the incident energy appears as ionization or excitation in the medium—hence the term "calorimeter". Such devices are essential in recording the energy of neutral hadrons; and since the fractional energy resolution varies as $E^{-1/2}$, calorimeters provide, even for charged hadrons, a precision at high energies (10–100 GeV) comparable with or better than what can be achieved by magnetic deflection. Just as important is the fact that total-absorption calorimeters provide fast (100-ns) "total energy" signals useful for quick decisions on event selection.

2.5.1. Electromagnetic Shower Detectors

For electrons and photons of high energy, a dramatic result of the combined phenomena of bremsstrahlung and pair production is the occurrence of cascade showers. A parent electron will radiate photons, which convert to pairs, which radiate and produce fresh pairs in turn, the number of particles increasing exponentially with depth in the medium. The development of such a shower can be discussed according to the following very simplified model. Starting off with a primary electron of energy E_0, suppose that, in traversing one radiation length, it radiates half its energy, $E_0/2$, as one photon. Assume that, in the next radiation length, the photon converts to a pair, the electron and positron each receiving half the energy (that is, $E_0/4$), and that the original electron radiates a further photon carrying half the remaining energy, $E_0/4$. Thus, after two radiation lengths, we shall have a photon of energy $E_0/4$ and two electrons and one positron, each of $E_0/4$. By proceeding in this way, it is easily seen that after t radiation lengths, there will be $N = 2^t$ particles, with photons, electrons, and positrons approximately equal in number. We have here neglected ionization loss and the dependence of radiation and pair-production cross-sections on energy. The energy per particle at depth t will then be $E(t) = E_0/2^t$. This process continues until $E(t) = E_c$, when we suppose that ionization loss suddenly becomes important and no further radiation is possible. The shower will thus reach a maximum and then cease abruptly. The maximum will occur at

$$t = t_{max} = \frac{\ln(E_0/E_c)}{\ln 2}, \qquad (2.22)$$

the number of particles at the maximum being

$$N_{max} = \exp[t_{max} \ln 2] = \frac{E_0}{E_c}. \qquad (2.23)$$

The number of particles of energy exceeding E will be

$$N(>E) = \int_0^{t(E)} N \, dt = \int_0^{t(E)} e^{t\ln 2} \, dt$$

$$\simeq \frac{e^{t(E)\ln 2}}{\ln 2} = \frac{E_0/E}{\ln 2},$$

where $t(E)$ is the depth at which the particle energy has fallen to E. Thus, the differential energy spectrum of particles $dN/dE \propto 1/E^2$. The total integral track length of *charged* particles (in radiation lengths) in the whole shower will be

$$L = \frac{2}{3} \int_0^{t_{max}} N \, dt = \frac{2}{3 \ln 2} \frac{E_0}{E_c} \simeq \frac{E_0}{E_c}. \qquad (2.24)$$

The last result also follows from the definition of E_c and conservation of energy; nearly all the energy of the shower must eventually appear in the form of ionization loss of charged particles in the medium.

In practice, the development of a shower consists of an initial exponential rise, a broad maximum, and a gradual decline. Nevertheless, the above equations indicate correctly the main qualitative features, which are:

(a) a maximum at a depth increasing logarithmically with primary energy E_0;
(b) the number of shower particles at the maximum being proportional to E_0; and
(c) a total track-length integral being proportional to E_0.

The observed longitudinal development of a 6-GeV electron-initiated shower in different absorbers is given in Fig. 2.19 together with the prediction of a Monte Carlo program incorporating the known energy-dependent cross-sections for radiation and ionization loss by electrons and for absorption of photons by pair production and other processes.

Because of Coulomb scattering, a shower spreads out laterally. The radial spread is determined by the radiation length in the medium and the angular deflection per radiation length at the critical energy. In all materials, this spread is of order one Moliere unit $R_m = 21(X_0/E_c)$, with E_c in MeV (see Fig. 2.20).

Electromagnetic shower detectors are built from high-Z materials of small X_0, so as to contain the shower in a small volume. In lead-loaded glass (55% PbO, 45% SiO_2) detectors, the Čerenkov light from relativistic

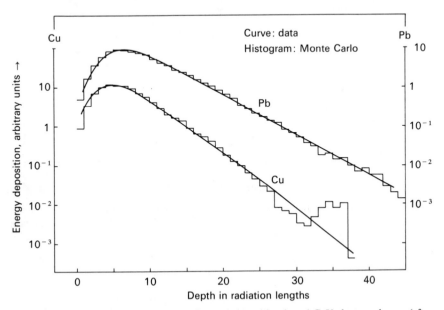

Figure 2.19 Longitudinal distribution of energy deposition in a 6-GeV electron shower (after Bathow et al. 1970).

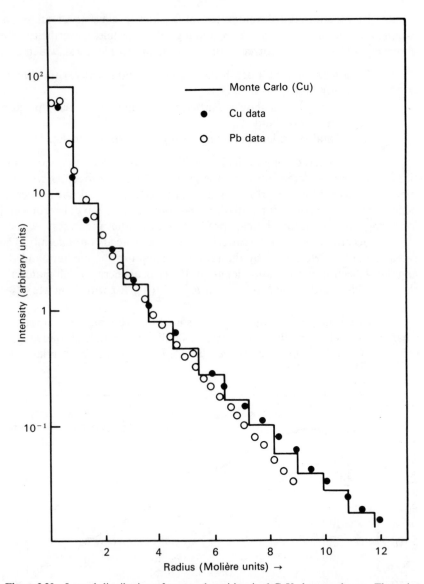

Figure 2.20 Lateral distribution of energy deposition in 6-GeV electron shower. The points indicate data, and the histogram the Monte Carlo predictions. The horizontal scale is in Molière units, $R_m = 21(X_0/E_c)$, with X_0 the radiation length and E_c the critical energy in MeV (after Bathow *et al.* 1970).

electrons is used to measure the shower energy. The resolution is typically $\Delta E/E = 0.05/\sqrt{E(\text{GeV})}$ and is determined by the fluctuation $\sqrt{N}$ in the number N of particles at maximum, where $N \propto E$. Calorimeters built from alternate sheets of lead and plastic scintillator are also used; the resolution depends on the sampling frequency but is comparable to that from lead glass.

2.4.2. Hadron-Shower Calorimeters

A hadron shower results when an incident hadron undergoes an inelastic nuclear collision with production of secondary hadrons, which again interact inelastically to produce a further hadron generation, and so on. The scale for longitudinal development is set by the nuclear absorption length λ, varying from $80 \, \text{g cm}^{-2}$ (C) through $130 \, \text{g cm}^{-2}$ (Fe) to $210 \, \text{g cm}^{-2}$ (Pb). This scale is big compared with the radiation length X_0 in heavy elements, so that, in comparison with electromagnetic shower detectors, hadron calorimeters are large. For an iron-scintillator sandwich, for example, the longitudinal and transverse dimensions are of order 2 m and 0.5 m, respectively.

In an electromagnetic cascade, the bulk of the incident energy appears eventually in the form of ionization. However, in a hadron cascade roughly 30% of the incident hadron energy is lost by the breakup of nuclei, nuclear excitation, and evaporation neutrons (and protons), and does not give an observable signal. One successful method of compensating for this is the use of ^{238}U as the cascade medium, the extra energy released by fast neutron and photon fission of ^{238}U making up for the "invisible" energy losses from nuclear breakup. The importance of obtaining this compensation is clear: In a hadronic cascade, containing both charged and neutral pions, which generate hadronic and electromagnetic cascades, respectively, the integrated scintillator pulse height should be proportional to the total incoming energy, irrespective of fluctuations in the neutral/charged pion ratio. Figure 2.21 shows results on the hadronic cascade development in a sandwich of iron and scintillator.

Typically the energy resolution of hadron calorimeters is $\Delta E/E \simeq 0.5/\sqrt{E}$ with E in GeV. Various types of hadron calorimeter are in current use. The iron-plus-scintillator sandwich has already been mentioned, but proportional tubes, flash tubes, and drift chambers have also been used for sampling.

2.4.3. Examples of Large Hybrid Detectors

Typical experiments in high-energy physics involve the simultaneous detection, measurement, and identification of many particles, both charged and neutral, from each interaction which occurs. They therefore usually incorporate several types of detection technique in a single detector array.

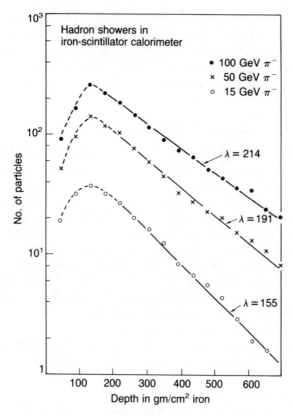

Figure 2.21 Average longitudinal development of hadronic showers due to negative pions in a sandwich array of 5-cm iron plates and scintillator. The ordinate gives the equivalent number of minimum ionizing particles at each depth. The rms fluctuation about the mean, in individual showers, is of order 100%. (After Holder *et al.* 1978.)

As one example, Fig. 2.22 shows the large spectrometer detector employed by the JADE collaboration at the e^+e^- storage ring PETRA at DESY, Hamburg. A vertical section containing the beam axis is shown. Charged particles from the beam intersection region are recorded by beam pipe counters, by an array of cylindrical drift chambers ("jet chambers"), and by time-of-flight counters. The drift chambers provide ionization information as well as track positions. All these are inside a solenoid magnet providing a field of 0.5 T parallel to the beam axis, extending over 3.5-m length by 2-m diameter. Outside the coil, lead-glass shower counters record electrons and photons, and the outermost layers of drift chambers ("muon chambers") identify and record muons by their penetration through iron and concrete absorber. Tagging counters placed upstream and downstream around the beam pipe record particles at small angles (see Fig. 2.23 for example of event).

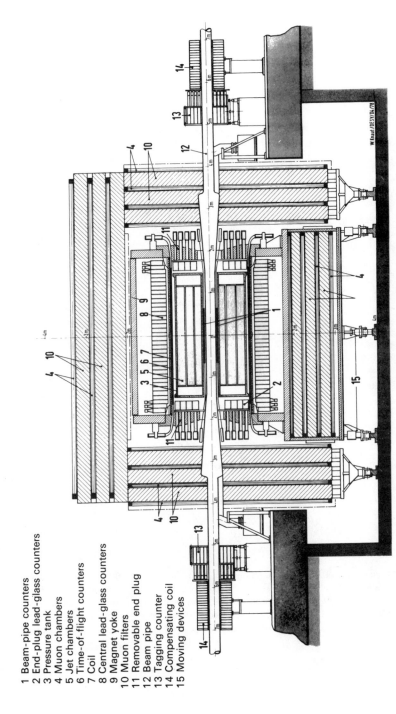

1 Beam-pipe counters
2 End-plug lead-glass counters
3 Pressure tank
4 Muon chambers
5 Jet chambers
6 Time-of-flight counters
7 Coil
8 Central lead-glass counters
9 Magnet yoke
10 Muon filters
11 Removable end plug
12 Beam pipe
13 Tagging counter
14 Compensating coil
15 Moving devices

Figure 2.22 The spectrometer detector JADE used at the PETRA e^+e^- storage ring at DESY, Hamburg. A vertical section through the beam pipe is shown. (Courtesy DESY.)

63

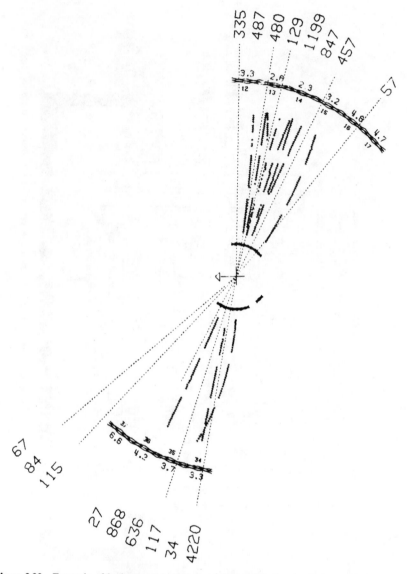

Figure 2.23 Example of hadron production in e^+e^- annihilation in PETRA at 30-GeV CMS energy. The computer reconstruction shows trajectories of charged particles as lines of crosses where they traverse drift chambers, while γ-rays from neutral-pion production and decay, recorded in lead-glass shower counters, are indicated by dotted lines. Projection in plane normal to beam pipe. The concentration of hadrons into two oppositely directed jets is well displayed. Event from JADE detector of Fig. 2.22.

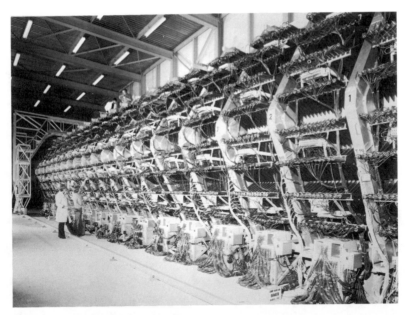

Figure 2-24 Magnetized calorimeter employed by CDHS collaboration in study of neutrino interactions at the CERN SPS. It is instrumented with magnetized iron toroids, scintillation counters, and drift chambers, for the identification and momentum measurement of secondary muons, and to measure the nuclear cascade energy of secondary hadrons. Total mass is 1400 tons. (Photograph courtesy CERN.)

Figure 2.24 shows the neutrino detector used by the CDHS collaboration at the CERN SPS. It is essentially a magnetized calorimeter of mass 1400 tons, and consists of a succession of iron plates 3.5 m in diameter and 15 cm thick, separated by scintillation counters and drift chambers. The latter record the trajectories of charged secondaries of neutrino reactions, while the integrated scintillator pulse height indicates the hadron (shower) energy. High-energy muon secondaries are identified by their penetration, and their momentum is determined from their deflection in the magnetized iron. Another example of a large detector is the UA1 calorimeter (see Fig. 7.18) employed in the observation of the W-and Z^0-bosons of the weak interactions.

PROBLEMS

2.1. The average number $\bar{n}$ of ionizing collisions suffered by a fast particle of charge ze in traversing an interval dx ($\mathrm{g\,cm^{-2}}$) of a medium, and resulting in energy transfers $E' \to E' + dE'$, is

$$\bar{n} = f(E')\,dE'\,dx = \frac{2\pi z^2 e^4 N_0 Z}{mv^2 A}\frac{dE'}{(E')^2}\left(1 - \frac{v^2}{c^2}\frac{E'}{E'_{\max}}\right)dx,$$

where the symbols are as in Eq. (2.7) and the maximum transferable energy is $E'_{max} = 2mv^2/(1 - \beta^2)$, with $\beta = v/c$. For individual particles, the distribution in number of collisions n follows the Poisson law, so that $\langle (n - \bar{n})^2 \rangle = \bar{n}$. If we multiply the above equation by $(E')^2$ and integrate, we obtain the mean squared deviation in energy loss, $\varepsilon^2 = \langle (\Delta E - \overline{\Delta E})^2 \rangle$, about the mean $\overline{\Delta E}$. Show that

$$\varepsilon^2 = 0.6 \frac{Z}{A} (mc^2)^2 \gamma^2 \left(1 - \frac{\beta^2}{2}\right) \Delta x.$$

Calculate the fractional rms deviation $\varepsilon/\overline{\Delta E}$ in energy loss for protons of kinetic energy 500 MeV traversing (a) 0.1, (b) 1.0, and (c) 10 g cm^{-2} of plastic scintillator $(Z/A = \frac{1}{2})$. Take dE/dx as 3 MeV g^{-1} cm^2.

2.2. A narrow pencil beam of singly charged particles of momentum p, traveling along the x-axis, traverses a slab of material s radiation lengths in thickness. If ionization loss in the slab may be neglected, calculate the rms lateral spread of the beam in the y-direction, as it emerges from the slab. [*Hint*: Consider an element of slab of thickness dx at depth x, and find the contribution $(dy)^2$ which this element makes to the mean squared lateral deflection; then integrate over the slab thickness.] Use the formula you derive to compute the rms lateral spread of a beam of 10-GeV/c muons in traversing a 100-m pipe filled with (a) air, (b) helium, at NTP.

2.3. Extensive air showers in cosmic rays contain a "soft" component of electrons and photons, and a "hard" component of muons. Suppose the central core of a shower, at sea level, contains a narrow, vertical, parallel beam of muons of energy 1000 GeV, which penetrate underground. Assume the ionization loss in rock is constant at 2 MeV g^{-1} cm^2. Find the depth in rock at which the muons come to rest, assuming the rock density to be 3.0. Using the formula of the preceding problem, estimate their radial spread in meters, taking account of the change in energy of the muons as they traverse the rock. (Radiation length in rock = 25 g cm^{-2}.)

2.4. Show that in the head-on collision of a beam of relativistic particles of energy E_1 with one of energy E_2, the square of the energy in the center-of-momentum frame is $4E_1 E_2$, and that for a crossing angle θ between the beams, this is reduced by a factor $(1 + \cos\theta)/2$. Show that the available kinetic energy in the head-on collision of two 25-GeV protons is equal to that in the collision of a 1300-GeV proton with a stationary nucleon.

2.5. A high-energy electron collides with an atomic electron. What is the threshold energy for production of an $e^+ e^-$ pair?

2.6. A proton of momentum **p**, large compared with its rest mass M, collides with a proton inside a target nucleus, with Fermi momentum $\mathbf{p}_f$. Find the available kinetic energy in the collision, as compared with that for a free-nucleon target, when $\mathbf{p}$ and $\mathbf{p}_f$ are (a) parallel, (b) antiparallel, (c) orthogonal.

2.7. It is sometimes possible to differentiate between the tracks due to relativistic pions, protons, and kaons in a bubble chamber by virtue of the high-energy δ-rays which are produced. For a pion of momentum 5 GeV/c, what is the minimum energy of a δ-ray which must be observed to prove it is not produced by a kaon or proton? What is the probability of observing such a knock-on electron in 1 m of liquid hydrogen (density 0.06)? Refer to Problem 2.1 for formulae.

2.8. An experiment on proton decay is to be carried out using a large cubical tank of water as the proton source, and the possible decay mode $p \to e^+ + \pi^0$ is to be detected by the Čerenkov light emitted when the electromagnetic showers from the decay products traverse the water. How big should the water tank be in order to contain such showers? Estimate the total track-length integral (TLI) of the showers in a decay event and hence the total number of photons emitted in the visible region ($\lambda = 400$–700 nm). The light is to be detected by an array of photomultipliers placed at the water surfaces. If the optical transmission of the water is 20% and the photocathode efficiency is 15%, what fraction of the surface must be covered by photocathode to give an energy resolution of 10%?

BIBLIOGRAPHY

Allison, W. M., and J. H. Cobb, "Relativistic charged particle identification by energy loss," *Ann. Rev. Nucl. Part. Science* **30**, 253 (1980).

Bethe, A. A., and J. Ashkin, "Passage of radiations through matter", in Segre (ed.), *Experimental Nuclear Physics*, John Wiley, New York, 1953, Vol. 1, p. 166.

Blewett, M. H., "Characteristics of typical accelerators", *Ann. Rev. Nucl. Science* **17**, 427 (1967).

Bradner, H., "Bubble chambers", *Ann. Rev. Nucl. Science* **10**, 109 (1960).

Charpak, G., "Evolution of automatic spark chambers", *Ann. Rev. Nucl. Science* **20**, 195 (1970).

Charpak, G., and F. Sauli, "High resolution electronic particle detectors," *Ann. Rev. Nucl. Part. Science* **34**, 285 (1984).

Courant, E. D., "Accelerators for high intensities and high energies", *Ann. Rev. Nucl. Science* **18**, 435 (1968).

Fabjan, C. W., and H. G. Fischer, "Particle detectors", *Rep. Prog. Physics* **43**, 1003 (1980).

Fabjan, C. W., and T. Ludlam, "Calorimetry in high energy physics", *Ann. Rev. Nucl. Part. Science* **32**, 335 (1982).

Glaser, D., "The bubble chamber", in S. Fluegge (ed.), *Encyclopaedia of Physics*, Springer (Berlin), 1955, Vol. 45.

Hutchinson, G., "Čerenkov detectors", *Prog. Nucl. Phys.* **8**, 195 (1960).

Kleinknecht, K., "Particle detectors", Proc. Advanced Study Inst. on Techniques and Concepts in High Energy Physics, St Croix USVI (ed. T. Ferbel), 1980.

Kohaupt, R. D., and G. A. Voss, "Progress and performance of e^+e^- storage rings", *Ann. Rev. Nucl. Part. Science* **33**, 67 (1983).

Lawson, J. D., and M. Tigner, "The physics of particle accelerators", *Ann. Rev. Nucl. Part. Science* **34**, 99 (1984).

Litt, J., and P. Meunier, "Čerenkov counter techniques in high energy physics", *Ann. Rev. Nucl. Part. Science* **23**, 1 (1973).

Livingstone, M. S., and J. P. Blewett, *Particle Accelerators*, McGraw-Hill, New York, 1962.

McMillan, E. M., "Particle accelerators", in Segre (ed.), *Experimental Nuclear Physics*, John Wiley, New York, 1959, Vol. 3, p. 639.

Palmer, R., and A. V. Tollestrup, "Superconducting magnet technology for accelerators", *Ann. Rev. Nucl. Part. Science* **34**, 247 (1984).

Panofsky, W. K. H., "High energy physics horizons", *Physics Today*, **26**, (June 1973).

Pelligrini, C., "Colliding beam accelerators", *Ann, Rev, Nucl. Science* **22**, 1 (1972).

Price, W. J., *Nuclear Radiation Detection*, McGraw-Hill, New York, 1964.

Rossi, B., *High Energy Particles*, Prentice-Hall, New Jersey, 1952.

Sanford, J. R., "The Fermi National Accelerator Laboratory", *Ann. Rev. Nucl. Part. Science* **26**, 151 (1976).

Segre, E., *Nuclei and Particles*, Benjamin, New York, 1977, Chapters 2–4.

Sternheimer, R. M., "Interaction of radiation with matter", in Yuan and Wu (eds.), *Methods of Experimental Physics*, Academic Press, 1961, Vol. 5A, p. 1. (This volume also contains articles on the different types of particle detector.)

Van der Meer, S., "Stochastic cooling and the accumulation of antiprotons", *Rev. Mod. Phys.* **57**, 699 (1985).

CHAPTER 3

Invariance Principles and Conservation Laws

One of the most important concepts in physics is that of symmetry or invariance of the equations describing a system under an operation—which might be, for example, a translation or rotation in space. Intimately connected with such invariance properties are conservation laws—in the above cases, conservation of linear and angular momentum. A particular type of interaction is generally observed to obey many different conservation laws, so that the mathematical description of the interaction has to fulfil several invariance requirements, which severely restricts the possible forms of the interaction, and furthermore allows one for example to obtain relations between cross-sections for different processes mediated by that interaction. Thus, the conservation of isospin in strong interactions is equivalent to invariance under rotations in "isospin space", and leads to relations between cross-sections for the various possible charge states in pion-nucleon scattering.

The transformations to be considered can be either continuous or discrete. A translation or rotation in space is an example of a continuous transformation, while spatial reflection through the origin of coordinates (the parity operation) is a discrete transformation. The associated conservation laws in the two cases are additive and multiplicative, respectively.

In this chapter, we discuss the more important examples of conservation rules and invariance in particle physics. Invariance under space-time transformations in special relativity is discussed in Appendix A.

3.1. INVARIANCE AND OPERATORS IN QUANTUM MECHANICS

In quantum mechanics a system of particles is represented by a wavefunction or state vector ψ. The Schrödinger and Heisenberg equations of motion describe the temporal development of such a system. The result of a physical measurement on the system corresponds to the expectation value of some operator Q acting on the wavefunction and has the value $q = \int \psi^* Q \psi \, dV$. The development of q with time can be attributed either to the time dependence of ψ (Schrödinger representation) or equivalently to that of the operator Q (Heisenberg representation). For the former we write $\psi = \psi_s(t)$ in obvious notation. The Schrödinger equation of motion for ψ is

$$i\hbar \frac{\partial}{\partial t} \psi_s(t) = H\psi_s(t), \tag{3.1}$$

where H is the total energy, or Hamiltonian, operator: it gives the energy eigenvalues E of a stationary state, $H\psi = E\psi$. The time dependence $\psi_s(t)$ is clearly

$$\psi_s(t) = T(t, t_0)\psi_s(t_0), \tag{3.2}$$

with

$$T(t, t_0) = \exp[-i(t - t_0)H/\hbar] \tag{3.3}$$

and where it has been assumed that H does not depend on t explicitly. The operator T preserves the norm of the wavefunction and is unitary, with $T^{-1} \equiv T^* = \exp[i(t - t_0)H/\hbar]$ and $T^{-1}T = 1$. Thus, the complex conjugate wavefunction satisfies

$$\psi_s(t)^* = \psi_s(t_0)^* T^{-1}(t, t_0). \tag{3.4}$$

The Heisenberg description attributes the time dependence to the operator Q rather than to ψ. Now the physically observable expectation value must be the same in either picture, so that

$$q = \int \underbrace{\psi_s(t_0)^* Q \psi_s(t_0) \, dV}_{\text{Heisenberg}} \equiv \int \underbrace{\psi_s(t)^* Q_0 \psi_s(t) \, dV}_{\text{Schrödinger}},$$

or from (3.2) and (3.4)

$$\psi_s(t_0)^* Q \psi_s(t_0) = \psi_s(t_0)^* T^{-1} Q_0 T \psi_s(t_0),$$

or

$$Q = T^{-1} Q_0 T, \tag{3.5}$$

giving an expression for the time dependence of the operator Q. So

$$i\hbar \frac{dQ}{dt} = i\hbar \frac{dT^{-1}}{dt} Q_0 T + i\hbar T^{-1} Q_0 \frac{dT}{dt}$$

$$= -HT^{-1}Q_0 T + T^{-1}Q_0 TH$$

$$= -HQ + QH = [Q, H], \tag{3.6}$$

using (3.3) and (3.5). Here, $[Q, H]$ is the commutator of Q and H. If Q depends explicitly on the time, $\partial Q/\partial t \neq 0$, then (3.6) generalizes to

$$i\hbar \frac{dQ}{dt} = i\hbar \frac{\partial Q}{\partial t} + [Q, H], \tag{3.7}$$

which is the Heisenberg equation of motion for the operator Q. When $\partial Q/\partial t = 0$, we have $dQ/dt = 0$ if $[Q, H] = 0$. Thus an operator, not depending explicitly on the time, will be a constant of the motion it is commutes with the Hamiltonian operator. Generally, *conserved quantum numbers are associated with operators commuting with the Hamiltonian.*

3.2. TRANSLATIONS AND ROTATIONS

The transformations produced by quantum-mechanical operators are broadly of two types: continuous transformations (e.g., translations in space and time) and discrete transformations (e.g., spatial inversion through the origin). As an example of a continuous transformation we first discuss space translations. The effect of an infinitesimal translation δr in space on a wavefunction ψ will be

$$\psi' = \psi(r + \delta r) = \psi(r) + \delta r \frac{\partial \psi(r)}{\partial r} = \left(1 + \delta r \frac{\partial}{\partial r}\right)\psi = D\psi,$$

where

$$D = \left(1 + \delta r \frac{\partial}{\partial r}\right) \tag{3.8}$$

is an infinitesimal space translation operator. Since the momentum operator is $p = i\hbar \, \partial/\partial r$, we can write this as

$$D = (1 + ip \, \delta r/\hbar). \tag{3.9}$$

A finite translation Δr can be obtained by making n steps in succession ($\Delta r = n \, \delta r$), giving

$$D = \lim_{n \to \infty} \left(1 + \frac{i}{\hbar} p \, \delta r\right)^n = \exp\left(\frac{i}{\hbar} p \, \Delta r\right) \tag{3.10}$$

Thus D is a unitary operator, with $D^*D = D^{-1}D = 1$. The momentum operator p is called the generator of the operator D of space translations.

If the Hamiltonian H is independent of such space translations, then $[D, H] = 0$. From the form (3.9) it is clear that if D commutes with H, so also does the generator p:

$$[p, H] = 0. \tag{3.11}$$

Thus, if the Hamiltonian is invariant under space translations, then the momentum operator p (which generates these translations) commutes with the Hamiltonian, and the expectation value of p (the momentum of the system) is conserved. So the following statements are all equivalent:

(a) Momentum is conserved in an isolated system.
(b) The Hamiltonian is invariant under space translations.
(c) The momentum operator commutes with the Hamiltonian.

In complete analogy with the operator D of space translations (3.8), the generator of infinitesimal rotations about some axis may be written

$$R = 1 + \delta\phi \, \frac{\partial}{\partial\phi}. \tag{3.12}$$

The operator of the z-component of angular momentum is (see Appendix C)

$$J_z = -i\hbar\left(x\,\frac{\partial}{\partial y} - y\,\frac{\partial}{\partial x}\right) = -i\hbar\,\frac{\partial}{\partial\phi}, \tag{3.13}$$

where ϕ measures the azimuthal angle about the z-axis. So

$$R = 1 + \frac{i}{\hbar} J_z \, \delta\phi.$$

A finite rotation $\Delta\phi$ is obtained by repeating the infinitesimal rotation n times: $\Delta\phi = n\,\delta\phi$, where $n \to \infty$ as $\delta\phi \to 0$. Then

$$R = \lim_{n\to\infty}\left(1 + \frac{i}{\hbar} J_z\,\delta\phi\right)^n = \exp\left(\frac{i}{\hbar} J_z\,\Delta\phi\right). \tag{3.14}$$

Conservation of angular momentum about an axis corresponds to invariance of the Hamiltonian under rotations about that axis, and is also expressed by the commutation relation $[J_z, H] = 0$.

3.3. PARITY

The operation of spatial inversion of coordinates $(x, y, z \to -x, -y, -z)$ is an example of a discrete transformation. This transformation is produced by the parity operator P, where

$$P\psi(\mathbf{r}) \to \psi(-\mathbf{r}).$$

Repetition of this operation clearly implies $P^2 = 1$, so that P is a unitary operator. The eigenvalue of the operator (if there is one) will be ± 1, and this is also called the parity P of the system. A wavefunction may or may not have a well-defined parity, which can be even $(P = +1)$ or odd $(P = -1)$. For example, for

$$\psi = \cos x, \quad P\psi \rightarrow \cos(-x) = \cos x = +\psi: \quad \text{even} \quad (P = +1),$$
$$\psi = \sin x, \quad P\psi \rightarrow \sin(-x) = -\sin x = -\psi: \quad \text{odd} \quad (P = -1),$$

while for

$$\psi = \cos x + \sin x, \quad P\psi \rightarrow \cos x - \sin x \neq \pm \psi,$$

so the last function has no definite parity eigenvalue. As usual, the parity of a system will be a conserved quantum number if $[H, P] = 0$. For example, any spherically symmetric potential has the property that $H(-\mathbf{r}) = H(\mathbf{r}) = H(r)$, so that $[P, H] = 0$: the bound states of the system have definite parity. A familiar example is provided by the hydrogen-atom wavefunctions (neglecting spin effects), where the angular solutions are spherical harmonics (see Appendix, Table II):

$$\psi(r, \theta, \phi) = \chi(r) Y_l^m(\theta, \phi)$$

$$= \chi(r) \sqrt{\frac{(2l + 1)(l - m)!}{4\pi(l + m)!}} P_l^m(\cos \theta) e^{im\phi}. \tag{3.15}$$

The spatial inversion $\mathbf{r} \rightarrow -\mathbf{r}$ is equivalent to

$$\theta \rightarrow \pi - \theta,$$

$$\phi \rightarrow \pi + \phi,$$

as in Fig. 3.1, with the result

$$e^{im\phi} \rightarrow e^{im(\pi + \phi)} = (-1)^m e^{im\phi},$$

$$P_l^m(\cos \theta) \rightarrow P_l^m(\cos(\pi - \theta)) = (-1)^{l + m} P_l^m(\cos \theta).$$

or

$$Y_l^m(\theta, \phi) \rightarrow Y_l^m(\pi - \theta, \pi + \phi) = (-1)^l Y_l^m(\theta, \phi). \tag{3.16}$$

Thus, the spherical harmonic functions have parity $(-1)^l$. So $s, d, g, \ldots$ atomic states have even parity, while $p, f, h, \ldots$ have odd parity. Electric dipole transitions between states are characterized by the selection rule $\Delta l = \pm 1$, so that, as a result of the transition, the parity of the atomic state must change. The parity of the electromagnetic $(E1)$ radiation (photons) emitted in this case must be -1, so that the parity of the whole system (atom + photon) is conserved.

Parity is a multiplicative quantum number, so that the parity of a composite system $\psi = \phi_a \phi_b \cdots$ is equal to the product of the parities of the parts.

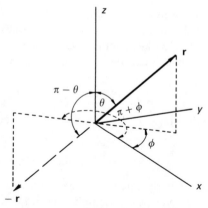

Figure 3.1

In strong as well as electromagnetic interactions, parity is conserved. This is true, for example, in the reaction $p + p \to \pi^+ + p + n$ in which a single boson (pion) is created. In such a case, it is necessary to assign an *intrinsic parity* to the pion in order to ensure the same parity in initial and final states, in just the same way that one has assigned a charge to the pion in order to ensure charge conservation in the same reaction.

3.3.1. Pion Parity

As will be shown in Chapter 4, the pion is a boson with zero intrinsic spin ($J = 0$). Using this fact and the observation that the absorption of slow negative pions in deuterium leads to the following reactions:

$$\pi^- + d \to n + n \tag{3.17}$$

$$\to 2n + \gamma, \tag{3.18}$$

we can prove that the pion must be assigned odd parity. This follows from the fact that capture of the pion by the deuteron takes place from an atomic S-state, as is proved by studies of mesonic X-rays as well as by direct calculations. Since $s_d = 1$ and $s_\pi = 0$, the total angular momentum is then $J = 1$ on either side of the reaction. Here $\mathbf{J} = \mathbf{L} + \mathbf{S}$, where L is the orbital angular momentum of the two final-state neutrons, and S their total spin. As indicated in (3.40) below, the triplet spin state of $S = 1$ is symmetric, while the singlet state $S = 0$ is antisymmetric. Hence the overall symmetry under interchange of the two identical neutrons will be $(-1)^{L+S+1}$, and this must be negative, from (1.2), so that $L + S$ is even. The overall angular momentum $J = 1$ requires either $L = 0$, $S = 1$, or $L = 1$, $S = 0$ or 1, or $L = 2$, $S = 1$. Of these, only $L = S = 1$ has $L + S$ even. So the only possibility is the $^3P_1$ state of two neutrons, with parity $(-1)^L = -1$, which requires negative parity also for the initial state.

Neutrons and protons are, by convention, assigned the same intrinsic parity $+ 1$. (Since baryon number is conserved, the actual assignment is immaterial; the nucleon parities cancel in any reaction.) Thus the deuteron parity is even, and negative parity must be assigned to the pion.

The parity of the neutral pion has been established from observations of the γ-ray polarization in the decay $\pi^0 \to 2\gamma$. Using arguments identical to those in Section 3.10 for the 2γ-decay of positronium, odd parity for the pion would predict plane polarization vectors (E-vectors) of the two photons to be preferentially orthogonal. Observations were actually made on the "double Dalitz decay"

$$\pi^0 \to (e^+ + e^-) + (e^+ + e^-),$$

in which each photon internally converts to a pair. The branching ratio, as compared with $\pi^0 \to 2\gamma$, is $\alpha^2 \sim 10^{-4}$. Since the plane of each electron-positron pair lies predominantly in the plane of the E-vector, measurement of the angular distribution between the plane of the pairs allows one to demonstrate the odd parity of the neutral pion (see Fig. 3.2).

In summary, the pion has zero spin and odd intrinsic parity, that is, $J^P = 0^-$. It is represented by a wavefunction which has the space transformation properties under inversions and rotations of a pseudoscalar. Such mesons are therefore called *pseudoscalar* mesons. By similar reasoning, particles of $J^P = 0^+$ are referred to as *scalar*, $J^P = 1^-$ as *vector*, and $J^P = 1^+$ as *axial-vector*.

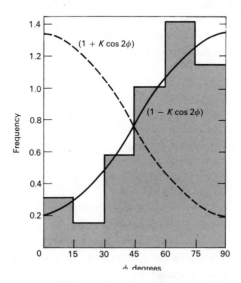

Figure 3.2 Plot of weighted frequency distribution of angle ϕ between planes of polarization of the pairs in the "double Dalitz decay," $\pi^0 \to (e^+ + e^-) + (e^+ + e^-)$. For a scalar π^0, this should have the form $1 + K \cos 2\phi$, and for a pseudoscalar π^0, $1 - K \cos 2\phi$. (After Plano *et al.* 1959.)

3.4. PARITY OF PARTICLES AND ANTIPARTICLES

While the intrinsic parity of the proton is a matter of convention, the relative parity of proton and antiproton (or of any fermion and antifermion) is not. For example, one can produce a proton-antiproton pair in the reaction

$$p + p \rightarrow p + p + (p + \bar{p}),$$

so that, just in the case of production of a single pion, the intrinsic parity of a nucleon-antinucleon pair is a measurable quantity. The Dirac theory of relativistic fermions (see Appendix D) predicts *opposite parity for fermion and antifermion*. This prediction has been verified experimentally, for example by observations on positronium (Section 3.10). For bosons, on the contrary, particles and antiparticles have the same intrinsic parity.

　　While the intrinsic parity assignment for the pion arises because pions can be created or destroyed singly, strange mesons must be created in association, for example in a reaction such as $p + p \rightarrow K^+ + \Lambda + p$. Thus, only the parity of a ΛK pair (relative to the nucleon) can be measured, and it is found to be odd. By convention, the Λ-hyperon is assigned the same (even) parity as the nucleon, so that of the kaon is odd.

3.5. TESTS OF PARITY CONSERVATION

While the purely strong and electromagnetic interactions are observed to be parity-conserving, the weak interactions are not. In such interactions, the matrix elements contain superpositions of amplitudes of even and odd parities. Thus nuclear β-decay is described by the so-called $V - A$ theory, in which the odd and even parity amplitudes have approximately the same magnitude. This is called the principal of maximal parity violation. A fuller discussion is given in Chapter 7.

　　In experimental studies of both strong and electromagnetic interactions, tiny degrees of parity violation are in fact observed. These arise, not from the breakdown of parity conservation in these interactions as such, but because the Hamiltonian describing the interaction inevitably contains contributions as well from the weak interactions between the particles concerned:

$$H = H_{\text{strong}} + H_{\text{electromagnetic}} + H_{\text{weak}}. \tag{3.19}$$

　　In nuclear transitions, the degree of parity violation will obviously be of the order of the ratio of weak to strong couplings, that is, $\sim 10^{-7}$ typically. As one example, we quote the observation of a fore-aft asymmetry in the γ-decay of polarized ^{19}F nuclei:

$$\begin{array}{ccc} ^{19}\text{F*} & & ^{19}\text{F} \\ J^P = \frac{1}{2}^- & \rightarrow & J^P = \frac{1}{2}^+ \end{array} + \gamma(110\,\text{keV}), \tag{3.20}$$

where there is parity mixing between the states. The observed asymmetry, $\Delta = -(18 \pm 9) \times 10^{-5}$, is in good accord with that expected from weak neutral-current effects (Adelberger *et al.* 1975). As a second example, we cite the α-decay of the excited state of ^{16}O at 8.87 MeV:

$$
\begin{array}{ccc}
^{16}O^* & & ^{12}C + \alpha, \\
J^P = 2^- & \rightarrow & J^P = 2^+
\end{array}
\tag{3.21}
$$

where the initial state is known to have odd parity, and the final state, even parity. The extremely narrow partial width for this decay, $\Gamma_\alpha = (1.0 \pm 0.3) \times 10^{-10}$ eV (Neubeck *et al.* (1974)), is consistent with the magnitude expected from the parity-violating (weak-interaction) contribution and may be contrasted with the width for γ-decay, $^{16}O^* \rightarrow {}^{16}O + \gamma$, of 3×10^{-3} eV. For more examples of parity violation in nuclear reactions, see the review of Tadic (1980).

Very small parity-violating effects have also been observed in atomic transitions. For example, a small (10^{-7} radian) rotation is observed to the plane of polarization of light traversing, and inducing optical transitions in, bismuth vapor. The parity violation arises through the interference between the weak-neutral-current and the purely electromagnetic contributions to the transition amplitude, and its magnitude is roughly in accord with that expected from the neutral-current couplings in the Weinberg-Salam-Glashow theory. For a recent review, see Fortson and Lewis (1984). Neutral-current effects are discussed in detail in Chapter 9.

3.6. CHARGE CONSERVATION, GAUGE INVARIANCE, AND PHOTONS

An important class of continuous transformations in particle physics are the gauge transformations, which are connected with conservation of electric charge. Electric charge is known to be very accurately conserved. For example, in the decay of the neutron, $n \rightarrow p + e^- + \bar{\nu}$, charge is conserved to an accuracy of better than 1 part in 10^{19} (see Problem 1.7).

3.6.1. Charge Conservation and Gauge Invariance in Classical and Quantum Physics

The concept of gauge invariance and charge conservation may be introduced using an argument of Wigner (1949). In electrostatics, the potential ϕ of a system is arbitrary. The equations are always concerned with *changes* of potential, and independent of the absolute value of ϕ at any point in space. Suppose now that charge is not conserved, that it can be created or destroyed by some magic process. To create a charge Q, say, will require work W, which can be recovered when the charge is destroyed. Let the charge be created at a point where the potential on the chosen scale is ϕ. The work done will be W, independent of ϕ, since by hypothesis no physical process can depend on the absolute potential scale. If the charge is now moved to a point where the potential is ϕ', the energy change will be $Q(\phi - \phi')$. Thus, when the charge is destroyed, we recover the original system but have gained a net energy $W - W + Q(\phi - \phi')$. So, conservation of energy implies that we cannot create or destroy charge if the scale of electrostatic potential is arbitrary. In other words, conservation of electric charge allows us the freedom to choose potential scales at will.

In quantum mechanics, the link between gauge invariance and charge conservation can be illustrated by a simple example. Figure 3.3 depicts a beam of low-energy electrons incident on a double slit AB and detected at the plane C, where one observes the characteristic diffraction pattern. The electron wave is described by a wavefunction of the form $\psi = \exp i(\mathbf{k} \cdot \mathbf{x} - \omega t)$, or, in units $\hbar = c = 1$, by

$$\psi = \exp i(\mathbf{p} \cdot \mathbf{x} - Et) = \exp(ipx), \qquad (3.22)$$

where $p = (\mathbf{p}, iE)$ is the 4-momentum of the electron and $x = (\mathbf{x}, it)$ is the space-time coordinate. In general we can include a constant phase factor, which—for reasons that will become clear later—we write as $-e\alpha$, where e is the electron charge:

$$\psi = \exp i(px - e\alpha). \qquad (3.23)$$

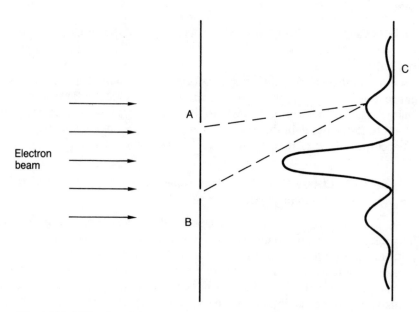

Figure 3.3   Diffraction of electron beam by a double slit AB, observed in the plane C.

The intensity at C depends on the *difference* of the phase from A and B, and is independent of the *global* (same over all space) phase $e\alpha$, which is arbitrary and unobservable. Suppose however that the phase is chosen differently at different points: $e\alpha = e\alpha(x)$. Then the space-time gradient of the overall phase is

$$\frac{\partial}{\partial x} i(px - e\alpha) = i\left(p - e\frac{\partial \alpha}{\partial x}\right) \tag{3.24}$$

Since α depends on x, the relative phase difference from A and B will change and the predicted pattern at C will be altered. So the expected physical result is not invariant under arbitrary *local* phase transformations. The electrons however are charged and they interact via an electromagnetic potential, which we write as the 4-vector $A = (\mathbf{A}, i\phi)$ where $\mathbf{A}$ is the vector and ϕ the scalar potential. The effect of the potential is to change the phase of an electron by the substitution in (3.22) of the form*

$$p \rightarrow p + eA, \tag{3.25}$$

so the derivative (3.24) now becomes

$$\frac{\partial}{\partial x} i(px + eAx - e\alpha) = i\left(p + eA - e\frac{\partial \alpha}{\partial x}\right). \tag{3.26}$$

The potential scale, or gauge, is also arbitrary and, as discussed later, one is free to change it—make a so-called local gauge transformation—by adding to A the gradient of any scalar function. Hence, if we make the transformation

$$A \rightarrow A + \frac{\partial \alpha}{\partial x}, \tag{3.27}$$

the derivative (3.26) becomes simply ip, independent of α. In other words, the effect of the original local phase transformation and the gauge transformation (3.27) cancel exactly, and the pattern at C is unchanged.

This cancellation depends on two things. First, there must exist a long-range field (the electromagnetic field in this example), acting on the electrons and changing their phases. It has to be long-range because our slit separation can be arbitrarily large. Second, the cancellation will only work if electric charge is conserved; it would not work if an electron were suddenly to lose, and later regain, its charge.

In summary, it is possible to describe charged particles by wavefunctions with phases chosen quite arbitrarily at different times and places, and this makes no difference to physical results, provided the charges are coupled to a long-range field (the electromagnetic field) to which are applied

---

* See, for example, J. D. Jackson, *Classical Electrodynamics* 2nd ed., John Wiley, New York, 1975, p. 254.

simultaneous local gauge transformations *and* provided the charge is conserved. Theories of particle interactions with this property of local gauge invariance have a high degree of symmetry and turn out to be renormalizable—that is, the predicted effects of the interaction, such as lifetimes or cross-sections, are finite and calculable to all energies and to all orders in the coupling constant.

In field theories with local gauge symmetry, an absolutely conserved quantity like electric charge implies the existence of a long-range field coupled to the charge. Similarly, if baryon number were absolutely conserved, there should exist a long-range field coupled to it (Lee and Yang (1955)). There is no evidence for such a field, and the limit on the coupling strength is less than 10^{-9} of the gravitational coupling (see, for example, Problem 9.5). For this reason, it is considered that protons might decay, albeit with a very long lifetime (see Chapter 9).

3.6.2. The Electromagnetic Field and Photons

In electromagnetism, the fields $\mathbf{B}$, $\mathbf{E}$ can be expressed as derivatives of vector and scalar potentials $\mathbf{A}$, ϕ, where

$$\mathbf{B} = \mathbf{V} \times \mathbf{A}, \tag{3.28}$$

$$\mathbf{E} = -\mathbf{V}\phi - \frac{1}{c}\frac{\partial \mathbf{A}}{\partial t}. \tag{3.29}$$

$\mathbf{B}$ and $\mathbf{E}$ satisfy Maxwell's equations, which incorporate the conservation of electric charge via a relation between charge and current densities: $\partial \rho / \partial t = -\mathbf{V} \cdot \mathbf{j}$. The values of $\mathbf{B}$ and $\mathbf{E}$ are invariant under a gauge transformation of the form

$$\mathbf{A} \rightarrow \mathbf{A}' = \mathbf{A} + \mathbf{V}\alpha, \tag{3.30}$$

$$\phi \rightarrow \phi' = \phi - \frac{1}{c}\frac{\partial \alpha}{\partial t}, \tag{3.31}$$

where α is *any* scalar function of space and time. This freedom to choose the value of α, called gauge invariance, means that $\mathbf{A}$, ϕ can be defined in various ways. For example, the Coulomb gauge is chosen to fulfil

$$\mathbf{V} \cdot \mathbf{A} = 0. \tag{3.32}$$

Using (3.29) and the Maxwell equation $\mathbf{V} \cdot \mathbf{E} = 4\pi\rho$, one obtains Poisson's equation

$$\nabla^2 \phi = -4\pi\rho,$$

so that in this gauge ϕ is determined just from the static charge distribution ρ.

In free space, the vector potential **A** obeys the wave equation

$$\nabla^2 \mathbf{A} - \frac{1}{c^2}\frac{\partial^2 \mathbf{A}}{\partial t^2} = 0, \tag{3.33}$$

corresponding to the propagation of free, massless photons. This has a plane-wave solution

$$\mathbf{A} = \mathbf{e}A_0 \exp i(\mathbf{k}\cdot\mathbf{r} - \omega t),$$

where **k** is the propagation vector and **e** is a unit vector (polarization vector) giving the direction of the E-field. The x-component of **A** is

$$A_x = e_x A_0 \exp i(\mathbf{k}\cdot\mathbf{r} - \omega t)$$
$$= e_x A_0 \exp i(k_x x + k_y y + k_z z - \omega t),$$

and

$$\frac{\partial A_x}{\partial x} = ie_x k_x A_0 \exp i(\mathbf{k}\cdot\mathbf{r} - \omega t),$$

with similar expressions for the y- and z-components. Using (3.32),

$$\mathbf{V}\cdot\mathbf{A} = \frac{\partial A_x}{\partial x} + \frac{\partial A_y}{\partial y} + \frac{\partial A_z}{\partial z} = 0,$$

we therefore obtain the relation in the Coulomb gauge

$$\mathbf{e}\cdot\mathbf{k} = 0. \tag{3.34}$$

Thus, the field **E** (and **B**) associated with a plane electromagnetic wave in free space is transverse to the propagation vector (Fig. 3.4). If one takes **k** along the z-axis, then e_z and A_z are zero, and one is left with two linearly independent components

$$A_x = e_x A_0 \exp i(kz - \omega t + \delta),$$

$$A_y = e_y A_0 \exp i(kz - \omega t),$$

where $e^2 = e_x^2 + e_y^2 = 1$, and where the amplitudes and the relative phase δ between the components are arbitrary. For example, if $\delta = 0$ one obtains plane polarization, and for $\delta = \frac{1}{2}\pi$ and $e_x = e_y$, circular polarization. The

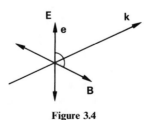

Figure 3.4

latter can be interpreted in terms of a rotating **e**-vector. For right and left circulatory polarized waves, the combinations are

$$e_R = \frac{1}{\sqrt{2}} (e_x + ie_y),$$

$$e_L = \frac{1}{\sqrt{2}} (e_x - ie_y). \tag{3.35}$$

Note that $|e_R|^2 = |e_L|^2$ for all values of e_x and e_y, as required by parity conservation. The polarization vectors of the classical field can be associated with the spin states of the free photons of the field. For an infinite plane wave, propagating in the z-direction, it is obvious that since the momentum components $p_x = p_y = 0$, the z-component of orbital angular momentum $L_z = xp_y - yp_x = 0$. Thus the total angular momentum about the z-axis, J_z, must refer to the spin of the associated photons.

Consider now a rotation through angle θ about the z-axis, which can be obtained via the rotation operator (3.14), $R = \exp(iJ_z\theta)$—where we have dropped the factor $\hbar$ for brevity. The transformed values of the components e_x, e_y, and e_z are

$$e'_x = e_x \cos \theta - e_y \sin \theta,$$

$$e'_y = e_x \sin \theta + e_y \cos \theta, \tag{3.36}$$

$$e'_z = e_z.$$

The first two equations can be rearranged as

$$e'_R = \frac{1}{\sqrt{2}} (e'_x + ie'_y) = \frac{1}{\sqrt{2}} (e_x + ie_y) \exp(i\theta)$$

$$e'_L = \frac{1}{\sqrt{2}} (e'_x - ie'_y) = \frac{1}{\sqrt{2}} (e_x - ie_y) \exp(-i\theta) \tag{3.37}$$

Thus, the states e_R, e_L, and e_z are eigenstates of the rotation operator $\exp(iJ_z\theta)$ with $J_z = +1$, -1, and 0, respectively, and correspond to the $2J + 1 = 3$ possible substates of a spin-1 photon. Because of the transversality condition (3.34), however, we have $e_z = 0$, and the $J_z = 0$ substrate of a free photon does not exist. In general, Lorentz invariance allows only two substates for a massless particle of spin $J: J_z = \pm J$.

Real photons travel at the speed of light and are therefore strictly massless (the experimental limit on the photon mass is $< 10^{-47}$ g (Goldhaber and Nieto 1971)). They are called *transverse* photons, meaning that the associated **E** and **B** fields are perpendicular to the propagation vector **k**. However, electromagnetic disturbances can also travel at $v < c$ (for example, in a waveguide), and both longitudinal and transverse field components are then possible. In such a situation, the photons cannot be exactly massless.

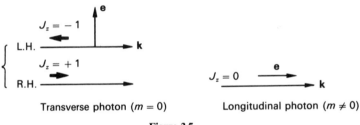

Figure 3.5

They have a spin component $J_z = 0$ as well as ± 1 (Fig. 3.5). A photon with $J_z = 0$ is called *longitudinal* (or scalar). An example of a longitudinal photon is the virtual photon mediating the static interaction between two charges, where the **e** vector lies along the line joining the charges.

The spin polarization of particles is often referred to by the term *helicity*. A particle with spin vector **σ**, and momentum **p**, has helicity

$$H = \frac{\boldsymbol{\sigma} \cdot \mathbf{p}}{|\boldsymbol{\sigma}||\mathbf{p}|} \tag{3.38}$$

Thus right-handed, left-handed, and scalar photons have $H = +1, -1, 0$ respectively. Since, under space inversion, $\boldsymbol{\sigma} \cdot \mathbf{p}$ changes sign (see Table 3.2), the net helicity of photons associated with the parity- conserving electromagnetic interactions must be zero: So right-handed and left-handed photons always occur with equal amplitudes, and the term "transverse" refers to both helicity states taken together.

To summarize this section: Gauge invariance is a property of electromagnetic fields associated with conservation of electric charge. It leads to the transversality condition for plane electromagnetic waves in free space. As a result, free photons—spin-1 carriers of a vector field—can exist only in two substates and must be massless. The masslessness of the photon follows from (1.7) and the fact that, as discussed above, the field must be of infinite range.

More complicated gauge transformations, of a type introduced by Yang and Mills (1954), are considered when both charged and neutral fields (bosons) are involved. This is the case for the unified gauge theories of electromagnetic and weak interactions, discussed in Chapter 9.

3.7. CHARGE-CONJUGATION INVARIANCE

As the name implies, the operation of charge conjugation reverses the sign of charge and magnetic moment of a particle (leaving all other coordinates unchanged). Symmetry under charge conjugation in classical physics is evidenced by the invariance of Maxwell's equations under change in sign of the charge and current density and also of **E** and **H**. In relativistic quantum mechanics the term "charge conjugation" also implies the interchange of particle and antiparticle. For baryons and leptons, a reversal of charge entails a change in sign of the baryon number or lepton number. As examples, we show in Table 3.1 the effects of charge conjugation on electron and proton. Note that μ positive (negative) means that the magnetic moment is parallel (antiparallel) to the spin vector.

Experimental evidence for invariance of the strong and electromagnetic interactions under the C-operation is discussed in Section 3.11. On the contrary, weak interactions violate charge-conjugation invariance just as they violate invariance under the parity operation. These effects are discussed in detail in Chapter 7 but it is perhaps worthwhile mentioning them briefly at this point. The noninvariance of weak interactions under the parity and charge-conjugation operations is exemplified by the longitudinal polarization of neutrinos (ν) and antineutrinos $(\bar{\nu})$ emitted in β-decay, in company with positrons and electrons respectively. Neutrinos have spin-$\frac{1}{2}$ and zero (or almost zero) mass, so that, as pointed out below (3.37), the possible spin eigenstates are of $J_z = \pm\frac{1}{2}$, where z denotes the direction of the momentum vector **p**. Experimentally, it is found that for neutrinos $J_z = -\frac{1}{2}$ only, and for anti-neutrinos $J_z = +\frac{1}{2}$ only. In other words, neutrinos are "left-handed" and antineutrinos "right-handed". This is shown in Fig. 3.6, where **p** is along the negative z-axis in (a). Under the parity operation, which in this case corresponds to the inversion $z \rightarrow -z$, the polar vector **p** changes sign, whereas, as Table 3.2 indicates, the axial vector **σ** remains unchanged. This situation is shown in diagram (b). It corresponds to a right-handed neutrino, which does not exist in nature. This tells us that the weak interaction is not invariant under space inversions. One can contrast this with photons emitted in electromagnetic processes. Both right-handed and left-handed photons

TABLE 3.1   Charge conjugation

	Proton	Antiproton
Q	$+e$	$-e$
B	$+1$	-1
μ	$+2.79(e\hbar/2Mc)$	$-2.79(e\hbar/2Mc)$
σ	$\frac{1}{2}\hbar$	$\frac{1}{2}\hbar$

	Electron	Positron
Q	$-e$	$+e$
L_e	$+1$	-1
μ	$-e\hbar/2mc$	$+e\hbar/2mc$
σ	$\frac{1}{2}\hbar$	$\frac{1}{2}\hbar$

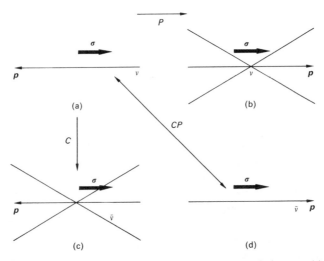

Figure 3.6 Results of the C- and P-operations on neutrino states. Only states (a) and (d) are observed in nature.

exist, but they occur with equal probability, so the interaction is invariant under a spatial inversion.

Similarly, if we apply the charge-conjugation operation to the ν-state (a), we obtain a left-handed $\bar{\nu}$, as in (c). This also does not exist. However, if, in addition, we make the spatial inversion of this state, we end up with a right-handed antineutrino (d), which *is* observed. Thus, the weak interactions are not invariant under C or P separately, but do exhibit CP-invariance. Although we come to this conclusion by considering neutrino states, it may be remarked that it is a general property of all weak interactions, whether they involve neutrinos or not.†

3.8. EIGENSTATES OF THE CHARGE-CONJUGATION OPERATOR

Consider the operation of charge conjugation performed on a charged-pion wave function, which we write as $|\pi^\pm\rangle$:

$$C|\pi^+\rangle \rightarrow |\pi^-\rangle \neq \pm|\pi^+\rangle.$$

In this operation, an arbitrary phase may enter; this is not important for the present discussion. We note that $|\pi^+\rangle$ and $|\pi^-\rangle$ are *not* C-eigenstates.

---

† The CP-invariance of weak interactions is not in fact exact. The small departures from CP-invariance are discussed in Chapter 7.

However, for a neutral system, the charge-conjugation operator may have a definite eigenvalue. Thus, for the neutral pion,

$$C|\pi^0\rangle = \eta|\pi^0\rangle$$

since the π^0 transforms into itself; η is a constant. Clearly, repeating the operation gives us $\eta^2 = 1$, so

$$C|\pi^0\rangle = \pm 1|\pi^0\rangle.$$

To find the sign, we note that electromagnetic fields are produced by moving charges (currents) which change sign under charge conjugation. As a consequence, the photon has $C = -1$. Since the charge-conjugation quantum number is multiplicative, this means that a system of n photons has C-eigenvalue $(-1)^n$. The neutral pion undergoes the decay

$$\pi^0 \to 2\gamma,$$

and thus has even C-parity:

$$C|\pi^0\rangle = +|\pi^0\rangle.$$

The decay

$$\pi^0 \to 3\gamma$$

will therefore be forbidden if the electromagnetic interactions are invariant under C. Experimentally, the branching ratio

$$\frac{\pi^0 \to 3\gamma}{\pi^0 \to 2\gamma} < 4 \times 10^{-7}.$$

3.9. POSITRONIUM DECAY

We now consider the restrictions imposed by C-invariance on the states of *positronium*, which undergoes annihilation in the modes

$$e^+e^- \to 2\gamma, 3\gamma.$$

The bound state of electron and positron possesses energy levels similar to the hydrogen atom (but with about half the spacing, because of the factor 2 in the reduced mass). We write down the total wave function of the positronium state as the product of three wave functions depending on the spin, space, and charge coordinates:

$$\psi(\text{total}) = \Phi(\text{space})\alpha(\text{spin})\chi(\text{charge}) \tag{3.39}$$

and consider how these functions behave under particle interchange. If electron and positron had been identical fermions, rather than fermion and antifermion, we should have been able to conclude that $\psi(\text{total})$ was antisymmetric, as in Eq. (1.1).

The spin functions for a combination of two spin-$\frac{1}{2}$ particles are written as follows. Denoting these as $\psi_1(s, s_z)$ and $\psi_2(s, s_z)$ for particles 1 and 2, and that of the combination by $\alpha(S, S_z)$ (where s, S refer to spins and s_z, S_z to third components), the four possible combinations are

$$\alpha(1, 1) = \psi_1(\tfrac{1}{2}, \tfrac{1}{2})\psi_2(\tfrac{1}{2}, \tfrac{1}{2}), \tag{3.40a}$$

$$\alpha(1, 0) = \frac{1}{\sqrt{2}} [\psi_1(\tfrac{1}{2}, \tfrac{1}{2})\psi_2(\tfrac{1}{2}, -\tfrac{1}{2}) + \psi_2(\tfrac{1}{2}, \tfrac{1}{2})\psi_1(\tfrac{1}{2}, -\tfrac{1}{2})], \tag{3.40b}$$

$$\alpha(1, -1) = \psi_1(\tfrac{1}{2}, -\tfrac{1}{2})\psi_2(\tfrac{1}{2}, -\tfrac{1}{2}), \tag{3.40c}$$

$$\alpha(0, 0) = \frac{1}{\sqrt{2}} [\psi_1(\tfrac{1}{2}, \tfrac{1}{2})\psi_2(\tfrac{1}{2}, -\tfrac{1}{2}) - \psi_2(\tfrac{1}{2}, \tfrac{1}{2})\psi_1(\tfrac{1}{2}, -\tfrac{1}{2})], \tag{3.40d}$$

where the first three form a spin triplet of $S = 1$ and $S_z = 1, 0, -1$, and the last is a singlet of $S = S_z = 0$. The triplet states have the property that they are *symmetric* under particle label interchange $1 \leftrightarrow 2$ (α does not change sign), while the singlet state is *antisymmetric* (α changes sign). Thus, the symmetry of the spin function α under particle interchange is $(-1)^{S+1}$, where S is the total spin.

The decomposition (3.40) can be formally expressed in group theory language. These states are representations of the group SU(2), where the letters stand for the special unitary group[‡] in two dimensions (corresponding to combining two basis states of spin-$\frac{1}{2}$). Symbolically this result is written as

$$2 \otimes 2 = 1 \oplus 3, \tag{3.41}$$

meaning that combining two of the fundamental "2" representations of SU(2), we obtain a "3" representation ($S = 1$ triplet) and a "1" representation ($S = 0$ singlet).

The space wavefunction Φ is expressed as a spherical harmonic $Y_l^m(\theta, \phi)$ as in Eq. (3.16). Particle interchange is equivalent to space inversion, introducing a factor $(-1)^l$, where l is the orbital angular momentum of the system.

Finally, let the charge wavefunction χ acquire a factor C under interchange. The product of the factors applying to the separate spin, space, and charge functions must then be that of the total wavefunction ψ, which we denote by K, so that

$$K = C(-1)^{S+1}(-1)^l. \tag{3.42}$$

To find C and K, we must appeal to additional information. As indicated before, two separate decay modes, into two and three γ-rays

---

‡ The matrix operator producing the transformations between members of the group has determinant $+1$ (making it special) and is unitary, preserving the norm of the wavefunction.

respectively, are observed for positronium annihilation from the ground state ($l = 0$). These modes obviously must correspond to the two possible spin states, the singlet ($J = 0$) and triplet ($J = 1$). By appealing either to the Bose symmetry of the two-photon system or to the masslessness of the photon (Fig. 3.5), we know that the 2γ-decay must have $J = 0$, so the 3γ-decay has to be assigned $J = 1$. Thus, from the formula $C = (-1)^n$ for a system of n photons, we find $C = +1$ for the $J = 0$ state and $C = -1$ for the $J = 1$ state. Inserting the appropriate factors in (3.36), we then obtain

Decay	$S = J$	l	C	K	Lifetime, s
Singlet ($^1S_0$) 2γ	0	0	$+1$	-1	1.25×10^{-10}
Triplet ($^3S_1$) 3γ	1	0	-1	-1	1.4×10^{-7}

We note that the factor acquired by the total wavefunction ψ is $K = -1$ under interchange; i.e., if we take account of the charge-conjugation operation, the total wavefunction is antisymmetric, just as for two identical fermions.

The annihilation rate to two γ-rays, expressed as a width $\Gamma = \hbar/\tau$, where τ is the lifetime, is calculated to be

$$\Gamma(2\gamma) = 4\pi r_e^2 \hbar c |\psi(0)|^2, \tag{3.43}$$

where $r_e = e^2/4\pi mc^2$ is the classical electron radius, $\psi(0)$ is the amplitude of the electron-positron radial wavefunction at the origin. From the solution of the Schrödinger equation for the ground state of the hydrogen atom, we know that

$$|\psi(0)|^2 = \frac{1}{\pi a^3}, \tag{3.44}$$

where a is the Bohr radius. Remembering the factor 2 for the reduced mass effect, the Bohr radius in positronium will have a value

$$a = \frac{2r_e}{\alpha^2}, \tag{3.45}$$

so

$$\Gamma(2\gamma) = \tfrac{1}{2}mc^2\alpha^5$$

and

$$\tag{3.46}$$

$$\tau(2\gamma) = 1.252 \times 10^{-10} \text{ s},$$

compared with the observed value of $(1.252 \pm 0.017) \times 10^{-10}$ s. For the 3γ-decay, the annihilation rate is slower by a factor of order α, with

$$\Gamma(3\gamma) = \left(\frac{2}{9\pi}\right)(\pi^2 - 9)\alpha^6 mc^2,$$

$$\tau(3\gamma) = 1.374 \times 10^{-7} \text{ s} \qquad \text{(observed } 1.377 \pm 0.004\text{).}$$

$$\tag{3.47}$$

Both the 3γ- and 2γ-periods were first detected in the work of Deutsch (1953), who measured the annihilation rate of positrons stopping in gases. Calculated and observed values are in excellent agreement. For comparison with the energy levels of the heavy quark-antiquark systems, see Chapter 5.

3.10. PHOTON POLARIZATION IN POSITRONIUM DECAY

To complete the discussion of the quantum numbers of positronium, we discuss the polarization of the two γ-rays emitted in the decay of the singlet state ($J = 0$). The electron and positron are fermion and antifermion, and as such must have opposite intrinsic parities (see Appendix D). The relative parity of e^+ and e^- will not affect the foregoing analysis, since under particle interchange, the total parity stays the same. However, it does imply that the two γ-rays emitted must have odd parity.

Let $\mathbf{k}$, $-\mathbf{k}$ be the momentum vectors of the two photons, and $\mathbf{e}_1$ and $\mathbf{e}_2$ their polarization vectors (E-vectors). The initial state has $J = 0$, and the simplest linear combinations one can form to include both E-vectors and satisfy requirements of exchange symmetry for identical bosons are

$$\psi(2\gamma) = a(\mathbf{e}_1 \cdot \mathbf{e}_2) \tag{3.48a}$$

and

$$\psi(2\gamma) = b(\mathbf{e}_1 \times \mathbf{e}_2) \cdot \mathbf{k}, \tag{3.48b}$$

where a and b are constants. The first product is a scalar and is therefore even under space inversions. The second is a pseudoscalar (the product of a polar vector with an axial vector) and therefore has the odd parity required. If $\psi(2\gamma)$ is to be finite, $\mathbf{e}_1$ and $\mathbf{e}_2$ cannot be parallel. Thus the planes of polarization of the two photons should be preferentially at right angles, the probability of observing an angle ϕ between the planes being from (3.48b), $|\psi|^2 \propto \sin^2 \phi$. Experimentally, this can be investigated by observing the angular distribution of the Compton scattering of the γ-rays, which depends strongly on polarization, being more probable in a plane normal to the electric vector. This is clear if one considers that the incident photon sets up oscillations of the target electron in the direction of the E-vector. In the subsequent dipole radiation of the scattered photon, the radiated intensity is greatest normal to E.

The experimental setup used by Wu and Shaknov (1950) is shown in Fig. 3.7. The coincidence rate of γ-rays scattered by aluminum blocks was measured in the anthracene counters $S1$ and $S2$ as a function of their relative azimuthal angle ϕ. The expected ϕ-dependence depends on the (polar) angle of scattering, θ, of the γ-rays by the cubes, and is maximum for $\theta \simeq 81°$ (Pryce and Ward 1947, Snyder et al. 1948). The observed ratio was

$$\frac{\text{rate}(\phi = 90°)}{\text{rate}(\phi = 0°)} = 2.04 \pm 0.08, \tag{3.49}$$

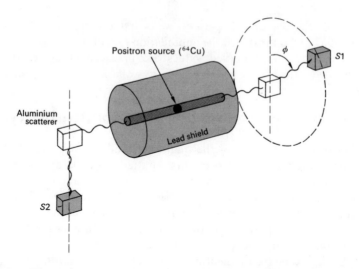

Figure 3.7 Sketch of the method used by Wu and Shaknov (1950) to measure the relative orientation of the polarization vectors of the two photons emitted in decay of $^1S_0$-positronium. S1 and S2 are anthracene counters, recording the γ-rays after Compton scattering by aluminum cubes. The results proved that fermion and antifermion have opposite intrinsic parity, as predicted by the Dirac theory.

compared with a theoretically expected value of 2.00. Such experiments therefore bear out the prediction of preferentially orthogonal polarizations of the γ-rays, and thus demonstrate the correctness of the assumption that fermions and antifermions have opposite intrinsic parities. Similar arguments have been made to demonstrate that the neutral pion has odd intrinsic parity (Fig. 3.2).

3.11. EXPERIMENTAL TESTS OF *C*-INVARIANCE

Experimental tests of charge conjugation invariance compare reactions in which particles are replaced by their antiparticles. For example, in strong interactions comparisons have been made of the rates and spectra of positive and negative mesons in the reactions

$$p + \bar{p} \to \pi^+ + \pi^- + \cdots$$
$$\to K^+ + K^- + \cdots,$$

and any possible violation of *C*-invariance has been found to be $\ll 1\%$.

The search for possible *C*-violation in electromagnetic interactions has an interesting history. In 1964, a small (0.1 %) violation of *CP* invariance was observed in the decay of neutral kaons (see Section 7.14.3). The origin was unknown, but clearly such an effect might imply C-violation in electromagnetic (or strong) interactions, since in these reactions parity is exactly conserved. An intensive analysis was therefore made of the decay products of the η-meson (mass 550 MeV/c^2), which decays through electromagnetic transitions. Some possible decay modes are

$$\eta \to \gamma\gamma \qquad\qquad (3.50a)$$

$$\to \pi^+\pi^-\pi^0 \qquad\qquad (3.50b)$$

$$\to \pi^+\pi^-\gamma \qquad\qquad (3.50c)$$

$$\to \pi^0 e^+ e^-. \qquad\qquad (3.50d)$$

The decay $\eta \to \gamma\gamma$, with a branching ratio of 38 %, establishes the C-parity of the η to be $+1$. Hence, the decay mode $\eta \to \pi^0 e^+ e^-$ should be forbidden, if we interpret it as $\eta \to \pi^0\gamma$ with internal conversion of the γ-ray to an e^+e^- pair, since $C_\gamma = -1$ and $C_{\pi^0} = +1$. The existence of this decay mode would therefore be definite evidence for C-violation. Present limits for the branching ratio are $<5 \times 10^{-5}$.

Unfortunately, the nonexistence of the decay (3.50d) does not unambiguously prove the absence of C-violation, since the interpretation of the limit is model-dependent. A better test is the comparison of the π^+ and π^- spectra in the decays $\eta \to \pi^+\pi^-\gamma$ and $\eta \to \pi^+\pi^-\pi^0$. Indeed, early experiments showed an apparently significant asymmetry, but subsequent and more careful work verified the equivalence of π^+ and π^- spectra down to the 0.5 % level.

In summary, during the later 1960s a great deal of effort went into investigation of possible C-violation in strong and electromagnetic interactions, but no such violations have been detected.

3.12. TIME-REVERSAL INVARIANCE

Both invariance and noninvariance under time reversal are familiar in classical physics. For example, Newton's law $F = m\,d^2x/dt^2$ is invariant under change of sign of the time coordinate: a film of the trajectory of a projectile in the earth's gravitational field looks equally realistic whether we run it backward or forward (if we neglect air resistance). The situation is different for the laws of heat conduction or diffusion, which depend on the first derivative of the time coordinate. In this case, the elementary collisions between molecules do obey the principle of microscopic reversibility. For each collision, there exists a time-reversed collision as in Fig. 3.8. For a gas in equilibrium, the two types of collision occur with equal probability and the entropy is constant. A film of an order-disorder transition, such as of gas at very high pressure expanding through a nozzle to a region of low pressure, does specify an arrow of time and looks unreal if run in the reverse direction.

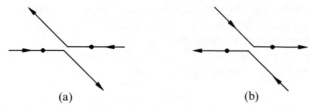

Figure 3.8   (a) Collision between two molecules. (b) Time-reversed collision.

This is simply a consequence however of the initial conditions and has nothing to do with time-reversal invariance in elementary collisions.

The transformations of common quantities in classical physics under space inversion P and time reversal T are given in Table 3.2. We note that an "elementary" particle of spin σ is not expected to possess a static electric dipole moment, if the interaction of the particle with the electromagnetic field is to be invariant under the T- or P-operations. At first sight this may seem strange, in view of the fact that molecules and even atoms may possess large electric dipole moments, while the interactions involved are surely invariant under the T- and P-operations. In the case of molecular or atomic systems, electric dipole moments are always associated with the existence of two degenerate, or nearly degenerate energy eigenstates. As an example, the first excited state of the hydrogen atom ($n = 2$) contains $s_{1/2}$ and $p_{1/2}$ levels which are degenerate, in energy (apart from a small effect associated with the Lamb shift). These states have opposite parity, but, under an applied electric field E, a Stark-effect mixing occurs. The new energy eigenstates are linear combinations of the $s(l = 0)$ and $p(l = 1)$ states and clearly do not have a definite parity. They are associated with an asymmetric (i.e., pear-shaped) electron charge distribution, and hence an electric dipole moment μ_e and a corresponding energy perturbation $\pm \mu_e E$. (No states are *exactly* degenerate: what we mean here is that the original energy separation between the levels is small compared with $\mu_e E$).

In the case of a neutron, for example, an electric dipole moment could only arise, in the absence of P, T violation, if there existed a "ghost"

TABLE 3.2

Quantity	T	P	
$\mathbf{r}$	$\mathbf{r}$	$-\mathbf{r}$	
$\mathbf{p}$	$-\mathbf{p}$	$-\mathbf{p}$	Polar vector
σ (spin)	$-\sigma$	σ	Axial vector ($\mathbf{r} \times \mathbf{p}$)
$\mathbf{E}$ (electric field)	$\mathbf{E}$	$-\mathbf{E}$	($\mathbf{E} = -\partial V/\partial \mathbf{r}$)
$\mathbf{B}$ (magnetic field)	$-\mathbf{B}$	$\mathbf{B}$	(As for σ; e.g., consider a ring current)
$\sigma \cdot \mathbf{B}$	$\sigma \cdot \mathbf{B}$	$\sigma \cdot \mathbf{B}$	Magnetic dipole moment
$\sigma \cdot \mathbf{E}$	$-\sigma \cdot \mathbf{E}$	$-\sigma \cdot \mathbf{E}$	Electric dipole moment
$\sigma \cdot \mathbf{p}$	$\sigma \cdot \mathbf{p}$	$-\sigma \cdot \mathbf{p}$	Longitudinal polarization
$\sigma \cdot (\mathbf{p}_1 \times \mathbf{p}_2)$	$-\sigma \cdot (\mathbf{p}_1 \times \mathbf{p}_2)$	$\sigma \cdot (\mathbf{p}_1 \times \mathbf{p}_2)$	Transverse polarization

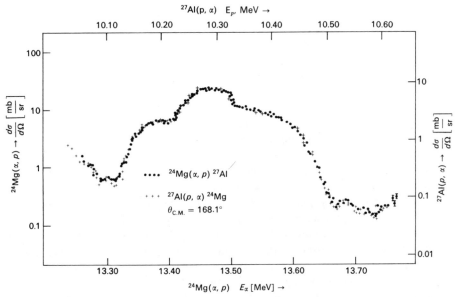

Figure 3.9 The differential cross-sections for the reaction $^{24}\text{Mg}(\alpha, p)^{27}\text{Al}$ and its inverse, as measured by Von Witsch *et al.* (1968).

neutron state of opposite parity degenerate with the first. A superposition of these states could result in a "handedness" and allow an asymmetric charge distribution and thus an electric dipole moment (see Fig. 3.10). There is absolutely no evidence for two types of neutron state; if such existed, the whole of nuclear structure physics would be quite different. We consider therefore that for the neutron we are dealing with one state of fixed parity. The spin vector σ prescribes the only possible direction in space of any electric (or magnetic) dipole moment (i.e., μ_e or μ_m must be parallel or antiparallel to σ). The measurement of any neutron electric dipole moment is therefore an important test of T-invariance; it is discussed in Section 3.14.

Experimentally, time reversal has been verified in strong interactions by application of the principle of detailed balance, as explained in Section 4.1. Figure 3.9 shows the results of a test of detailed balance in the reaction

$$p + {}^{27}\text{Al} \rightleftarrows \alpha + {}^{24}\text{Mg},$$

from which it was concluded that the T-violating amplitude was $<0.3\%$ of the T-conserving amplitude.

3.13. *CP*-VIOLATION AND THE *CPT* THEOREM

Experimental tests of T-invariance have acquired considerable importance in recent years, because of the observed breakdown of CP-invariance in K^0-decay in weak interactions. The reason for this is that the T- and

CP-transformations are connected by the famous *CPT theorem*, which is one of the most important principles of quantum field theory. This theorem states that *all* interactions are invariant under the succession of the three operations C, P, and T taken in any order. The proof of this theorem is based on very general assumptions and is cherished by theorists in the sense that it is difficult to formulate field theories which are not automatically *CPT*-invariant. However, the CPT theorem is not on quite such experimentally solid ground as conservation of energy, for example, and bearing in mind the history of long-respected conservation laws which have fallen by the wayside, we must make experimental checks.

Some consequences of the CPT theorem which may be verified experimentally relate to the properties of particles and antiparticles, which should have the same mass and lifetime, and magnetic moments equal in magnitude but opposite in sign. These results would follow from charge-conjugation invariance alone, if it held universally. However, weak interactions are not C-invariant, so that the prediction rests on the more general theorem.

The experimental consequences of the CPT theorem seem to be well verified. Current results are as shown in Table 3.3. The best experimental limit comes from the K^0-$\bar{K}^0$ comparison.

Another consequence of the CPT theorem is that particles would be expected to obey the "normal" spin-statistics relationship, i.e., integral spin and half-integral spin particles should follow Bose and Fermi statistics, respectively.

Until 1964, it was believed that all types of interaction were invariant under the combined operation CP. Weak interactions were known to violate C- and P-invariance separately, but to respect the CP-symmetry. In that year, however, it was discovered by Christenson *et al.* (1964) that the long-lived neutral K-particle, which normally decays by a weak interaction into three

TABLE 3.3   Tests of *CPT* theorem

		Limit on fractional difference
Lifetime	$\tau_{\pi^+} - \tau_{\pi^-}$	$< 10^{-3}$
	$\tau_{\mu^+} - \tau_{\mu^-}$	$< 10^{-4}$
	$\tau_{K^+} - \tau_{K^-}$	$< 10^{-3}$
Magnetic moment	$|\mu_{\mu^+}| - |\mu_{\mu^-}|$	$< 10^{-8}$
	$|\mu_{e^+}| - |\mu_{e^-}|$	$< 10^{-10}$
Mass	$M_{\pi^+} - M_{\pi^-}$	$< 10^{-3}$
	$M_{\bar{p}} - M_p$	$< 10^{-4}$
	$M_{K^+} - M_{K^-}$	$< 10^{-4}$
	$M_{K^0} - M_{\bar{K}^0}$	$< 10^{-14}$

pions of CP eigenvalue -1, could occasionally (with probability 2×10^{-3}) decay into two pions, of $CP = +1$. These results are discussed in detail in Section 7.14.

The origin of CP-violation is not at present established, although it is a feature of theories of fundamental interactions incorporating at least six quark flavors. If one believes in the "big-bang" theory of the universe with no initial asymmetry, the results of CP-violation (or T-violation) are all around us: the universe has evolved with time in such a way as to result in a huge preponderance of matter over antimatter. But the discussion of these questions has to be deferred until Chapter 9. As indicated above, the magnitude of the violation of CP-symmetry is very small and so far has only been (and possibly will only be) observed in the laboratory using the very precise "interferometer" provided by the $K^0 \bar{K}^0$-system. Nevertheless, attempts have been made to find evidence of T-violation in other processes, and we next discuss an experiment on the neutron electric dipole moment.

3.14. ELECTRIC DIPOLE MOMENT OF THE NEUTRON

We previously noted that the existence of an electric dipole moment implies violation of both T- and P-invariance. Since a dipole moment can be measured with great precision, this provides a sensitive test of T-invariance. Before describing the experiment on the electric dipole moment (EDM) of the neutron, let us try to make a guess at the magnitude of the effect, just from dimensional arguments. One can write

EDM $=$ charge $(e) \times$ a length $(l) \times T$-violation parameter (f).

The neutron is uncharged, so that the dipole moment could result from an asymmetry between positive and negative charge clouds, of net value zero, relative to the spin direction σ (Fig. 3.10). Since P-invariance is also violated, we must somehow bring in the weak interactions; the natural length

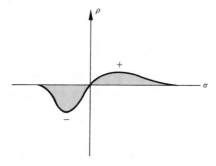

Figure 3.10 An asymmetric distribution of positive and negative charge density ρ in the neutron would give rise to an electric dipole moment.

involving the weak interaction is $l = GM$, where M is some chosen mass—the obvious one being the nucleon mass M — and $G = 10^{-5}/M^2$ is the weak-interaction coupling constant. We have used units $\hbar = c = 1$ here. Then

$$\text{EDM} = 10^{-5} \frac{ef}{M} \sim 10^{-19} fe \text{ cm}, \qquad (3.51)$$

where for $1/M$ we put the proton Compton wavelength $\hbar/Mc = 2 \times 10^{-14}$ cm. What do we take for f? From the parity-conserving electromagnetic interactions, the CPT theorem tells us that if T-invariance is violated, so is C-invariance, and hence the η-decay results (Section 3.11) suggest $f < 10^{-2}$. Or we may take the T-violation parameter from K^0-decay, giving $f < 10^{-3}$. Either way, one can expect a maximum value of the EDM of order $10^{-22} e$ cm. Note that the effective length of the dipole is very small compared with the "size" of an elementary particle, of about 10^{-13} cm.

In the experiment by Dress et al. (1968), a reactor was used as a source of (predominantly) thermal neutrons. In order to make the experiment more sensitive, these neutrons were "cooled" by passage through a narrow, curved tube of highly polished nickel of 1-m radius of curvature (see Fig. 3.11). The critical angle for total internal reflection of neutrons by the tube is inversely proportional to velocity, so that for a beam of finite divergence, only low-velocity neutrons are transmitted with high intensity. The beam emerging from the tube then falls on a polarizing magnet, consisting of a polished, magnetized mirror of cobalt-iron alloy, the direction of the field $\mathbf{B}$ being normal to the surface. Total internal reflection of neutrons will occur for an angle of incidence (relative to the surface) less than the critical angle θ_c, where

$$\sin^2 \theta_c = 1 - n^2 = \frac{\lambda^2 Na}{\pi} \pm \frac{\mu B}{T}. \qquad (3.52)$$

Here λ, T, and μ are the wavelength, kinetic energy, and magnetic moment of the neutrons, n is the refractive index of the mirror, N is the number of scattering nuclei per unit volume, and a is the coherent nuclear scattering length. Because of the second term in (3.52), θ_c depends on the sign of μ and therefore the neutron spin direction. So the reflection angle can be chosen to provide a transversely spin-polarized beam. In a typical case, for neutrons of velocity $v = 100 \text{ m} \cdot \text{s}^{-1}$ (temperature 1°K), $\theta = 2°$ gives a beam with 70% polarization.

After traversing a spectrometer, the neutrons are reflected from the analyzing magnet (similar to the polarizer), and recorded in the detector, consisting of a ^{6}Li-loaded glass scintillator, sensitive to neutrons. The transmitted intensity I is a maximum for neutrons which do not suffer depolarization in the spectrometer.

The spectrometer consists first of a uniform magnetic field $H(\simeq 10 \text{ G})$, which causes the neutrons to precess with the Larmor frequency $\nu_L = \mu H/h$, where μ is the neutron magnetic moment ($\nu_L \simeq 25 \text{ kHz}$), Second,

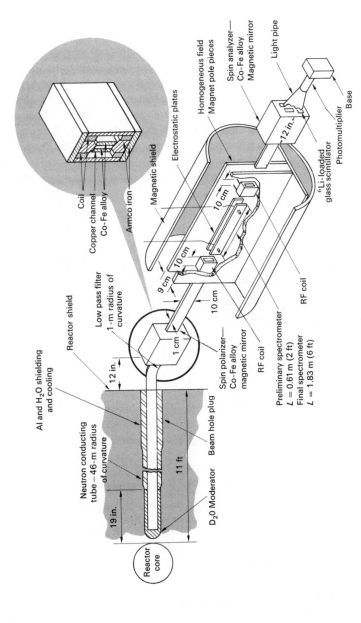

Figure 3.11 Sketch of the apparatus used by Dress *et al.* (1968) to measure the electric dipole moment of the neutron.

a RF field of frequency v is applied by means of two coils, so that at resonance, when $v = v_L$, spin-flip transitions occur, the neutron beam is partly depolarized, and the transmitted intensity I changes. Two coils are used to give an interferometer effect, producing several maxima and minima in the resonance curve (Fig. 3.12), and providing a rapid change in counting rate with RF frequency. Finally, a reversible electric field E of 100 kV/cm is applied in the direction of the steady magnetic field H.

The experiment consists essentially of sitting in a region of the resonance curve where dI/dv is large, and observing the change in I when the electrostatic field E is reversed. If the neutron possesses an electric dipole moment in the direction of the spin, the field E will produce an additional small precession and consequently a change in I when the frequency v is held constant. dI/dv is proportional to the time spent by the neutron between the RF coils and is thus greatest for large coil separations and low velocities— hence the advantage of using "cold" neutrons. The final result of the most

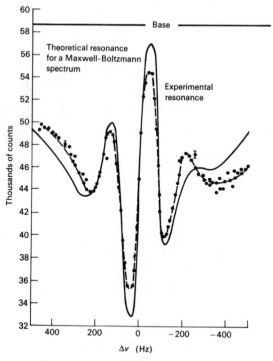

Figure 3.12 Resonance curve obtained with the apparatus of Fig. 3.11 as the RF frequency is varied about the resonance value. At the steepest part of the curve, the counting rate varies by 1% for a change of frequency $\Delta v = 1$ Hz. The full-line curve is that calculated assuming a simple Maxwellian distribution in the velocity of the neutron beam.

recent version of the experiment (Dress *et al.* (1978)) sets a limit on the electric dipole moment of

$$\text{EDM} = (0.4 \pm 1.5) \times 10^{-24} \, e \, \text{cm}. \tag{3.53}$$

It may be remarked that one limitation of the experiment is the difficulty of obtaining **E** and **H** exactly parallel. If there is a small component $E_\perp$ perpendicular to **H**, this will induce an extra magnetic field $\Delta H = (v/c)E_\perp$ in the direction of **H** and hence a spurious effect.

From (3.52) we note that for ultracold neutrons ($T \simeq 0.002°\text{K}, v \simeq 6 \, \text{m s}^{-1}$), θ_c can exceed $\pi/2$, so that neutrons can be reflected at normal incidence. Such neutrons can be captured and stored in magnetic "bottles", allowing longer observation times and more precise measurement of the EDM. The most recent results using this technique are those of Altarev *et al.* (1981):

$$\text{EDM} = (2.3 \pm 2.3) \times 10^{-25} \, e \, \text{cm}$$

and of Pendlebury *et al.* (1984):

$$\text{EDM} = (0.3 \pm 4.8) \times 10^{-25} \, e \, \text{cm}. \tag{3.54}$$

For a recent review of measurements of the electric dipole moments of the neutron and other particles, see Ramsey (1982).

Theoretical estimates of the EDM vary over many orders of magnitude. Since there is no direct evidence for *C*-violation in electromagnetism, it is usual to assume that any EDM of the neutron is associated with *CP*-violation, such as is observed in K^0-decay (see Section 7.14). This leads to predictions in the region of $10^{-30} \, e \, \text{cm}$ for the so-called "standard model" with six quark flavors, but higher values are possible with other assumptions. It is clear that future experiments on the neutron electron dipole moment could be of fundamental significance for our understanding of *CP*-violation.

PROBLEMS

3.1. Λ-hyperons are produced by a pion beam in the reaction $\pi^- + p \to K^0 + \Lambda$, and observed via their decay $\Lambda \to p + \pi^-$. Let J denote the spin of the Λ, z the beam direction, and θ the angle of a decay product, relative to z, measured in the Λ rest frame. (a) In the case where the Λ is produced exactly along the z-direction, what are the possible values of J_z? (b) Show that, for unpolarized protons, the decay angular distributions for the forward-produced Λs as a function of Λ spin will be as follows:

$$J_\Lambda = \tfrac{1}{2} \quad \text{(isotropic)},$$
$$J_\Lambda = \tfrac{3}{2} \quad (3 \cos^2 \theta + 1),$$
$$J_\Lambda = \tfrac{5}{2} \quad (5 \cos^4 \theta - 2 \cos^2 \theta + 1).$$

[This method to determine the Λ spin was first proposed by Adair (1955). For a discussion see Sakurai (1964) and Tripp (1965).] (c) State how one might determine the spin of the $\Sigma^{\pm}$ from the (s-state) capture of negative kaons in hydrogen, $K^{-} + p \to \Sigma^{\pm} + \pi^{\mp}$.

3.2. The Σ^0-hyperon decays electromagnetically in the mode $\Sigma^0 \to \Lambda + \gamma$. Show how the relative parity of Σ^0 and Λ determines the multipolarity of the γ-ray emitted. From the polarization vector ε of the photon, and the propagation vector $\mathbf{k}$ and spin $\boldsymbol{\sigma}$ of the Λ, deduce the simplest forms for the matrix element for even or odd relative parity. The experimental determination of the Σ-Λ parity has been based on the analysis of the Dalitz decay $\Sigma \to \Lambda e^{+} e^{-}$. Which of the parity assignments has the steeper distribution in the invariant mass of the $e^{+} e^{-}$ pair?

3.3. Show that a scalar meson cannot decay to three pseudoscalar mesons in a parity-conserving process.

3.4. The intrinsic parity of the hyperon Ξ^{-}, of strangeness -2, can in principle be determined from observations on capture in hydrogen from an S-orbit:

$$\Xi^{-} + p \to \Lambda + \Lambda.$$

The polarization of the Λ-hyperons can be determined from the asymmetry in the weak decay $\Lambda \to p + \pi^{-}$ (see Section 7.7). State what is the polarization (if any) of the Λs produced in the above reaction and how the relative polarizations are determined by the Ξ-parity.

3.5. Capture of negative kaons in helium sometimes leads to the formation of a hypernucleus (a nucleus in which a neutron is replaced by a Λ-hyperon) according to the reaction

$$K^{-} + {}^{4}\text{He} \to {}^{4}\text{H}_{\Lambda} + \pi^0.$$

Study of the decay branching ratios of ${}^{4}\text{H}_{\Lambda}$, and the isotropy of decay products establishes that $J({}^{4}\text{H}_{\Lambda}) = 0$. Show that this implies negative parity for the K^{-}, independent of the orbital angular momentum of the state from which the K^{-} is captured.

3.6. Show that the reaction $\pi^{-} + d \to n + n + \pi^0$ cannot occur for pions at rest.

3.7. Show that, for pions with zero relative orbital angular momentum, the combination $\pi^{+}\pi^{-}$ is an eigenstate of $CP = +1$, and $\pi^{+}\pi^{-}\pi^0$ is an eigenstate of $CP = -1$.

3.8. What restrictions does the decay mode $K_1^0 \to 2\pi^0$ place on (a) the kaon spin, (b) the kaon parity?

BIBLIOGRAPHY

Fraser, W. R., *Elementary Particles*, Prentice-Hall, Englewood Cliffs, New Jersey, 1966.

Hamilton, W. D., "Parity violation in electromagnetic and strong interaction processes", *Prog. Nucl. Phys.* **10**, 1 (1969).

Henley, E. M., "Parity and time-reversal invariance in nuclear physics", *Ann. Rev. Nucl. Science* **19**, 367 (1969).

Jackson, J. D., *The Physics of Elementary Particles*, Princeton University Press, Princeton, New Jersey, 1958.

Kemmer, N., J. C. Polkinghorne, and D. Pursey, "Invariance in elementary particle physics", *Rep. Prog. Phys.* 22, 368 (1959).

Muirhead, A., *The Physics of Elementary Particles*, Pergamon, London, 1965, Chapter 5.

Ramsey, N. F., "Dipole moments and spin rotations of the neutron", *Phys. Reports* **43**, 409 (1978).

Ramsey, N. F., "Electric dipole moments of particles", *Ann. Rev. Nucl. Part. Science* **32**, 211 (1982).

Rowe, E. G., and E. J. Squires, "Present status of *C-*, *P-*, *and T*-invariance", *Rep. Prog. Phys.* **32**, 273 (1969).

Sakurai, J. J., *Invariance Principles and Elementary Particles*, Princeton University Press, Princeton, New Jersey, 1964.

Tadic, D., "Parity non-conservation in nuclei", *Rep. Prog. Phys.* **43**, 67 (1980).

Tripp, R. D., "Spin and parity determination of elementary particles", *Ann. Rev. Nucl. Science* **15**, 325 (1965).

Wick, G. C., "Invariance principles of nuclear physics", *Ann. Rev. Nucl. Science* **8**, 1 (1958).

Williams, W. S., *An Introduction to Elementary Particles*, Academic Press, New York, and London, 1971.

CHAPTER 4

Hadron-Hadron Interactions

In this chapter, we discuss the characteristics of the strong interactions between hadrons. As indicated in Chapter 5, the fundamental strong interactions are considered to take place between the elementary quark constituents of hadrons, and there is indeed a plausible field theory of the interactions between a pair of quarks, called quantum chromodynamics, which is however only calculable for high momentum transfers between the quarks. Hadron-hadron collisions, on the contrary, involve many quark (or antiquark) constituents simultaneously—thus, an insoluble many-body problem—and the individual momentum transfers are generally small.

The situation is somewhat similar to that of collisions between atoms. The fundamental (electromagnetic) interaction between the individual electrons and nuclei is well understood, but the description of atom-atom interactions is very complex because many electrons are involved, and a largely empirical approach to the problem is required. The treatment of hadron-hadron interactions antedates the quark model by many years. One very general approach—the S-matrix theory—discusses hadron-hadron collisions in terms of amplitudes and phases of matter waves, in analogy with optics, and using concepts and methods (such as that of resonances) first developed in nuclear structure physics.

The description of hadron-hadron interactions has been of great importance in establishing the existence of the different hadron states, and in measuring their masses and quantum numbers, and this accumulation of data was the essential input required for the quark concept to emerge.

4.1. CROSS-SECTIONS AND DECAY RATES

Much of our information about properties of particles and their interactions comes from experimental data on collision cross-sections and rates for spontaneous decay. As an example we consider a reaction of the form

$$\underbrace{a + b}_{i} \rightarrow \underbrace{c + d}_{f} \tag{4.1}$$

with two particles in the initial state i, and two in the final state f. If we regard b as the target and a as the projectile particle—usually in a well-collimated beam—then the *cross-section* for the above reaction is defined as the transition rate W per unit incident flux per target particle. If n_a is the density of particles in the incident beam, and v_i the relative velocity of a and b, then the flux of particles per unit time through unit area normal to the beam is

$$\phi = n_a v_i. \tag{4.2}$$

If there are n_b particles in the target per unit area, each of effective cross-section σ, the probability that any incident particle will hit a target is σn_b, and the number of interactions per unit area per second will be $n_a n_b \sigma v_i$. Per target particle, the transition rate is therefore

$$W = \sigma\phi = \sigma n_a v_i. \tag{4.3}$$

The value of W is given by the product of the square of a matrix element M_{if} and a density of final states, or phase-space factor, ρ_f. M_{if} contains various dynamical features of the interaction: coupling strength, energy dependence, angular distribution, and so forth. The formula for W is (see Appendix E)

$$W = \frac{2\pi}{\hbar} |M_{if}|^2 \rho_f. \tag{4.4}$$

This formula holds equally for a decay process, in which the initial state i consists of an unstable particle decaying to the final state f (i.e., $X \rightarrow c + d$). In first-order perturbation theory, where the interaction is assumed to be "weak," M_{if} can be interpreted as the overlap integral over volume, $\int \psi_f^* H' \psi_i \, d\tau$, where H' is the interaction potential, and ψ_i and ψ_f the initial- and final-state wavefunctions. This interpretation holds provided $H' \ll H_0$, the unperturbed energy operator (Hamiltonian). When the interaction is strong, however, M_{if} cannot be calculated explicitly, and (4.4) can then be regarded as a definition of M_{if}.

To calculate the phase space factor ρ_f we restrict the particles to some arbitrary volume, which cancels in the calculation and which we shall take as unity. Then

$$\rho_f = \frac{dn}{dE_0} = \frac{d\Omega}{h^3} p_f^2 \frac{dp_f}{dE_0} g_f, \tag{4.5}$$

where E_0 is the total energy in the center-of-mass frame, p_f is the final-state momentum in this frame ($|\mathbf{p}_c| = |\mathbf{p}_d| = p_f$), and $d\Omega$ is the solid angle containing the final-state particles; g_f is a spin-multiplicity factor. Equation (4.1) refers to one particle of each type in the normalization volume. With $n_a = 1$, and integrating over all angles of the products c and d, one obtains; using (4.3),

$$\sigma = \frac{W}{v_i} = \frac{1}{\pi \hbar^4} |M_{if}|^2 \frac{p_f^2 \, g_f}{v_i v_f} \tag{4.6}$$

where conservation of energy gives

$$\sqrt{p_f^2 + m_c^2} + \sqrt{p_f^2 + m_d^2} = E_0$$

and thus

$$\frac{dp_f}{dE_0} = \frac{E_c E_d}{E_0 p_f} = \frac{1}{v_f}, \tag{4.7}$$

where E_c, E_d are total energies, and $v_f = v_c + v_d$ is the relative velocity of c and d. If s_c and s_d are the spins of particles c and d, the numbers of possible substates for them are $2s_c + 1$ and $2s_d + 1$, and thus $g_f = (2s_c + 1)(2s_d + 1)$. Equation (4.6) then becomes

$$\sigma(a + b \rightarrow c + d) = \frac{1}{\pi \hbar^4} |M_{if}|^2 \frac{(2s_c + 1)(2s_d + 1)}{v_i v_f} p_f^2. \tag{4.8}$$

Implied in this expression, integrated over angles, is that $|M_{if}|^2$ has been averaged over all possible spin states of a and b, and summed over all orbital-angular-momentum states involved.

As indicated above, the precise form of M_{if} is usually not known. However, if we compare the forward and backward reactions $a + b \rightleftharpoons c + d$ at the same center-of-mass energy, this information is not necessary. We rely on the principle of detailed balance, which states that

$$|M_{if}|^2 = |M_{fi}|^2. \tag{4.9}$$

This relation follows by assuming invariance of the interaction under time reversal and space inversion, both of which hold good for the strong (hadronic) interactions (see Chapter 3). Time reversal interchanges final and initial states, but reverses all momenta and spins; space inversion then changes back the signs of momenta but leaves spins unchanged. Thus, writing $M_{if} = \langle f | T | i \rangle$, where T is a suitable transition matrix operator, we have

$$\langle f(p_c, p_d, s_c, s_d) | T | i(p_a \cdot p_b, s_a, s_b) \rangle$$

$$\xrightarrow[\text{+ space inversion}]{\text{time reversal}} \langle i(p_a, p_b, -s_a, -s_b) | T | f(p_c, p_d, -s_c, -s_d) \rangle. \tag{4.10}$$

Summing over all $2s + 1$ spin projections, running from $-s$ to $+s$, we obtain $|M_{if}|^2 = |M_{fi}|^2$ as in (4.9).

Invariance under time reversal and space inversion does not hold good in the weak interactions. In these, however, first-order perturbation theory can be applied ($H' \ll H_0$), and H' is a Hermitian operator (that is, $\langle f|H'|i \rangle = \langle i|H'|f \rangle^*$), so that the detailed balance relation (4.9) is again valid. The relations (4.8) and (4.9) will now be used in determining the spin of the pion.

4.1.1. Pion Spin

For charged pions, the spin was determined by applying detailed balance to the reversible reaction

$$p + p \rightleftharpoons \pi^+ + d. \tag{4.11}$$

Applying Eq. (4.8) to the forward reaction, we obtain

$$\sigma_{pp \to \pi^+ d} = |M_{if}|^2 \frac{(2s_\pi + 1)(2s_d + 1)}{v_i v_f} p_\pi^2 \tag{4.12}$$

where p_π stands for the absolute value of the momentum $\mathbf{p}_\pi = -\mathbf{p}_d$ in the center-of-momentum system (CMS) and $s_d = 1$. For the backward reaction

$$\sigma_{\pi^+ d \to pp} = \tfrac{1}{2}|M_{fi}|^2 \frac{(2s_p + 1)^2}{v_f v_i} p_p^2. \tag{4.13}$$

At a given energy in the CMS, detailed balance ensures that the factors $|M_{if}|^2/v_i v_f$ for the forward and backward reactions will be the same. The factor $\tfrac{1}{2}$ in (4.13) arises from the fact that the integration of the phase-space factor (4.5) over 2π (rather than 4π) solid angle gives all the states when a pair of identical protons is involved. Thus, the cross-section ratio

$$\frac{\sigma(pp \to \pi^+ d)}{\sigma(\pi^+ d \to pp)} = 2 \frac{(2s_\pi + 1)(2s_d + 1)}{(2s_p + 1)^2} \frac{p_\pi^2}{p_p^2}. \tag{4.14}$$

The differential cross-section $d\sigma/d\Omega$ for the reaction $p + p \to \pi^+ + d$ was first measured by Cartwright et al. (1953), using 340-MeV incident protons, corresponding to a pion CMS kinetic energy $T_\pi = 21.4$ MeV. The total cross-section for the inverse reaction $\pi^+ + d \to p + p$ had been measured by Clark et al. (1951, 1952), with $T_\pi = 23$ MeV, giving $\sigma = 4.5 \pm 0.8$ mb, and by Durbin et al. (1951), with $T_\pi = 25$ MeV, giving $\sigma = 3.1 \pm 0.3$ mb. Integration of the Cartwright measurements predicted for this reaction $\sigma = 3.0 \pm 1.0$ mb for $s_\pi = 0$, and $\sigma = 1.0 \pm 0.3$ mb for $s_\pi = 1$. These data clearly showed that $s_\pi = 0$, and later experiments confirmed this.

For neutral pions, the existence of the decay

$$\pi^0 \to 2\gamma$$

proves that s_π must be integral (since $s_\gamma = 1$) and that $s_\pi \neq 1$. A photon has only two possible spin states, either parallel or antiparallel to the direction of motion (see Section 3.6.2). Taking the common line of flight of the photons in the π^0 rest frame as the quantization axis, the z-component of total photon spin can have the values $s_z = 0$ or 2. Suppose $s_\pi = 1$, then only $s_z = 0$ is possible. In this case the two-photon amplitude must behave under spatial rotations like the polynomial $P_1^m(\cos \theta)$ with $m = 0$, where θ is the angle relative to the z-axis. Under a $180°$ rotation about an axis normal to z, $\theta \rightarrow \pi - \theta$; since $P_1^0 \propto \cos \theta$, it therefore changes sign. For $s_z = 0$, the situation corresponds to two right circularly polarized (or two left circularly polarized) photons, traveling in opposite directions. The above rotation is therefore equivalent to interchange of the two photons, for which the wavefunction must be symmetric. Hence, the neutral-pion spin cannot be unity, and $s_\pi = 0$, or ≥ 2. In high-energy nucleon-nucleon collisions, it is observed that neutral, positive and negative pions are produced in equal numbers, indicating that the neutral pion has spin zero, with the same spin multiplicity as the charged pions.

4.2. ISOSPIN

Heisenberg suggested in 1932 that neutron and proton might be treated as different charge substates of one particle, the nucleon. A nucleon is ascribed a quantum number, isospin, denoted by the symbol I, with a value $I = \frac{1}{2}$, and two substates with I_z, or I_3, equal to $\pm\frac{1}{2}$. The charge is then given by $Q/e = \frac{1}{2} + I_3$, if we assign $I_3 = +\frac{1}{2}$ to the proton and $I_3 = -\frac{1}{2}$ to the neutron. This purely formal description is in complete analogy with that of a particle of ordinary spin $\frac{1}{2}$, with substates $J_z = \pm\frac{1}{2}$ (in units of $h/2\pi$).

Isospin is a useful concept because it is a conserved quantum number in strong interactions. Consequently these depend on I and not on the third component, I_3. The specifically strong interactions between nucleons, for example, are specified by I and we do not distinguish between neutron and proton—they are degenerate states. A pictorial way to visualize isospin is as a vector, $\mathbf{I}$, in a three-dimensional "isospin space," with Cartesian coordinates I_x, I_y, I_z—or, as they are more usually denoted, I_1, I_2, and I_3. Isospin conservation corresponds to invariance of the length of this vector under rotation of the coordinate axes in isospin space. Electromagnetic interactions do not conserve $\mathbf{I}$ and are not invariant under such rotation. Because they couple to charge Q, they single out the I_3-axis in isospin space. (see Fig. 4.1).

The earliest evidence for isospin conservation in strong interactions came from the observation of charge symmetry and charge independence of nuclear forces, that is, the equivalence of n-p, p-p, and n-n forces in the same angular-momentum states, once Coulomb effects had been subtracted. This equality followed from the remarkable similarity in the level schemes of

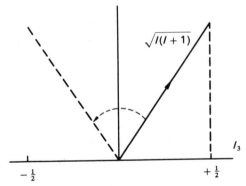

Figure 4.1 The conservation of isospin I in strong interactions can be interpreted as the invariance, under rotations of the axes, of an isospin vector of length $\sqrt{I(I+1)}$ in a three-dimensional isospin space. For $I = \frac{1}{2}$, the z-component can have eigenvalues $I_3 = \pm\frac{1}{2}$. In an electric field, these are differentiated as proton and neutron.

mirror nuclei, that is nuclei with similar configurations of nucleons, but with a neutron replaced by a proton or vice versa.

4.3. ISOSPIN IN THE TWO-NUCLEON SYSTEM

The isospin states of the two-nucleon system can be written down in direct analogy with the combination of two objects of spin-$\frac{1}{2}$, as in (3.40). Writing the wavefunctions n and p to denote neutron and proton states, we therefore obtain

$$
S\begin{cases}
\chi(1, 1) = p(1)p(2), & \text{(4.15a)} \\
\chi(1, 0) = \dfrac{1}{\sqrt{2}}\left[p(1)n(2) + n(1)p(2)\right], & \text{(4.15b)} \\
\chi(1, -1) = n(1)n(2), & \text{(4.15c)}
\end{cases}
$$

$$
A \quad \chi(0, 0) = \frac{1}{\sqrt{2}}\left[p(1)n(2) - n(1)p(2)\right], \quad \text{(4.15d)}
$$

where the first three states are members of an $I = 1$ triplet, symmetric under label interchange $1 \leftrightarrow 2$, and the last is an $I = 0$ singlet, which is anti-symmetric.

The total wavefunction for a two-nucleon state may be written

$$
\psi(\text{total}) = \phi(\text{space})\,\alpha(\text{spin})\,\chi(\text{isospin}) \quad \text{(4.16)}
$$

provided orbital and spin angular momentum can be separately quantized (i.e. the system is nonrelativistic). Applying (4.16) to a deuteron, which has spin 1, we see that α is symmetric under interchange of the two nucleons. The space wavefunction ϕ has symmetry $(-1)^l$ under interchange. The two nucleons in the deuteron are known to be in an $l = 0$ state (with a few percent $l = 2$ admixture). Thus ϕ is symmetric, and χ must be antisymmetric in order to satisfy overall antisymmetry of the total wavefunction ψ. From (4.15) it follows that $I = 0$: The deuteron is an isosinglet.

As an example, consider the reactions

$$\text{(i) } p + p \rightarrow d + \pi^+, \quad \text{(ii) } p + n \rightarrow d + \pi^0.$$
$$I \qquad 1 \qquad 0 \qquad 1 \qquad\quad 0 \text{ or } 1 \quad 0 \quad 1$$

In each case the final state is of $I = 1$. On the left-hand side, we have a pure $I = 1$ state in reaction (i), but 50% $I = 0$ and 50% $I = 1$ in reaction (ii). Conservation of isospin means that either reaction can only proceed through the $I = 1$ channel. Consequently, $\sigma(\text{ii})/\sigma(\text{i}) = \frac{1}{2}$, as is observed.

4.4. ISOSPIN IN THE PION-NUCLEON SYSTEM

The pion exists in three charge states of roughly the same mass: π^+, π^-, and π^0. Consequently it is assigned $I = 1$, with the charge given by $Q/e = I_3$ simply. This formula is different from that used for the nucleon: $Q/e = I_3 + \frac{1}{2}$. Both can be accommodated in the formula by introducing baryon number:

$$\frac{Q}{e} = I_3 + \frac{B}{2}. \tag{4.17}$$

An important application of isospin conservation arises in the strong interactions of nonidentical particles, which will generally consist of mixtures of different isospin states. The classical example of this is pion-nucleon scattering. Since $I_\pi = 1$ and $I_N = \frac{1}{2}$, one can have $I_{\text{total}} = \frac{1}{2}$ or $\frac{3}{2}$. If the strong interactions depend only on I and not on I_3, then the $3 \times 2 = 6$ pion-nucleon scattering processes can all be described in terms of two isospin amplitudes.

Of the six elastic scattering processes,

$$\pi^+ p \rightarrow \pi^+ p \tag{4.18a}$$

and

$$\pi^- n \rightarrow \pi^- n \tag{4.18b}$$

have $I_3 = \pm\frac{3}{2}$, and are therefore described by a pure $I = \frac{3}{2}$ amplitude. Clearly, at a given bombarding energy, (4.18a) and (4.18b) will have identical cross-sections, since they differ only in the sign of I_3.

The remaining interactions,

$$\pi^- p \to \pi^- p, \tag{4.18c}$$

$$\pi^- p \to \pi^0 n, \tag{4.18d}$$

$$\pi^+ n \to \pi^+ n, \tag{4.18e}$$

$$\pi^+ n \to \pi^0 p, \tag{4.18f}$$

have $I_3 = \pm \frac{1}{2}$ and therefore $I = \frac{1}{2}$ or $\frac{3}{2}$. The weights of the two amplitudes in the mixture are given by Clebsch-Gordan coefficients (alternatively known as vector-coupling or Wigner coefficients). Their derivation is given in Appendix C.

An alternative method, which can be applied to the present case, has been given by Feynman. The force between nucleons is charge independent. This force may be thought of as being due to exchange of virtual particles between the nucleons (just as the electrostatic force between two charges can be thought of in terms of exchange of virtual photons). However, for hadron-hadron interactions, many processes, such as pion exchange, ρ-meson exchange, $K\bar{K}$ exchange, etc., may contribute. It is very plausible to assume that the *contribution* from single pion exchange alone is charge independent—if it were not, charge independence would be an accident depending on chance cancellation of forces due to all possible exchange mechanisms, which individually were not charge-independent. This possibility is discounted as too remote.

The pp, pn, and nn interactions contributed by single pion exchange are represented diagrammatically in Fig. 4.2, where a, b, and c are unknown coupling constants. In the last diagram, the virtual π^- can of course be

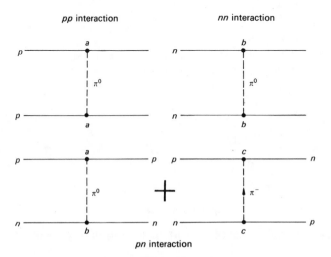

Figure 4.2

replaced by a π^+ with arrow reversed—we do not distinguish the two. If the pp, nn, and np forces are equal, we have

$$a^2 = b^2 = ab + c^2.$$

Thus, $a = \pm b$. Furthermore, c cannot be zero, since it is known that the last diagram exists and leads to charge-exchange pn scattering. One must then have

$$b = -a,$$

$$c = \pm\sqrt{2}a,$$

where the last sign is arbitrary.

Because of its strong interaction, a nucleon can virtually dissociate into a nucleon plus one or more pions, a hyperon-kaon pair, and so on. The possible virtual states of the neutron involving a nucleon plus one pion are clearly those in the bottom part of Fig. 4.2: we can denote them $|n\pi^0\rangle$ and $|p\pi^-\rangle$, respectively. The initial state (the neutron) has $I = \frac{1}{2}$, $I_3 = -\frac{1}{2}$. Hence the decomposition of this state is

$$|\chi(\tfrac{1}{2}, -\tfrac{1}{2})\rangle = b|n\pi^0\rangle + c|p\pi^-\rangle$$
$$= b(|n\pi^0\rangle \pm \sqrt{2}|p\pi^-\rangle).$$

With the normalization of the states $\langle\chi|\chi\rangle = \int \chi^*\chi\, dV = 1$, $\langle n\pi^0|n\pi^0\rangle = \langle p\pi^-|p\pi^-\rangle = 1$ we therefore obtain

$$|\chi(\tfrac{1}{2}, -\tfrac{1}{2})\rangle = \sqrt{\tfrac{1}{3}}|n\pi^0\rangle - \sqrt{\tfrac{2}{3}}|p\pi^-\rangle. \tag{4.19}$$

Similarly, the proton state will be

$$|\chi(\tfrac{1}{2}, \tfrac{1}{2})\rangle = -\sqrt{\tfrac{1}{3}}|p\pi^0\rangle + \sqrt{\tfrac{2}{3}}|n\pi^+\rangle, \tag{4.20}$$

where the sign of c has been chosen to conform to the usual notation (Condon and Shortley 1951).

To find the coefficients for the $I = \frac{3}{2}$, $I_3 = \pm\frac{1}{2}$ states, we make use of the orthogonality and normalization conditions; for example

$$\langle\chi(\tfrac{3}{2}, \tfrac{1}{2})|\chi(\tfrac{1}{2}, \tfrac{1}{2})\rangle = 0, \qquad \langle n\pi^0|n\pi^0\rangle = 1, \qquad \langle p\pi^-|n\pi^0\rangle = 0.$$

Thus, if we set

$$|\chi(\tfrac{3}{2}, \tfrac{1}{2})\rangle = A|p\pi^0\rangle + B|n\pi^+\rangle, \tag{4.21}$$

$$|\chi(\tfrac{3}{2}, -\tfrac{1}{2})\rangle = C|p\pi^-\rangle + D|n\pi^0\rangle, \tag{4.22}$$

we clearly must have

$$A^2 + B^2 = C^2 + D^2 = 1,$$

and, taking the products of (4.21) with (4.20) and of (4.22) with (4.19),

$$A = B\sqrt{2}, \qquad D = C\sqrt{2}.$$

TABLE 4.1 Clebsch-Gordan coefficients in pion-nucleon scattering

Pion	Nucleon	$I = \frac{3}{2}$				$I = \frac{1}{2}$	
		$I_3 = \frac{3}{2}$	$\frac{1}{2}$	$-\frac{1}{2}$	$-\frac{3}{2}$	$\frac{1}{2}$	$-\frac{1}{2}$
π^+	p	1					
π^+	n		$\sqrt{\frac{1}{3}}$			$\sqrt{\frac{2}{3}}$	
π^0	p		$\sqrt{\frac{2}{3}}$			$-\sqrt{\frac{1}{3}}$	
π^0	n			$\sqrt{\frac{2}{3}}$			$\sqrt{\frac{1}{3}}$
π^-	p			$\sqrt{\frac{1}{3}}$			$-\sqrt{\frac{2}{3}}$
π^-	n				1		

Thus

$$B = C = \sqrt{\tfrac{1}{3}}, \qquad D = A = \sqrt{\tfrac{2}{3}}, \tag{4.23}$$

again with the usual convention on sign. The resulting list of Clebsch-Gordan coefficients is given in Table 4.1.

We can now calculate the relative cross sections for the following three processes, at a fixed energy:

$$\pi^+ p \to \pi^+ p \qquad \text{(elastic scattering)}, \tag{4.24a}$$

$$\pi^- p \to \pi^- p \qquad \text{(elastic scattering)}, \tag{4.24b}$$

$$\pi^- p \to \pi^0 n \qquad \text{(charge exchange)}. \tag{4.24c}$$

The cross-section is proportional to the square of the matrix element connecting initial and final states, i.e.

$$\sigma \propto \langle \psi_f | H | \psi_i \rangle^2 = M_{if}^2,$$

where H is an isospin operator, having a value $H = H_1$ if it operates on initial and final states of $I = \frac{1}{2}$, and $H = H_3$ for states of $I = \frac{3}{2}$. By conservation of isospin, there is no operator connecting initial and final states of different isospin. Let

$$M_1 = \langle \psi_f(\tfrac{1}{2}) | H_1 | \psi_i(\tfrac{1}{2}) \rangle,$$

$$M_3 = \langle \psi_f(\tfrac{3}{2}) | H_3 | \psi_i(\tfrac{3}{2}) \rangle.$$

The reaction (4.24a) involves a pure state of $I = \frac{3}{2}$, $I_3 = +\frac{3}{2}$. Therefore,

$$\sigma_a = K |M_3|^2,$$

where K is some constant.

Referring to our table, in the reaction (4.24b) we may write

$$|\psi_i\rangle = |\psi_f\rangle = \sqrt{\tfrac{1}{3}} |\chi(\tfrac{3}{2}, -\tfrac{1}{2})\rangle - \sqrt{\tfrac{2}{3}} |\chi(\tfrac{1}{2}, -\tfrac{1}{2})\rangle.$$

Therefore,

$$\sigma_b = K\langle\psi_f|H_1 + H_3|\psi_i\rangle^2$$
$$= K|\tfrac{1}{3}M_3 + \tfrac{2}{3}M_1|^2.$$

For the reaction (4.24c), one has

$$|\psi_i\rangle = \sqrt{\tfrac{1}{3}}|\chi(\tfrac{3}{2}, -\tfrac{1}{2})\rangle - \sqrt{\tfrac{2}{3}}|\chi(\tfrac{1}{2}, -\tfrac{1}{2})\rangle,$$
$$|\psi_f\rangle = \sqrt{\tfrac{2}{3}}|\chi(\tfrac{3}{2}, -\tfrac{1}{2})\rangle + \sqrt{\tfrac{1}{3}}|\chi(\tfrac{1}{2}, -\tfrac{1}{2})\rangle,$$

and thus,

$$\sigma_c = K|\sqrt{\tfrac{2}{9}}M_3 - \sqrt{\tfrac{2}{9}}M_1|^2.$$

The cross-section ratios are then

$$\sigma_a:\sigma_b:\sigma_c = |M_3|^2:\tfrac{1}{9}|M_3 + 2M_1|^2:\tfrac{2}{9}|M_3 - M_1|^2. \qquad (4.25)$$

The limiting situations, if one or other isospin amplitude dominates under the experimental conditions, are

$$M_3 \gg M_1, \qquad \sigma_a:\sigma_b:\sigma_c = 9:1:2,$$
$$M_1 \gg M_3, \qquad \sigma_a:\sigma_b:\sigma_c = 0:2:1.$$

Numerous experimental measurements have been made of the total and differential pion-nucleon cross-sections. The earliest and simplest experiments measured the attenuation of a collimated, monoenergetic $\pi^\pm$-beam in traversing a liquid hydrogen target. Thus, in the sketch of Fig. 4.3, one would measure, for a given number of coincidences in the counters $S1$ and $S2$, the change in rate of counters $S3$ and $S4$ with the target both full and empty. The results of such measurements are shown in Fig. 4.4. For both positive and negative pions, there is a strong peak in σ_{total} at a pion kinetic energy of 200 MeV. The ratio $(\sigma_{\pi^+p}/\sigma_{\pi^-p})_{\text{total}} = 3$, proving that the $I = \tfrac{3}{2}$ amplitude dominates this region. This bump is referred to as a *resonance*, the $\Delta(1236)$ (1236 MeV being the invariant pion-nucleon mass). The width of this state at half height is 120 MeV. Because the spin-parity turns out to be $J^P = \tfrac{3}{2}^+$, and $I = \tfrac{3}{2}$, it is often referred to as the (3, 3) resonance. A discussion of this pion-nucleon state is given in Section 4.9. As Fig. 4.4 indicates, more pion-nucleon resonances are observed—for example, there is an $I = \tfrac{1}{2}$ resonance at 1525 MeV. In general, in any one region of pion-nucleon invariant mass, several amplitudes will contribute, and one cannot simply interpret a bump in cross-section as signifying a unique resonant state. Only for the first (3, 3) resonance is such an interpretation unambiguous.

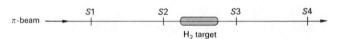

Figure 4.3 Schematic drawing of measurement of the total pion-proton cross-section.

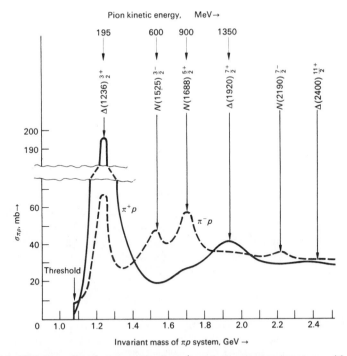

Figure 4.4 Variation of total cross section for π^+- and π^--mesons on protons, with incident pion energy. The symbol Δ refers to resonances of $I = \frac{3}{2}$; N refers to $I = \frac{1}{2}$. The positions of only a few of the known states, together with their spin-parity assignments, are given.

4.5. STRANGENESS AND ISOSPIN

As explained in Chapter 1, the strange particles were so named because of their long lifetime (for decay via the weak interactions) in contrast with their copious production (in strong interactions). It was argued that a new quantum number, called the strangeness S, was involved. Strangeness would be conserved in the associated strong production of particles of opposite strangeness, but violated in the weak decay of single strange particles into nonstrange particles.

The assignments of isospin and strangeness were obtained as follows. First we note that the Λ-hyperon has no charged counterpart, implying $I_\Lambda = 0$. The values of I and I_3 involved in Λ-decay are as follows:

$$\Lambda \rightarrow p + \pi^-.$$

I	0	$\frac{1}{2}$	1
I_3	0	$\frac{1}{2}$	-1

This is a *weak* decay process, and neither I nor I_3 is conserved on the two sides of the equation. From the fact that the Λ-hyperon and neutral kaon have been observed to be produced in association in strong interactions of pions with protons (in the early diffusion-cloud-chamber experiments), we can assign the kaon half-integral isospin. $I_K = \frac{1}{2}$ is the simplest choice:

$$\pi^- + p \rightarrow \Lambda + K^0.$$

| I | 1 | $\frac{1}{2}$ | 0 | $\frac{1}{2}$ |
| I_3 | -1 | $\frac{1}{2}$ | 0 | $-\frac{1}{2}$ |

The correct charge for the neutral kaon is obtained if we set

$$\frac{Q}{e} = I_3 + \frac{1}{2}, \tag{4.26}$$

implying that the K^+-meson, of $I_3 = +\frac{1}{2}$, is the charged member of a kaon doublet. This assignment also forbids the strong decay of a kaon according to the scheme $K^+ \rightarrow \pi^+ + \pi^- + \pi^+$; as already seen, this is a weak decay process. The K^--meson does not fit into this scheme. It was therefore necessary to postulate a second K-doublet, with

$$\frac{Q}{e} = I_3 - \frac{1}{2}, \tag{4.27}$$

which predicts a second neutral kaon of $I_3 = +\frac{1}{2}$—called $\bar{K}^0$. K^+ and K^- are considered as particle and antiparticle, as are K^0 and $\bar{K}^0$. The interesting phenomena associated with K^0 and $\bar{K}^0$ are discussed in Section 7.14.

Gell-Mann (1953) and Nishijima (1955) pointed out that the formulae (4.17), (4.26), and (4.27) could be more elegantly expressed by introducing the *strangeness* quantum number, according to the formula

$$\frac{Q}{e} = \frac{B}{2} + \frac{S}{2} + I_3. \tag{4.28}$$

The assignment of strangeness S follows from the isospin assignments. Nucleons and pions clearly must have $S = 0$, the Λ-hyperon $S = -1$, the K^0, K^+ doublet $S = +1$, and the K^-, $\bar{K}^0$ doublet $S = -1$. These assignments are shown in Table 4.2.

An example of conservation of strangeness in the reaction

$$K^- + p \rightarrow \Lambda + \pi^0$$

| S | -1 | 0 | -1 | 0 |
| I_3 | $-\frac{1}{2}$ | $+\frac{1}{2}$ | 0 | 0 |

is shown in Fig. 4.5.

The Σ-hyperons fit into a charge triple, $I = 1$. This assignment fits into the observed strong production reaction

$$\pi^\pm + p \rightarrow \Sigma^\pm + K^+$$

I	1	$\frac{1}{2}$	1	$\frac{1}{2}$
I_3	± 1	$\frac{1}{2}$	± 1	$\frac{1}{2}$
S	0	0	-1	1

TABLE 4.2 Isospin and strangeness assignments for particles decaying by weak or electromagnetic interactions

					I_3		
B	S	I	-1	$-\frac{1}{2}$	0	$+\frac{1}{2}$	$+1$
1	0	$\frac{1}{2}$		n		p	
1	-1	0			Λ		
0	0	1	π^-		π^0		π^+
0	$+1$	$\frac{1}{2}$		K^0		K^+	
0	-1	$\frac{1}{2}$		K^-		$\bar{K}^0$	
1	-1	1	Σ^-		Σ^0		Σ^+
1	-2	$\frac{1}{2}$		Ξ^-		Ξ^0	
1	-3	0			Ω^-		
0	0	0			η		

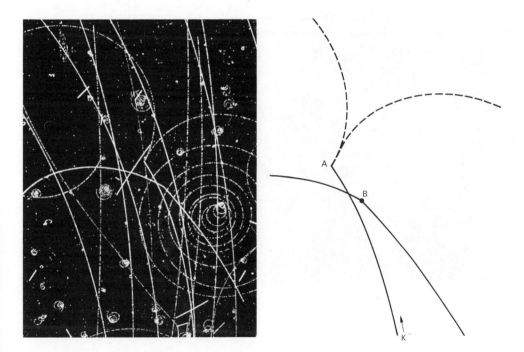

Figure 4.5 Example of the reaction $K^- + p \rightarrow \Lambda + \pi^0$ occurring when a K^--meson comes to rest in a hydrogen bubble chamber, at the point A. The neutral pion undergoes Dalitz decay, $\pi^0 \rightarrow e^+ e^- \gamma$. The Λ-hyperon decays ($\Lambda \rightarrow \pi^- + p$) at the point B. (Courtesy CERN.)

and accounts for the fact that the decay $\Sigma^+ \to n + \pi^+$ is weak. The predicted neutral member, Σ^0, was not finally identified until 1959. It undergoes the electromagnetic decay mode

$$\Sigma^0 \to \Lambda + \gamma.$$

I	1	0	0
I_3	0	0	0
S	-1	-1	0

The cascade or Ξ^--hyperon was first observed in the early cosmic-ray work with cloud chambers; the $S = -2$ assignment followed from the production in association with a pair of K^0-mesons, and from the fact that the decay

$$\Xi^- \to \Lambda + \pi^-$$

was weak. The expected neutral counterpart Ξ^0 was detected in 1959. An example of Ξ^--decay is shown in Fig. 4.6.

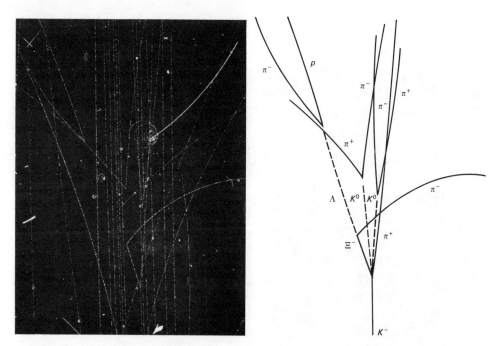

Figure 4.6 Example of the decay $\Xi^- \to \pi^- + \Lambda$ in a hydrogen bubble chamber. The reaction is produced by an incident 10-GeV/c K^--meson:

$$K^- + p \to \Xi^- + \pi^+ + K^0 + K^0 + \bar{K}^0$$

Strangeness S: -1 0 -2 0 $+1$ $+1$ -1.

The final $\bar{K}^0$ is not observed in this picture. Both K^0-mesons decay in the mode $K^0 \to \pi^+\pi^-$. (Courtesy CERN Information Service.)

TABLE 4.3 Mass differences in isospin multiplets

	Δm (MeV/c^2)	m_{av} (MeV/c^2)	$10^3 \, \Delta m/m$
$n - p$	1.3	939	1.4
$\Sigma^0 - \Sigma^+$	3.1	1190	2.6
$\Sigma^- - \Sigma^0$	4.9	1195	4.1
$\Xi^- - \Xi^0$	6.5	1318	4.9
$K^0 - K^\pm$	4.0	495	8.1
$\pi^\pm - \pi^0$	4.6	140	33

The existence of the Ω^--baryon of $S = -3$, as well as its mass and decay modes, was predicted before it was observed in 1964, on the basis of unitary symmetry and the quark model (Chapter 5). A picture of the first Ω^--event is given in Fig. 5.2.

The electromagnetic mass splittings between members of isospin multiplets are summarized in Table 4.3. As expected, they are of order $\Delta m/m \simeq \alpha \simeq 10^{-2}$. Note that particle and antiparticle must have identical masses, by the *CPT* theorem, and thus $m_{\pi^+} \equiv m_{\pi^-}$, $m_{K^+} \equiv m_{K^-}$. However, $m_{\Sigma^+} \neq m_{\Sigma^-}$, since Σ^+ and Σ^- are both baryons, rather than baryon and antibaryon.

4.5.1. *G*-parity

The charge-conjugation operator C can have eigenvalues only for neutral systems, such as the π^0, γ, η, and e^+e^-. It is useful, however, to be able to formulate selection rules for some charged systems; this can be done for strong interactions by combining the operation of charge conjugation with an isospin rotation. For this purpose, consider the operation

$$G = CR = C \exp(i\pi I_2). \tag{4.29}$$

The operation G consists of a rotation R of $180°$ about the y-axis in isospin space, followed by charge conjugation. Applied to a state with a z-component of isospin I_3, this amounts to first flipping $I_3 \to -I_3$ and then reversing the process, $-I_3 \to I_3$. It is therefore plausible that charged states may be eigenfunctions of the G-operator. In order to get the eigenvalues, consider an isospin state $\chi(I, I_3 = 0)$. Under isospin rotations, this state behaves precisely like the angular-momentum wavefunction $Y_l^{m=0}(\theta, \phi)$ of (3.16) under rotations in ordinary space. The R-operation $\exp(i\pi L_y)$ implies $\theta \to \pi - \theta$, $\phi \to \pi - \phi$, and hence

$$Y_1^0 \underset{R}{\to} (-1)^l Y_l^0.$$

Therefore,

$$\chi(I, 0) \underset{R}{\to} (-1)^I \chi(I, 0).$$

As an example, for a state of nucleon and antinucleon, of total spin s and orbital angular momentum l, the effect of the C-operation is to give a factor $(-1)^{l+s}$, just as in the case of positronium. Thus, the effect of the operation $G = CR$ on a neutral ($I_3 = 0$) nucleon-antinucleon system $|\psi\rangle$ will be

$$G|\psi\rangle = (-1)^{l+s+I}|\psi\rangle.$$

Since the strong interactions are invariant under isospin rotations, this must, in fact, be a general formula, and not limited to the case $I_3 = 0$ for which it was derived.

Now suppose the G-operator acts on a pion wave function, $|\pi^+\rangle$. R reverses I_3, thus converting $\pi^+ \to \pi^-$, and C flips the charge back, $\pi^- \to \pi^+$. We may then write

$$G|\pi^+\rangle = \pm|\pi^+\rangle,$$

$$G|\pi^-\rangle = \pm|\pi^-\rangle,$$

$$G|\pi^0\rangle = \pm|\pi^0\rangle.$$

The neutral pion must be an eigenstate of C, with eigenvalue $+1$, since it decays in the mode $\pi^0 \to 2\gamma$. Then the R-eigenvalue is $(-1)^I = -1$. Thus

$$G|\pi^0\rangle = -|\pi^0\rangle.$$

The eigenvalue of the G-operator is called the G-parity. While the G-parity of the neutral pion is unambiguous, that of the charged pions is not. They are not eigenstates of C, and in the process of charge conjugation, an arbitrary phase appears, which can be chosen at will. For convenience, however, it is the practice to define the phases so that all members of an isospin triplet have the same G-parity as the neutral member. In the present case we can then write

$$G|\pi\rangle = -|\pi\rangle \tag{4.30}$$

provided we define

$$C|\pi^\pm\rangle = -|\pi^\mp\rangle.$$

Since the C-operation reverses the sign of the baryon number, it will be apparent that eigenstates of G-parity must have baryon number zero. The charge-conjugation quantum number is multiplicative and isospin additive, so that G-parity is multiplicative. Thus, for a state of n pions,

$$G|\psi(n\pi)\rangle = (-1)^n|\psi(n\pi)\rangle. \tag{4.31}$$

The concept of G-parity introduces nothing that is not already known from the twin postulates of charge conjugation and isospin invariance; it simply allows some short cuts when we consider selection rules for the decay of meson resonances. The relevant quantum numbers are given in Table 4.4.

TABLE 4.4

Particle $\left(\begin{smallmatrix}\text{mass,}\\\text{MeV}\end{smallmatrix}\right)$	$\pi(140)$	$\rho(770)$	$\omega(783)$	$\phi(1020)$	$f(1270)$	$\eta(549)$	$\eta'(958)$
Spin parity J^P	0^-	1^-	1^-	1^-	2^+	0^-	0^-
Isospin I	1	1	0	0	0	0	0
G-parity	-1	$+1$	-1	-1	$+1$	$+1$	$+1$
Dominant pion decay mode	—	2π	3π	3π	2π	3π	5π

Note that the vector mesons ρ, ω, ϕ, and f decay by strong interactions, being resonant states with large widths (3 to 100 MeV), and that the multiplicity of the pion decay modes follows the rule $G = (-1)^n$. On the other hand, the existence of the $\gamma\gamma$-decay mode of the η and η' proves that these mesons decay by electromagnetic transitions, with total widths of 0.9 keV and 0.3 MeV, respectively. Since $I = 0$, and $C(2\gamma) = +1$, they must have $G = +1$. The strong decay into two pions is forbidden by parity conservation, leaving the three-pion, G-violating electromagnetic decay as the only possibility.

4.6. DALITZ PLOTS

In a reaction such as

$$\pi + p \rightarrow \pi + \pi + p \tag{4.32}$$

one wants to investigate possible interactions among the final-state particles —for example, the occurrence of resonances $(\pi p, \pi\pi)$—and thus how the observed distributions in secondary momentum and angle depart from the situation where the matrix element is constant. In the latter case, the distributions will be determined simply from the phase-space factors. The departure of such distributions from phase space can be assessed from a phase-space or *Dalitz* plot (Dalitz 1953, Fabri 1954).

4.6.1. Three-Body Phase Space

As indicated in Eq. (4.5), the number of quantum states available in phase space, per unit normalization volume, is

$$\frac{p^2 \, dp \, d\Omega}{h^3}$$

for a spinless particle of momentum $p \to p + dp$ inside the solid-angle element $d\Omega$. In a reaction such as (4.32) with three final-state particles, labeled 1, 2, and 3, and fixed initial energy, the number of states will be proportional to

$$p_1^2 \, dp_1 p_2^2 \, dp_2 \, d\Omega_1 \, d\Omega_2.$$

In the center-of-momentum system (CMS), $\mathbf{p}_3 = -(\mathbf{p}_1 + \mathbf{p}_2)$ is fixed, so there is no factor for particle 3. If the initial state is unpolarized, the overall orientation in space will be isotropic. The integral over all directions of particle 1 is then $\int d\Omega_1 = 4\pi$, while $d\Omega_2 = 2\pi \, d(\cos \theta_{12})$, where θ_{12} is the angle between particles 1 and 2. Thus, the number of states in the element $dp_1 \, dp_2 \, d(\cos \theta_{12})$ is

$$dN = \text{const } p_1^2 \, dp_1 p_2^2 \, dp_2 \, d(\cos \theta_{12}).$$

The matrix element for the interaction is usually cast in Lorentz-invariant form, by normalizing the wavefunctions of the particles in their individual rest frames. A volume V in the particle rest frame will be Lorentz-contracted to a value Vm/E in a reference frame (in this case the CMS) where the particle has total energy E. The phase-space expression can be made relativistically invariant by including a factor m/E or $1/E$ for each final-state particle (see Appendix A). This then has the form

$$dN = \text{const } \frac{p_1^2 \, dp_1 p_2^2 \, dp_2 \, d(\cos \theta_{12})}{E_1 E_2 E_3}. \tag{4.33}$$

Using the relations

$$E_1^2 = p_1^2 + m_1^2, \qquad E_2^2 = p_2^2 + m_2^2,$$

$$E_3^2 = p_3^2 + m_3^2 = p_1^2 + p_2^2 + m_3^2 + 2p_1 p_2 \cos \theta,$$

$$E_1 \, dE_1 = p_1 \, dp_1, \qquad E_2 \, dE_2 = p_2 \, dp_2,$$

$$(E_3 \, dE_3)_{p_1, \, p_2 \text{ fixed}} = p_1 p_2 \, d(\cos \theta),$$

one obtains

$$dN = \text{const } \frac{E_1 \, dE_1 E_2 \, dE_2 E_3 \, dE_3}{E_1 E_2 E_3}.$$

The density of final states is obtained by dividing by dE_f, where $E_f = E_1 + E_2 + E_3$ is the total energy and, for E_1 and E_2 fixed, $dE_f = dE_3$. So

$$\rho = \frac{dN}{dE_f} = \text{const } dE_1 \, dE_2. \tag{4.34}$$

If now one includes the matrix element M for the interaction, the Dalitz plot population in the interval $dE_1\ dE_2$ will be

$$\rho = |M(E_1, E_2)|^2\ dE_1\ dE_2, \tag{4.35}$$

so the density at any point is a measure of the square of the matrix element.

4.6.2. $K_{\pi 3}$-Decay

The original Dalitz plot was made for the decay

$$K^+ \rightarrow \pi^+ + \pi^+ + \pi^-, \qquad Q = 75 \text{ MeV} = \text{total kinetic energy}.$$

In this case, the three particles have equal mass and are nonrelativistic (or very nearly so). The plot was actually made of the kinetic energies of the pions along three axes at $120°$, as shown in Fig. 4.7(a). Again, the density of points should be uniform if the matrix element is constant, and they should lie inside a circle of radius $Q/3$, where Q is the total kinetic energy available (see Problem 4.10).

We now ask what the departures from phase space for various spin assignments are for the kaon, i.e., for the $\pi^+ \pi^+ \pi^-$ combination. Let l^+ be the

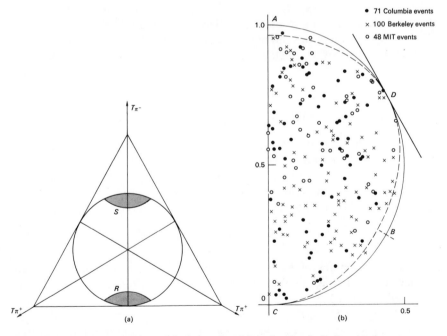

Figure 4.7 (a) Expected regions of depletion, R and S, in the $K \rightarrow 3\pi$ Dalitz plot, for a nonzero kaon spin, J_K. (b) Observed distribution of events in $K \rightarrow 3\pi$, showing uniform population, and thus $J_K = 0$. All events are plotted in one semicircle. The dashed curve is the boundary using relativistic kinematics, and departs slightly from a circle. (After Orear et al. 1956.)

angular momentum of the two *like* pions in the CMS—which must be even, by Bose symmetry—and let l^- be that of the negative pion relative to the dipion system. If $l^+ > 0$ (i.e., 2, 4,...) the matrix element and Dalitz-plot density should vanish in the region S in Fig. 4.7(a), where the two like pions are mutually at rest. If $l^- > 0$ (i.e., 1, 2, 3,...), the plot should be depleted in the region R, where the negative pion has very small energy relative to the dipion. Since the actual plot (Fig. 4.7(b)) is uniform everywhere, neither l^+ nor l^- can be nonzero, and hence $J_K = 0$ is the only possible assignment.

It must be emphasized that the above analysis gives no information about the kaon *parity*. Three pions in a relative S-state have parity $(-1)^3 = -1$; since, however, parity is not conserved in the weak decay $K \to 3\pi$, this tells us nothing about the initial state. The kaon parity may be determined by the study of *hypernuclei*. These are nuclei in which one neutron is replaced by a bound Λ-hyperon. For example, when negative kaons are brought to rest in helium, and undergo capture from an atomic S-state, a few percent of the events give rise to bound, rather than free, Λ-hyperons, according to the reaction

$$K^- + {}^4\text{He} \to {}^4\text{H}_\Lambda + \pi^0,$$

where the hypernucleus ${}^4\text{H}_\Lambda$ consists of a triton (${}^3\text{H}$) plus a bound Λ-particle. Detailed measurements of its (weak) decay modes establish that $J_{{}^4\text{H}_\Lambda} = 0$. Thus, $J = l = 0$ on both sides of the equation. If the Λ-parity is defined to be positive, like that of the nucleon, the kaon must then have $J^P = 0^-$, like the pion.

4.6.3. Dalitz Plots Involving Three Dissimilar Particles

As an example of a Dalitz plot involving three particles of unequal mass, we take the case of the hyperon resonance $Y_1^*(1385)$—also denoted $\Sigma(1385)$. This state was first observed by Alston *et al.* in 1960 in interactions of 1.15-GeV/c K^--mesons in a liquid-hydrogen bubble chamber at the Lawrence Radiation Laboratory, Berkeley. The reaction studied was

$$K^- + p \to \pi^+ + \pi^- + \Lambda. \tag{4.36}$$

The kinetic energies of π^+- and π^--mesons, as computed in the overall center-of-mass frame, are plotted along the x- and y-axes respectively. If Q is the total available kinetic energy of the three final-state particles in this frame, then $T_\Lambda = Q - (T_{\pi^+} + T_{\pi^-})$, and lines of constant T_Λ are thus inclined at $45°$ to the axes, as shown in Fig. 4.8. Momentum-energy conservation keeps the points representing individual events inside the distorted ellipse shown. If there are no strong correlations in the final state (4.36), the density of points should be uniform, as usual.

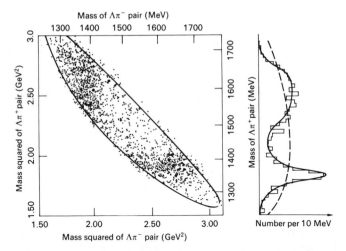

Figure 4.8 Dalitz plot of the $\Lambda\pi^+\pi^-$ events from reaction (4.36), as measured by Shafer *et al.* (1963), for 1.22-GeV/c incident momentum. The effective $\Lambda\pi^+$ mass spectrum is shown at right. The dashed curve is that expected for a phase-space distribution (ordinate equal to the interval in $M^2_{\Lambda\pi^-}$ within the Dalitz-plot boundary), while the full curve corresponds to a Breit-Wigner resonance expression fitted to the $\Lambda\pi^+$ and $\Lambda\pi^-$ systems.

Figure 4.8 shows clearly the strong departures from uniform density, as evidenced by horizontal and vertical bands on the plot, corresponding to favored values of T_{π^-} and T_{π^+}. This means that the reaction (4.36) is then proceeding as a two-body reaction, the two bodies consisting of one pion ($\pi 1$) and a resonant state of the Λ with the other pion ($\pi 2$), with (more or less) unique mass $M_{\Lambda\pi} = 1385$ MeV. Thus, if p is the CMS momentum of each, the total CMS energy will be

$$W = \sqrt{M^2_{\Lambda\pi} + p^2} + \sqrt{m^2_\pi + p^2}, \qquad (4.37)$$

so that, for fixed $M_{\Lambda\pi}$, the momentum p is unique. Obviously, if there is a broad resonance, the quantity T_π will have a spread in values, and thus form a band rather than a line on the Dalitz plot. Note that *both* vertical and horizontal bands are found, since either π^+ or π^- can resonate with the Λ-hyperon. Thus

$$\Sigma(1385) \to \Lambda + \pi^\pm,$$
$$I \quad 1 \quad 0 \quad 1$$

and this resonance must have isospin 1, since it decays by strong interaction and isospin is conserved.

In the Dalitz plot it is usual to display $M^2_{\Lambda\pi^-}$ instead of T_{π^+} along the x-axis (and $M^2_{\Lambda\pi^+}$ along the y-axis), so that one can read off the $\Lambda\pi$ invariant mass directly. It is readily shown that

$$M^2_{\Lambda\pi^-} = W^2 + m^2_\pi - 2WE_{\pi^+} = a + bT_{\pi^+}, \qquad (4.38)$$

so that $M^2_{\Lambda\pi^-}$ is proportional to T_{π^+} for fixed W, the total CMS energy. Figure 4.8 also includes the $\Lambda\pi^+$ mass spectrum. The pure phase-space distribution (4.34) is shown, together with a curve obtained by assuming that either $\Lambda\pi^+$ or $\Lambda\pi^-$ may resonate, the resonance being described by the Breit-Wigner formula (4.55) with a width $\Gamma = 40$ MeV.

Finally, the spin-parity of the $\Sigma(1385)$ has been determined to be $J^P = \frac{3}{2}^+$; this conclusion is arrived at by analysis of the polarization in the Λ-decay, relative to the production plane of reaction (4.36).

4.7. WAVE-OPTICAL DISCUSSION OF HADRON SCATTERING

Consider a beam of particles to be represented by a plane wave traveling in the z-direction and incident on a spinless target particle, or scattering center. Such an incident wave, which we take as of unit amplitude, is represented by

$$\psi_i = e^{ikz},$$

where $k = 1/\lambdabar$ and $2\pi\lambdabar$ is the de Broglie wavelength, and where the time dependence $e^{-i\omega t}$ has been omitted for brevity. A plane wave can be represented as a superposition of spherical waves, incoming and outgoing. At a radial distance r from the scattering center, such that $kr \gg 1$, the radial dependence of these spherical waves has the form $e^{\pm ikr}/kr$, so that the flux through a spherical shell is independent of r, as it must be to conserve probability. The angular dependence is determined by the Legendre polynomials $P_l(\cos\theta)$.

The expansion is $(kr \gg 1)$*

$$\psi_i = e^{ikz} = \frac{i}{2kr}\sum_i (2l + 1)[(-1)^l e^{-ikr} - e^{ikr}]P_l(\cos\theta), \qquad (4.39)$$

where the first term in square brackets denotes the incoming wave and the second denotes the outgoing wave as in Fig. 4.9 (the sense is clear if one includes the time-dependent term $e^{-i\omega t}$). In all problems with which we deal here, k is typically 10^{13} cm^{-1}, and r, the distance from the scattering center where the particle wave is observed, is many centimeters. Thus, the asymptotic form (4.39) is appropriate.

The scattering center or potential cannot affect the incoming waves, but can in general alter both phase and amplitude of the outgoing wave. The change of phase of the lth partial wave is denoted by $2\delta_l$, and its amplitude by η_l, where $1 > \eta_l > 0$.

---

* See, for example: F. Mandl, *Quantum Mechanics*, Butterworths, London, 1957, p. 166; L. I. Schiff, *Quantum Mechanics*, McGraw-Hill, New York, 1955, p. 103.

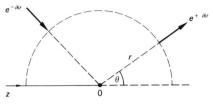

Figure 4.9

The total wave now has the asymptotic form

$$\psi_{\text{total}} = \frac{i}{2kr} \sum_l (2l + 1)[(-1)^l e^{-ikr} - \eta_l e^{2i\delta_l} e^{ikr}] P_l(\cos \theta). \qquad (4.40)$$

Thus, the scattered wave, representing the difference between the outgoing waves with and without the scattering potential, will be

$$\psi_{\text{scatt.}} = \psi_{\text{total}} - \psi_i = \frac{e^{ikr}}{kr} \sum_l (2l + 1) \frac{(\eta_l e^{2i\delta_l} - 1)}{2i} P_l(\cos \theta)$$

$$= \frac{e^{ikr}}{r} F(\theta), \qquad (4.41)$$

where the scattering amplitude

$$F(\theta) = \frac{1}{k} \sum_l (2l + 1) \left(\frac{\eta_l e^{2i\delta_l} - 1}{2i} \right) P_l(\cos \theta). \qquad (4.42)$$

We note that this corresponds to an *elastically* scattered wave, since the wavenumber k is taken to be the same before and after scattering. This can be true in the laboratory frame only if the scattering center is infinitely massive. In general, the target (scattering) particle will acquire both momentum and energy; thus, the quantities k and λ strictly refer to the properties of the wave in the center-of-momentum frame of the incident and target particles, so they do not change in an elastic collision. The scattered outgoing flux in solid angle $d\Omega$, through a sphere of radius r, is

$$v_0 \psi_{\text{scatt.}} \psi_{\text{scatt.}}^* r^2 \, d\Omega = v_0 |F(\theta)|^2 \, d\Omega, \qquad (4.43)$$

where v_0 is the velocity of the outgoing particles (relative to the scattering center). But (4.43) is, by definition, the product of the scattering cross-section and the incident flux $(= v_i \psi_i \psi_i^* = v_i)$. Since $v_i = v_0$ for elastic scattering,

$$v_0 \, d\sigma = v_0 |F(\theta)|^2 \, d\Omega,$$

or

$$\left(\frac{d\sigma}{d\Omega} \right)_{\text{el.}} = |F(\theta)|^2. \qquad (4.44)$$

The Legendre polynomials P_l obey the orthogonality condition

$$\int P_l P_{l'} \, d\Omega = \frac{4\pi\delta_{l,l'}}{2l+1},$$

where

$$\delta_{l,l'} = 1 \qquad \text{for} \quad l = l',$$
$$= 0 \qquad \text{for} \quad l \neq l'.$$

Thus, the total *elastic* scattering cross-section, integrated over angle, is, from (4.42) and (4.44),

$$\sigma_{\text{el.}} = 4\pi\lambdabar^2 \sum_l (2l+1) \left| \frac{\eta_l e^{2i\delta_l} - 1}{2i} \right|^2. \tag{4.45}$$

When $\eta = 1$, the case for no absorption of the incoming wave, this becomes

$$\sigma_{\text{el.}} = 4\pi\lambdabar^2 \sum_l (2l+1) \sin^2 \delta_l. \tag{4.46}$$

Obviously, $\sigma_{\text{el.}}$ is zero when $\delta_l = 0$, corresponding to zero scattering potential. If $\eta < 1$, the *reaction* cross-section, σ_r, is then obtained from conservation of probability:

$$\sigma_r = \int (|\psi_{\text{in}}|^2 - |\psi_{\text{out}}|^2) r^2 \, d\Omega,$$

where ψ_{in} is the first term of (4.39) and ψ_{out} is the second term of (4.40). This gives

$$\sigma_r = \pi\lambdabar^2 \sum_l (2l+1)(1 - \eta_l^2). \tag{4.47}$$

The *total* cross-section will be

$$\sigma_T = \sigma_r + \sigma_{\text{el.}} = \pi\lambdabar^2 \sum_l (2l+1)2(1 - \eta_l \cos 2\delta_l).$$

Since $P_l(1) = 1$ for all l, (4.42) therefore gives in the *forward direction* $\cos \theta = 1$, $\theta = 0$:

$$\text{Im } F(0) = \frac{1}{2k} \sum_l (2l+1)(1 - \eta_l \cos 2\delta_l).$$

Comparing the last two equations finally gives us the *optical theorem*,

$$\text{Im } F(0) = \frac{k}{4\pi} \sigma_T, \tag{4.48}$$

relating the total cross-section to the imaginary part of the forward elastic scattering amplitude.

The relations derived above describe the various cross-sections $\sigma_{el.}$, σ_r, and σ_T in terms of the parameters η and δ. They set bounds on the cross-sections imposed by the conservation of probability (often called the unitarity condition). For example, we see from (4.46) that the maximum elastic scattering cross-section for the lth partial wave occurs when $\delta_l = \pi/2$, having the value

$$\sigma_{el.}^{max} = 4\pi\lambda^2(2l + 1), \tag{4.49}$$

for $\eta_l = 1$, i.e., the case where there is pure scattering without absorption. Similarly, from (4.47) we obtain the maximum absorption or reaction cross-section by setting $\eta_l = 0$:

$$\sigma_r^{max} = \pi\lambda^2(2l + 1).$$

This last equation can be reproduced from a simple classical argument. An orbital angular momentum l corresponds to an "impact parameter" b given by $l\hbar = pb$, or $b = l\lambda$. Particles of angular momentum $l \rightarrow l + 1$ therefore impinge on, and are absorbed by, an annular ring of cross-sectional area

$$\sigma = \pi(b_{l+1}^2 - b_l^2) = \pi\lambda^2(2l + 1).$$

Note also that, for the case of complete absorption, $\eta = 0$, the elastic cross-section is also $\pi\lambda^2(2l + 1)$ — see Section 4.11.

The quantity

$$f(l) = \frac{\eta_l e^{2i\delta_l} - 1}{2i} = \frac{i}{2} - \frac{i\eta_l}{2}e^{2i\delta_l} \tag{4.50}$$

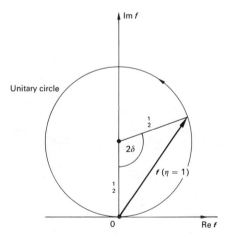

Figure 4.10 The scattering amplitude **f** plotted as a vector in the complex plane. Causality requires that as the energy increases, the vector will trace out the circle in an anticlockwise direction.

of (4.42) is the elastic scattering amplitude (for the lth partial wave). It is a complex quantity, and in Fig. 4.10 we show **f** plotted as a vector in the complex plane. For $\eta = 1$, the end of the vector describes a circle of radius $\frac{1}{2}$ and center $i/2$, as the phase shift varies from 0 to $\pi/2$. When $\delta = \pi/2$, **f** is purely imaginary and has magnitude unity, corresponding to (4.49). If $\eta < 1$, the end of the vector **f** lies within the circle shown, sometimes referred to as the *unitary circle*. It is so called because, as **f** is defined, its maximum modulus must be unity if probability is conserved—the intensity in a particular outgoing partial wave cannot exceed that in the corresponding incoming wave.

4.8. THE BREIT-WIGNER RESONANCE FORMULA

Let us now make the connection between the preceding wave-optical discussion and the scattering of two elementary particles—one corresponding to the incident wave and the other to the scattering center. To simplify matters let the two particles be spinless. If the elastic scattering amplitude $f(l)$ passes through a maximum for a particular value of l and for a particular CMS wavelength $\lambdabar$, the two particles are said to *resonate*. The resonant state is then characterized by a unique angular momentum or spin $J = l$, a unique parity and isospin, and a mass corresponding to the total CMS energy of the two particles. A criterion of resonance is that the phase shift δ_l of the lth partial wave should pass through $\pi/2$. The cross-section may also be described in terms of the width Γ or lifetime τ of the resonant state, as follows: dropping the subscript l in (4.50), and with $\eta = 1$, we may rewrite f in the form

$$f = \frac{e^{i\delta}(e^{i\delta} - e^{-i\delta})}{2i}$$

$$= e^{i\delta} \sin \delta = \frac{1}{\cot \delta - i}. \tag{4.51}$$

Near resonance $\delta \simeq \pi/2$, so that $\cot \delta \simeq 0$. If E is the total energy of the two-particle state in the CMS, and E_R is the value of E at resonance ($\delta = \pi/2$), then expanding by a Taylor series

$$\cot \delta(E) = \cot \delta(E_R) + (E - E_R)\left[\frac{d}{dE} \cot \delta(E)\right]_{E=E_R} + \cdots$$

$$\simeq -(E - E_R)\frac{2}{\Gamma},$$

where $\cot \delta(E_R) = 0$ and we have defined $2/\Gamma = -[d(\cot \delta(E))/dE]_{E=E_R}$. Neglecting further terms in the series is justified provided $|E - E_R| \simeq \Gamma \ll$

E_R. Then the resonance is symmetric and $[d^2(\cot \delta(E))/dE^2]_{E=E_R} = 0$. (If the resonance is broad however, the phase-space factor for decay varies appreciably over the width, and the resonance is asymmetric). From (4.51)

$$f(E) = \frac{1}{\cot \delta - i} = \frac{\Gamma/2}{(E_R - E) - i\Gamma/2}. \qquad (4.52)$$

From (4.45) and (4.50), we obtain for the elastic scattering cross-section

$$\sigma_{\text{el.}}(E) = 4\pi \lambdabar^2(2l + 1) \frac{\Gamma^2/4}{(E - E_R)^2 + \Gamma^2/4}. \qquad (4.53)$$

This is known as the *Breit-Wigner formula*. The resonance curve of $\sigma(E)$ is shown in Fig. 4.11. The width Γ is defined so that the elastic cross section $\sigma_{\text{el.}}$ falls by a factor 2 from the peak value when $|E - E_R| = \pm \Gamma/2$.

As pointed out in Chapter 1, the width Γ and lifetime τ of the resonant state are connected by the relation $\tau = \hbar/\Gamma$. The energy dependence of the amplitude (4.53) is simply the Fourier transform of an exponential time pulse, corresponding to the radioactive decay of the resonance. The wavefunction of a nonstationary decaying state of central angular frequency $\omega_R = E_R/\hbar$ and lifetime $\tau = \hbar/\Gamma$ can be written

$$\psi(t) = \psi(0)e^{-i\omega_R t}e^{-t/2\tau}$$
$$= \psi(0)e^{-t(iE_R + \Gamma/2)} \qquad (4.54)$$

in units $\hbar = c = 1$. The intensity $I(t) = \psi\psi^* = I(0)e^{-t/\tau}$ obeys the normal exponential law of radioactive decay. The Fourier transform of this expression is

$$g(\omega) = \int_0^\infty \psi(t)e^{i\omega t} \, dt,$$

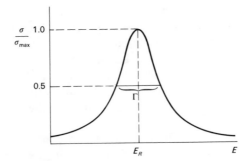

Figure 4.11 Breit-Wigner resonance curve.

with $\omega = E/\hbar = E$. The amplitude as a function of E is then

$$\chi(E) = \int \psi(t) e^{iEt}\, dt = \psi(0) \int e^{-t[(\Gamma/2) + iE_R - iE]}\, dt$$

$$= \frac{K}{(E_R - E) - i\Gamma/2},$$

where K is some constant.

The final step consists of making the connection between the probability, integrated over time, that the resonant state will decay with total energy E, and the cross-section $\sigma_{el.}(E)$. If the resonance is purely elastic, it is clear from conservation of probability that the cross-section—which measures the probability of forming the resonant state—must be proportional to the probability of decay. Thus, we may write $\sigma(l, E) \propto \chi^*\chi$, where $\chi = \chi_l(E)$ and l is the value of the orbital angular momentum involved in forming the resonant state in question. Both of these quantities have a maximum when $E = E_R$ and $\delta_l = \pi/2$, so that

$$(\sigma_{el.})_{max} = 4\pi\lambda^2(2l + 1) \quad \text{and} \quad (\chi^*\chi)_{max} = 4K^2/\Gamma^2.$$

Thus, in terms of Γ, measuring the width of a decay process, rather than δ, measuring the phase-shift in a scattering process, we have

$$\sigma_{el.} = 4\pi\lambda^2(2l + 1) \frac{\Gamma^2/4}{(E - E_R)^2 + \Gamma^2/4},$$

as in (4.53).

For a spinless projectile hitting a spinless target, $l = J$, the total angular momentum of the resonant state. Therefore, $(2l + 1) \to (2J + 1)$. For the formation and decay of a resonance of angular momentum J in the collision of particles of spins s_a and s_b, it being understood that the cross-section is averaged over the spin states of a and b, we get

$$\sigma_{el.}(E) = \frac{4\pi\lambda^2(2J + 1)\Gamma^2/4}{(2s_a + 1)(2s_b + 1)[(E_R - E)^2 + \Gamma^2/4]}. \tag{4.55}$$

The spin-multiplicity factors follow from considerations of the quantum states available in the reversible reaction $a + b \rightleftharpoons c$, as in (4.8) and (4.14).

In the foregoing, it is assumed that the resonance can only decay elastically, e.g., $\pi + n \to \Delta \to \pi + n$, but not $\Delta \to 2\pi + n$. In general, $\Gamma = \Gamma_{el.} + \Gamma_r$, where $\Gamma_{el.}$ and Γ_r are partial widths for decay in the elastic channel and inelastic channel. Then for $\sigma_{el.}(E)$, the Γ^2 numerator in (4.55) should be replaced by $\Gamma_{el.}^2$. The inelastic cross section $\sigma_r(E)$ is the same as (4.55), with the Γ^2 numerator replaced by $\Gamma_{el.}\Gamma_r$.

The scattering amplitude f, in the case of an elastic resonance ($\eta = 1$), traces out the unitary circle of Fig. 4.10. By applying the concept of causality—the outgoing wave cannot leave the scattering center before the incoming wave arrives—it may be proved that f traces out a circle in the *anticlockwise* direction (for an attractive potential).

4.9. AN EXAMPLE OF A BARYON RESONANCE—THE Δ(1232)

Figure 4.4 shows the $\pi^+ p$ and $\pi^- p$ total cross-sections as a function of kinetic energy of the incident pion. There is a very obvious $I = \frac{3}{2}$ resonance at $T_\pi = 195$ MeV, corresponding to a pion-proton mass of 1232 MeV. It was discovered by Fermi and Anderson in 1949. It is designated $P_{33}(1232)$, meaning that it is a p-wave ($l = 1$) pion-nucleon resonance, of $I = \frac{3}{2}$ and $J = \frac{3}{2}$. One can also distinguish other humps and bumps in σ(total). For example, in the $I = \frac{1}{2}$ channel one can see evidence for the states $D_{13}(1520)$ and $F_{15}(1688)$ as peaks in $\sigma(\pi^- p)$, and $F_{37}(1950)$ as a peak in $\sigma(\pi^+ p)$, i.e., $I = \frac{3}{2}$.

For the lowest-lying Δ(1232) πN resonance, the amplitude is almost purely elastic because of the low mass, and the "tails" of higher-lying resonances may be neglected. From (4.55) we expect $\sigma_{\text{el.}} = 2\pi \lambda^2 (2J + 1)$ at the peak. The limiting value for $J = \frac{3}{2}$, $\sigma_{\text{el.}} = 8\pi \lambda^2$, is shown dashed in Fig. 4.12, clearly proving the Δ(1236) to be a p-wave resonance of spin-parity $J^P = \frac{3}{2}^+$.

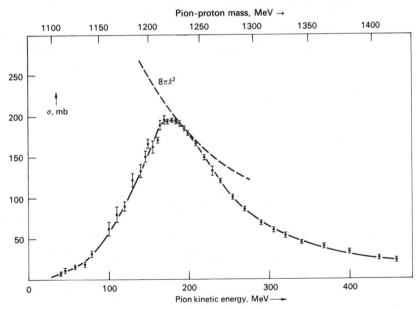

Figure 4.12 The $\pi^+ p$ total cross-section as a function of kinetic energy of the incident pion, or the $\pi^+ p$ mass, in the region of the 1232 MeV, $I = \frac{3}{2}$, $J^P = \frac{3}{2}^+$ resonance. Not all experimental points have been included. The maximum cross-section, $8\pi \lambda^2$, allowed by conservation of probability is shown dashed.

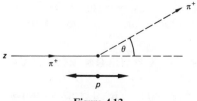

Figure 4.13

The assignment $J = \frac{3}{2}$ for the spin of the $\Delta(1232)$ may be confirmed from the angular distribution of the $\pi^+ p$ elastic scattering in the region of this resonance. Take the incident pion direction as the quantization axis (z-axis). The angular-momentum wave function of a p-state pion will be $\phi(l, m) = \phi(1, 0)$ since the pion is spinless. For the proton, we have $\alpha(\frac{1}{2}, \pm\frac{1}{2})$ corresponding to the two possible orientations of spin (Fig. 4.13). The product is the state

$$\psi(j, m) = \psi(\tfrac{3}{2}, \pm\tfrac{1}{2}).$$

When the Δ radiates a pion, the remaining proton may or may not have its spin "flipped". From Table 4.1 of Clebsch-Gordon coefficients, we can write for the final-state wave functions α' and ϕ'

$$\psi(\tfrac{3}{2}, \tfrac{1}{2}) = \sqrt{\tfrac{1}{3}}\, \phi'(1, 1)\alpha'(\tfrac{1}{2}, -\tfrac{1}{2}) + \sqrt{\tfrac{2}{3}}\, \phi'(1, 0)\alpha'(\tfrac{1}{2}, \tfrac{1}{2}).$$

Note that, since the scattered pion comes off at some angle θ to the z-direction, it is now possible for its orbital angular momentum to have a finite projection (1 or 0) on the old quantization axis. The ϕ' are simply the spherical harmonics:

$$\phi'(1, 1) = Y_1^1 = -\sqrt{\frac{3}{4\pi}} \sin\theta\, \frac{e^{i\phi}}{\sqrt{2}},$$

$$\phi'(1, 0) = Y_1^0 = \sqrt{\frac{3}{4\pi}} \cos\theta.$$

The angular distribution of the pions is therefore

$$I(\theta) = \psi\psi^* = \tfrac{1}{3}(Y_1^1)^2 + \tfrac{2}{3}(Y_1^0)^2,$$

the cross term being zero, since Y_1^1 and Y_1^0, as well as $\alpha'(\frac{1}{2}, -\frac{1}{2})$ and $\alpha'(\frac{1}{2}, \frac{1}{2})$, are orthogonal. Thus,

$$I(\theta) \propto \sin^2\theta + 4\cos^2\theta = 1 + 3\cos^2\theta. \tag{4.56}$$

The differential cross-section is plotted in Fig. 4.14 as a function of CMS angle θ for different values of T_π. At resonance, the dependence is in agreement with (4.56).

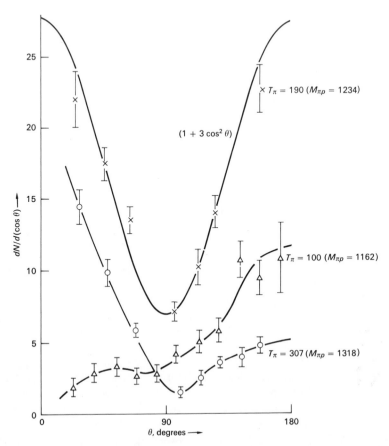

Figure 4.14 The angular distribution of the scattered pion, relative to the incident pion, in $\pi^+ p$ elastic scattering, as measured in the center-of-mass frame. In the region of the Δ-resonance of mass 1232 MeV ($T_\pi = 190$ MeV), the distribution has the form $1 + 3 \cos^2 \theta$, as in (4.56).

4.10. BOSON RESONANCES

Examples of *boson resonances* are given later in the text. Notable are the ψ/J resonance observed in $e^+ e^-$ annihilation (Section 5.13) and the W^+ and Z^0-resonances (Section 7.13 and 9.8).

4.11. TOTAL AND ELASTIC CROSS-SECTIONS AT HIGH ENERGY

Figure 4.15 shows total and elastic pp cross-sections at high energy, and Fig. 4.16 the total cross-section for various particles on protons. The elastic cross-section in Fig. 4.15 accounts for only a small fraction of the total cross-section

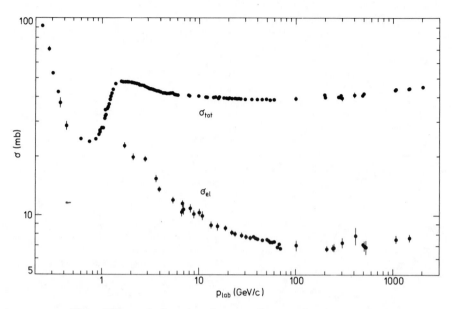

Figure 4.15 pp elastic and total cross-sections as a function of energy.

at high energy. The total cross-sections are constant within 10% above 5 GeV, with a slow decrease followed by an increase at the upper energies. The magnitude of the cross-section varies with the type of particle, but is in the region of 20 to 40 mb. If one equates this value to a "geometrical" cross-section πR^2, one obtains $R \simeq 10^{-13}$ cm = 1 fm as the "range" of the strong interaction. Secondly, especially in the lower energy range, one observes a large difference in the $\bar{p}p$ and pp cross sections, and this is not unexpected in view of the larger number of isospin channels open for the nucleon-antinucleon process, as well as the higher available energy following annihilation. Similar remarks apply in comparing $\pi^- p$ with $\pi^+ p$, and $K^- p$ with $K^+ p$. At very high energies there is a prediction from quantum field theory, known as the Pomerančuk theorem, that the cross-sections should become the same for particle and antiparticle, and moreover isospin-independent. Thus, cross-sections for $\pi^- p$ and $\pi^+ p (\equiv \pi^- n$ from charge independence) should become equal on both counts. The trends in the data (Fig. 4.17) support this.

The simplest possible model of absorption and scattering is that of a totally absorbing black disc of well-defined radius R. Setting $\eta_l = 0$ in (4.45) and (4.47) for this case, one obtains

$$\sigma_{\text{el.}} = \pi \lambdabar^2 \sum (2l + 1) = \pi R^2,$$

$$\sigma_{\text{inel.}} = \pi \lambdabar^2 \sum (2l + 1) = \pi R^2, \qquad (4.57)$$

$$\sigma_{\text{total}} = 2\pi \lambdabar^2 \sum (2l + 1) = 2\pi R^2,$$

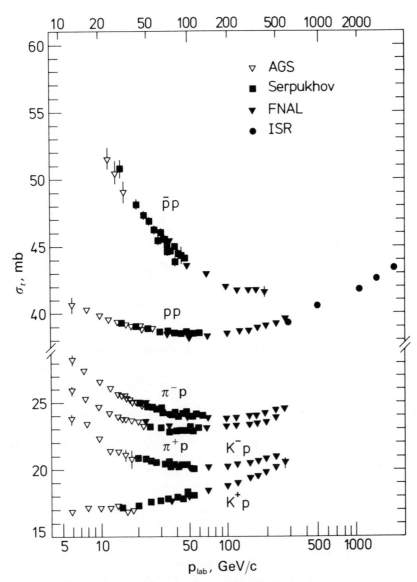

Figure 4.16   Total cross-sections for various particles on proton targets.

where $\lambdabar$ is the de Broglie wavelength of the colliding particles in their center-of-momentum system. In this model, the angular momentum contributed by the incident wave, l, varies from 0 to $l_{max} = R/\lambdabar$. For example, for $R = 1$ fm, and 20 GeV/c incident momentum, $\lambdabar = 0.01$ fm and $l_{max} = 100$. Then

$$\sum (2l + 1) = (l_{max} + 1)^2 \simeq l_{max}^2 = R^2/\lambdabar^2.$$

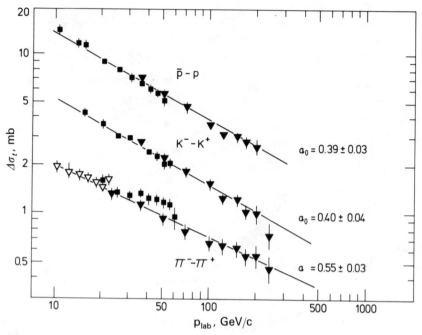

Figure 4.17 Difference of antiparticle and particle cross-sections on protons as a function of energy.

As expected, the inelastic or reaction or absorption cross-section is simply the geometrical area of the disc, πR^2. We note that $\sigma_{el.} = \sigma_{inel.}$ and represents the diffraction or shadow scattering, which is familiar in optics when a plane wave is interrupted by a completely absorbing obstacle. The angular distribution of the elastic scattering is the Fourier transform of the spatial distribution of the obstacle. For the lth partial wave, it is represented by the polynomial $P_l(\cos \theta)$. The sum over Legendre polynomials can be approximated, for small scattering angles, by a Bessel function of first order. In terms of the momentum transfer $q = 2p \sin(\theta/2)$, where p is the CMS momentum of the colliding particles

$$\frac{d\sigma_{el.} \text{ (black disc)}}{dq^2} = \pi R^4 \left| \frac{J_1(Rq)}{Rq} \right|^2 \tag{4.58}$$

$$\simeq \frac{\pi R^4}{4} \exp\left(-\frac{R^2 q^2}{4}\right), \tag{4.59}$$

where the second expression is accurate, with $R = 1$ fm, for values of $q^2 < 0.2$ (GeV/c)^2. For larger values of Rq, the function $J_1(Rq)/Rq$ undergoes maxima

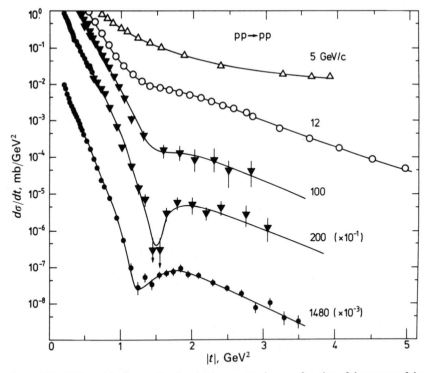

Figure 4.18 Differential cross-section for elastic pp scattering as a function of the square of the momentum transfer, $|t| = q^2$.

and minima characteristic of the diffraction phenomenon. The minima are at $Rq = 3.83$, 7.02, 10.17, etc., so that for $R = 0.7$ fm, the first zero is at $q^2 = 1.15$ GeV^2. Examples of data on pp scattering are shown in Fig. 4.18, indicating the exponential fall-off at small q^2, and diffraction-type minimum, especially at high incident momenta. It is noticeable that the minimum in q^2 decreases slowly with increasing energy, corresponding to a slow (logarithmic) increase in the effective size of the "disc". A similar dilation effect is apparent in the total cross-sections (Fig. 4.16).

While our black-disc model is qualitatively able to predict features of the elastic scattering of hadrons, it fails quantitatively. For example, it suggests $\sigma_{el.}/\sigma_{total} = 0.5$, while in practice the ratio is well below 0.5 and falls with increasing energy. The energy dependence of $d\sigma/dq^2$ is also not accounted for in simple models. Much better fits to the data are obtained with Regge pole theory, involving exchange of objects with complex angular momentum, but we shall not discuss these matters. The reader is referred to the bibliography for details.

4.12. PARTICLE PRODUCTION AT HIGH ENERGIES

As is clear from Fig. 4.15, inelastic processes dominate high-energy hadron
collisions. These are characterized by multiple production of secondary
mesons and of baryon-antibaryon pairs. The average multiplicity of the
different types of secondary particle in pp collisions is shown in Fig. 4.19,
indicating a slow increase in multiplicity at the highest incident energies. Very
roughly, this varies as

$$n = A + B \ln s, \qquad (4.60)$$

where s is the square of the CMS energy.

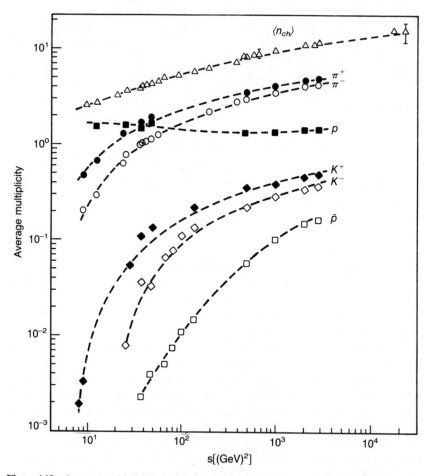

Figure 4.19   Average multiplicity of π^+, π^-, K^+, K^-, $\bar{p}$, and p secondaries in pp collisions as a
function of the square of the CMS energy, s. The total multiplicity, $\langle n_{ch} \rangle$, varies roughly as $\ln s$ at
high energy. (Antinucci *et al.* (1973).)

The momentum distribution of the secondary particles follows, approximately, simple scaling laws. If E, $\mathbf{p}$ represent the total energy and 3-momentum of a particular secondary particle, then the invariant cross-section (see Appendix A) has the form

$$\frac{E \, d^3\sigma}{dp^3} = \frac{E \, d^3\sigma}{dp_x \, dp_y \, dp_z} = \frac{d^2\sigma}{\pi d(p_T^2) \, d(p_L/E)} = F(x, p_T, s), \qquad (4.61)$$

where $p_T = \sqrt{p_x^2 + p_y^2}$ is the transverse momentum of the secondary, and $p_L = p_z$ its longitudinal CMS momentum, relative to the direction of the primary particle. The Feynman x-variable is $p_L/p_L(\text{max})$, where the maximum value of the longitudinal momentum is $p_L(\text{max}) \simeq \sqrt{s}/2$. At sufficiently large incident energies (above 10 GeV in pp collisions), the function F is observed to become nearly independent of s (i.e., incident energy) and factorizes into the product

$$F(x, p_T) = F_1(x) \cdot F_2(p_T). \qquad (4.62)$$

Thus, the transverse momentum distribution of the secondaries becomes almost independent of both s and p_L (a fact first observed in cosmic rays over 25 years ago). At fixed x, $d\sigma/dp_T^2$ approximates an exponential in p_T, with a mean value $\bar{p}_T \simeq 0.35$ GeV/c (see Fig. 4.20). The p_T of the secondaries is determined on average by the "size" of hadrons, that is, $p_T c \simeq \hbar c/R$, where $R \simeq 0.5$ fm.

Feynman postulated that for small x, $F_1(x) \simeq B$ (a constant), so that from (4.61) and (4.62),

$$d^2\sigma = \pi F_2(p_T) \cdot d(p_T^2) \cdot B \cdot (dp_L/E). \qquad (4.63)$$

The longitudinal momentum distribution is often discussed in terms of the rapidity, y, defined by

$$y = \frac{1}{2} \ln\left(\frac{E + p_L}{E - p_L}\right) = \ln\left(\frac{E + p_L}{\sqrt{p_T^2 + m^2}}\right), \qquad (4.64)$$

where m is the mass of the secondary and we have used the fact that $E^2 = p_T^2 + p_L^2 + m^2$. If E, p_L are measured in the CMS system, then $p_L = 0$ corresponds to $y = 0$, whereas

$$y_{\text{max}} = \frac{1}{2} \ln\left(\frac{s}{p_T^2 + m^2}\right),$$

$$y_{\text{min}} = -y_{\text{max}}. \qquad (4.65)$$

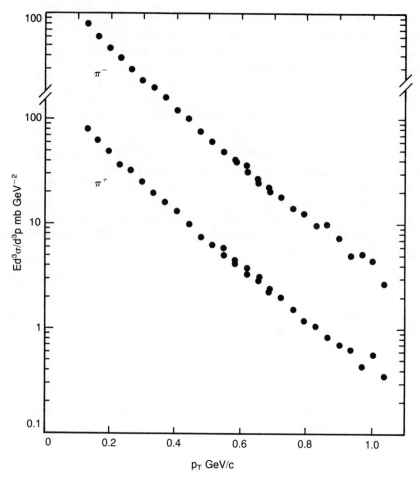

Figure 4.20 The transverse momentum distribution, dN/dp_T^2, of pions at ISR energies ($\sqrt{s} = 31$ GeV) and at 90° in the CMS, varies exponentially with p_T. (Alper *et al.* (1973).)

If a Lorentz transformation is made to another frame moving at velocity β along the z-axis (incident beam direction) then (see Appendix A)

$$y' = \ln\left(\frac{E' + p'_L}{\sqrt{p_T^2 + m^2}}\right) = \ln\left(\frac{\gamma(E - \beta p_L) + \gamma(p_L - \beta E)}{\sqrt{p_T^2 + m^2}}\right)$$

$$= \ln\frac{(E + p_L)\gamma(1 - \beta)}{\sqrt{p_T^2 + m^2}} = y + \frac{1}{2}\ln\frac{(1 - \beta)}{(1 + \beta)}, \qquad (4.66)$$

so that the y-distribution is invariant and a transformation to another reference frame simply amounts to a shift in the origin of y.

From (4.64)

$$dy = \frac{dp_L}{E},$$

and integrating (4.63) over p_T we get

$$\frac{d\sigma}{dy} = \text{constant}. \tag{4.67}$$

Figure 4.21 shows data from the CERN ISR on the rapidity distribution of $\pi^{\pm}$, p and $\bar{p}$ in pp collisions at high energy. The plateau at small y is apparent, in support of (4.67). It is here therefore that the bulk of produced secondaries are concentrated. For large y-values, $y \simeq y_{max}$, the density falls off rapidly. Neglecting this fall-off, the total multiplicity is given by the integral of $d\sigma/dy$ and thus varies as y_{max}, which is proportional to $\ln s$, from (4.65), in agreement with the empirical relation (4.60).

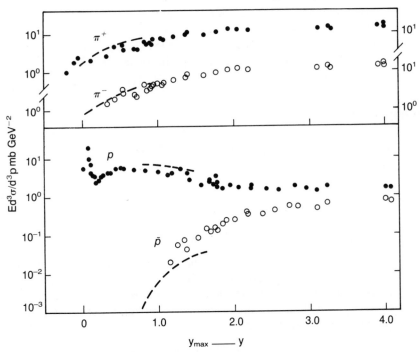

Figure 4.21 Rapidity distribution $d\sigma/dy$ at fixed $p_T = 0.4$ GeV/c, from pp collisions at the CERN ISR. The values of $\sqrt{s}$ varied from 22.5 to 53 GeV. The abscissa is $y_{max} - y$, where y is the rapidity in the CMS of the collision. Depending on s, $y = 0$ varies between 3 and 4 on this scale. The plateau at small y (on RHS) is apparent. The increase in $d\sigma/dy$ for protons near $y = y_{max}$ (on LHS) is associated with diffractive processes, where the protons emerge at small angle and with nearly the full energy.

Apart from the effect of the slow rise in multiplicity with incident energy, therefore, individual hadrons receive on average a rather constant fraction of the collision energy and thus a longitudinal momentum scaling with the incident particle momentum. On the contrary, the transverse momenta are limited and almost independent of either primary or secondary energy. Hence, the produced hadrons tend to be concentrated in a cone or jet of particles along the incident beam direction.

The features of hadron-hadron collisions at high energy are, in the large, dominated by the long-distance, confining behavior of the strong quark forces. It is virtually impossible, from such "soft" collisions, to gain any insight on the nature of the elementary quark/gluon constituents of the hadrons. This is why the analysis of such collisions has to be essentially phenomenological. However, it *is* possible, at extremely high collision energies, to select very rare events (typically one per billion of all collisions) whose characteristics are determined largely from the short-distance behavior of the quark-quark interaction. These events correspond to the occasional "head-on" or hard collision of quark on quark (or on gluon), with the scattered quarks recoiling at large angle and p_T up to 100 GeV/c or so, and "fragmenting" into two jets of hadrons. Such dramatic events have been observed at the CERN $p\bar{p}$ collider (see Fig. 8.18) and their analysis tells us very directly about the nature of the interquark forces (Section 8.9).

PROBLEMS

4.1. Find a relation between the total cross-sections (at a given energy) for the reactions

$$\pi^- p \to K^0 \Sigma^0,$$

$$\pi^- p \to K^+ \Sigma^-,$$

$$\pi^+ p \to K^+ \Sigma^+.$$

4.2. At a given center-of-mass energy, what is the ratio of the cross-sections for $p + d \to {}^3\text{He} + \pi^0$ and $p + d \to {}^3\text{H} + \pi^+$?

4.3. A hypernucleus is one in which a neutron is replaced by a bound Λ-hyperon. ${}^4\text{He}_\Lambda$ and ${}^4\text{H}_\Lambda$ are a doublet of mirror hypernuclei. Deduce the ratio of the reaction rates

$$K^- + {}^4\text{He} \to {}^4\text{He}_\Lambda + \pi^-$$

$$\to {}^4\text{H}_\Lambda + \pi^0.$$

4.4. State which of the following combinations can or cannot exist in a state of $I = 1$, and give the reasons: (a) $\pi^0 \pi^0$, (b) $\pi^+ \pi^-$, (c) $\pi^+ \pi^+$, (d) $\Sigma^0 \pi^0$, (e) $\Lambda \pi^0$.

4.5. In which isospin states can (a) $\pi^+\pi^-\pi^0$, (b) $\pi^0\pi^0\pi^0$ exist? [*Hint*: First write down the isospin functions for a pair, e.g., $\pi^0\pi^0$, and then combine with the third pion. Refer to Table III in the Appendix for any Clebsch-Gordan coefficients required.]

4.6. Deduce through which isospin channels the following reactions may proceed: (a) $K^+ + p \rightarrow \Sigma^0 + \pi^0$, (b) $K^- + p \rightarrow \Sigma^+ + \pi^-$. Find the ratio of cross-sections for (a) and (b), assuming that one or other channel dominates.

4.7. The $A1$ meson, of $I = 1$, is considered to be a resonant state of a ρ-meson ($I = 1$) and pion ($I = 1$). Thus decay $A1 \rightarrow \rho + \pi$ is dominant. Find the expected branching ratio

$$\frac{A1 \rightarrow \pi^0\pi^0\pi^+}{A1 \rightarrow \pi^+\pi^-\pi^+}.$$

4.8. The ω-meson has isospin $I = 0$, and the ρ-meson, $I = 1$. They have the same spin and parity (1^-). The ρ-meson has central mass 775 MeV and is a broad state with a width $\Gamma \simeq 120$ MeV, overlapping the ω-state (783 MeV, $\Gamma \sim 10$ MeV). Would you expect the ω- and ρ-states to interfere, and what qualitative effects would any interference have on the $\pi^+\pi^-$ and $\pi^+\pi^-\pi^0$ mass spectra in reactions where both ω and ρ can be produced?

4.9. As shown in Chapter 7, the neutral kaons decay from the states K_1^0 and K_2^0, of CP eigenvalues $+1$ and -1 respectively. If, as we believe, $p\bar{p}$ annihilation at rest takes place from an atomic S-state only, show that $p\bar{p} \rightarrow K_1^0 + K_2^0$ occurs, but that $p\bar{p} \rightarrow 2K_1^0$ or $p\bar{p} \rightarrow 2K_2^0$ does not.

4.10. In the Dalitz-plot analysis of the decay $K \rightarrow 3\pi$, the pion kinetic energies are plotted along three axes at $120°$, that is, normal to the sides of an equilateral triangle. The height of this triangle is Q, the total kinetic energy in the decay. Show each of the following if the pions are nonrelativistic: (a) conservation of energy requires all points to lie inside the triangle, (b) conservation of momentum constrains all points to lie inside the circle inscribed in the triangle, (c) for a constant-matrix element, the density of points inside this circle should be uniform. Show that in a decay involving three relativistic secondaries of equal mass, the boundary of the Dalitz plot would be deformed to an equilateral triangle inscribed in the above circle.

4.11. Show that in $K \rightarrow 3\pi$ decay, the relativistic factor $E_1E_2E_3$ is constant within a range of $\pm 1\%$ over different regions of the Dalitz plot.

4.12. In the decay of a resonance of mass M into three particles of momenta $\mathbf{p}_1, \mathbf{p}_2$, and $\mathbf{p}_3$, show that the boundary of the Dalitz plot is given by the condition $|\mathbf{p}_1| + |\mathbf{p}_2| - |\mathbf{p}_3| = 0$. Deduce the equation of the boundary in terms of E_1 and E_2 in the case where the three decay products have the same rest mass m. Show that if $m \ll M$, the boundary of the Dalitz plot becomes the inscribed triangle.

4.13. In a reaction $A + B \rightarrow C + D + E$, at a fixed bombarding energy, the quantities m_{CD}^2 and m_{DE}^2 are displayed along the x- and y-axes in a Dalitz plot (m_{CD} is the mass of particles C and D, etc.). If θ is the direction of particle E with respect to either C or D in the CD center-of-mass frame, show that when C and D resonate at a fixed mass, we have $m_{DE}^2 = \alpha - \beta \cos \theta$, where α and β are constants.

4.14. In the following reaction in hydrogen,

$$\pi^- + p \rightarrow X^- + p,$$

a boson resonance X is observed with mass 2.4 GeV. The incident pion momentum is 12 GeV/c. Calculate the maximum angle of emission of the recoil proton with respect to the beam direction, and its momentum. Calculate also the angle and momentum of the proton when the 4-momentum transfer is a maximum, and compute q_{max}^2.

4.15. The Breit-Wigner formula (4.55) describes a resonance of width $\Gamma \ll E_0$, the peak energy. For a broad resonance, this formula is not exact, since the phase-space available for two-body decay changes appreciably as one goes through the resonance. Show that for an S-wave resonance this effect may be accounted for by replacing Γ with $\Gamma_0 \cdot (pE_0/Ep_0)$, where p is the CMS momentum of either particle, and Γ_0, p_0, and E_0 refer to the resonance peak. What additional factors would you expect for a resonance decaying to two particles with orbital angular momentum l?

4.16. Show that if the value of m^2 of a relativistic secondary is neglected compared with p_T^2, then the rapidity becomes dependent only on the angle of emission of the secondary, and that $y = \ln(2 \cot \theta) \simeq \ln(\cot \theta^*/2) + \ln 2\gamma$, where θ and θ^* are the angles in the lab system and CMS system, respectively, and γ is the CMS Lorentz factor.

4.17. State which of the following decay modes of the ρ-meson ($J^P = 1^-$, $I = 1$) are allowed by the strong or electromagnetic interactions:

$$\rho^0 \rightarrow \pi^+ \pi^-$$
$$\rightarrow \pi^0 \pi^0$$
$$\rightarrow \eta^0 \pi^0$$
$$\rightarrow \pi^0 \gamma.$$

BIBLIOGRAPHY

Amaldi, U., M. Jacob, and G. Matthiae, "Diffraction of matter waves", *Ann. Rev. Nucl. Science* **26**, 385 (1976).

Blatt, J., and V. F. Weisskopf, *Theoretical Nuclear Physics*, John Wiley, New York, 1952.

Bøggild, H., and T. Ferbel, "Inclusive reactions", *Ann. Rev. Nucl. Science* **24**, 451 (1974).

Dalitz, R. H., "Strange particle resonant states", *Ann. Rev. Nucl. Science* **13**, 339 (1963).

Foa, L., "High energy hadron physics", *Riv. Nuovo Cim.* **3**, 283 (1973).

Giacomelli, G., "Total cross-sections and elastic scattering at high energies", *Phys. Reports* **23**, 1231 (1976).

Källen, G., *Elementary Particle Physics*, Addison-Wesley, Reading, Mass., 1964.

CHAPTER 5

Static Quark Model of Hadrons

5.1. INTRODUCTION

During the great accumulation of data on baryon and meson resonances in the 1960's, regularities or patterns were noted among these hadron states and interpreted in terms of an approximate higher symmetry, called unitary symmetry. This description has been superseded by one in which the patterns or multiplets of states could be simply accounted for in terms of quark constituents, a baryon consisting of three quarks, and a meson of a quark-antiquark pair. This evidence is especially compelling in the level systems of bound states formed from heavy quark-antiquark pairs. First however we discuss the construction of the lighter baryon and meson multiplets in terms of quark constituents.

5.2. THE BARYON DECUPLET

Figure 5.1 indicates the 10 baryon states of lowest mass and of spin-parity $J^P = \frac{3}{2}^+$, where we plot the strangeness S against the third component of isospin, I_3, for each of the 10 members. Working downward, these consist of an $S = 0$, $I = \frac{3}{2}$ isospin quadruplet, the $\Delta(1232)$, existing in the charge substates Δ^{++}, Δ^+, Δ^0, Δ^-. The number 1232 in parentheses indicates the central resonance mass in MeV. Next come an $I = 1$ isospin triplet of

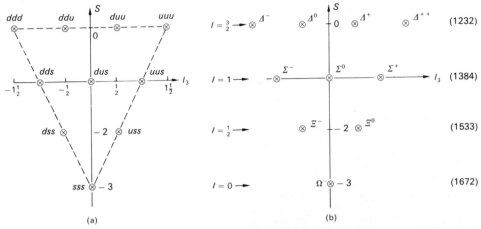

Figure 5.1 (a) Quark label assignments in the baryon decuplet. (b) The observed decuplet of baryon states of spin-parity $\frac{3}{2}^+$. The mean mass of each isospin multiplet is given in brackets.

$S = -1$, the $\Sigma(1384)$; an $S = -2$, $I = \frac{1}{2}$ isospin doublet, the $\Xi(1533)$; and finally an $I = 0$ singlet of $S = -3$, the $\Omega^-(1672)$. The members of each isospin multiplet have essentially the same central mass, differing only by a few MeV, characteristic of electromagnetic mass splittings in isospin multiplets. The states of different strangeness differ considerably in mass, but the mass difference for each increment of strangeness is roughly the same. This surely cannot be an accident; indeed, the Ω^- baryon was predicted on this basis three years before it was observed (see Fig. 5.2).

The regularities such as that in the decuplet can be accounted for by postulating three types of fermion constituent in a baryon, called *quarks* with the quantum numbers shown in Table 5.1. The quark hypothesis was put forward in 1964 by Gell-Mann and by Zweig. These quarks consist of an $S = 0$ isospin doublet, labeled u and d [standing for $I_3 = +\frac{1}{2}$ (up) and $I_3 = -\frac{1}{2}$ (down)], and a $S = -1$ isosinglet, labeled s (for strange). The assignments u, d, s are called the *flavor* of the quark. Baryons are assumed to

TABLE 5.1 Quark quantum numbers (as of 1964)[a]

Flavor	B	J	I	I_3	S	Q/e
u	$\frac{1}{3}$	$\frac{1}{2}$	$\frac{1}{2}$	$+\frac{1}{2}$	0	$+\frac{2}{3}$
d	$\frac{1}{3}$	$\frac{1}{2}$	$\frac{1}{2}$	$-\frac{1}{2}$	0	$-\frac{1}{3}$
s	$\frac{1}{3}$	$\frac{1}{2}$	0	0	-1	$-\frac{1}{3}$

[a] Antiquarks $\bar{u}$, $\bar{d}$, and $\bar{s}$ have the signs of B, I_3, S, and Q/e reversed.

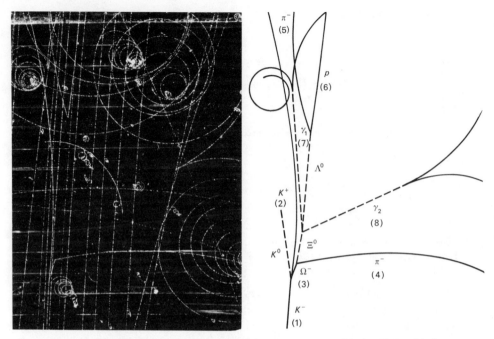

Figure 5.2 The first Ω^- event (Barnes *et al.* (1964).) (Courtesy Brookhaven National Laboratory.) It depicts the following chain of events:

$$K^- + p \to \Omega^- + K^+ + K^0$$
$$\quad\quad\quad\hookrightarrow \Xi^0 + \pi^- \ (\Delta S = 1 \text{ weak decay})$$
$$\quad\quad\quad\quad\hookrightarrow \pi^0 + \Lambda \ (\Delta S = 1 \text{ weak decay})$$
$$\quad\quad\quad\quad\quad\quad\hookrightarrow \pi^- + p \ (\Delta S = 1 \text{ weak decay})$$
$$\quad\quad\quad\hookrightarrow \ \gamma \ + \ \gamma \ \ (\text{e.m. decay})$$
$$\quad\quad\quad\quad\quad\downarrow \quad\quad \downarrow$$
$$\quad\quad\quad e^+e^- \quad e^+e^-.$$

be composed of three quarks, and the most democratic assignment is that each has a fractional baryon number, $B = \frac{1}{3}$. From the relation (4.28), i.e.,

$$Q/e = \tfrac{1}{2}(B + S) + I_3, \tag{5.1}$$

where the combination $Y = B + S$ is called the hypercharge, it follows that quarks must also carry fractional charges of $\frac{2}{3}$ and $-\frac{1}{3}$. The appropriate combinations of quarks indicated in Fig. 5.1 can then account for the quantum numbers I, I_3, S(or Y) of the members of the decuplet, and of course their electric charges. The progressive increase in mass of the decuplet members with decreasing S can then be simply ascribed to a difference in mass of s, as compared with u and d, quarks in the region $m_s - m_{u,d} \simeq 150$

MeV. The masses of u- and d-quarks are expected to be nearly equal, since any difference must be of the order of the electromagnetic mass differences among the members of an isospin multiplet (see Table 4.3).

Quarks are not observed as free particles and hence must be confined in hadrons by the interquark potential. Within this confining potential, the quark can be considered as a quasi-free particle of effective mass m^* (of order one-third of the baryon mass) with a momentum of order R_0^{-1}, where R_0 is the typical hadron size (about 1 fm). Early calculations assumed the quarks would be in nonrelativistic motion, that is, $m^* \gg R_0^{-1}$. This is a somewhat marginal assumption for the light quarks (u, d, s) but seems to work. The best tests of the nonrelativistic quark model in fact apply to hadrons containing the heavier (c, b) quarks, as described later.

Given that a baryon is to consist of 3 quarks chosen from any 3 flavors, 27 combinations are possible; so what is special about only 10 of them? We have to introduce some symmetry principle peculiar to members of a particular multiplet. Such symmetry will concern quark spin as well as quark flavor. First, with regard to flavor, we can require that the flavor part of the baryon wavefunction should have a definite symmetry under interchange of any pair of quarks. The corner states uuu, ddd, and sss of Fig. 5.1 are clearly symmetric under interchange, so it is natural to require the same symmetry for the other states. We have indicated these as ddu, duu, uss, etc., as shorthand for the properly symmetrized expressions. For example, the full form of the ddu state is

$$\frac{1}{\sqrt{3}}(ddu + udd + dud), \tag{5.2}$$

which is symmetric under interchange of any two quarks; the numerical factor is for normalization. Similarly, dus is shorthand for

$$\frac{1}{\sqrt{6}}(dsu + uds + sud + sdu + dus + usd) \tag{5.3}$$

etc. The different states in the multiplet can be obtained from each other, horizontally, by applying the isospin shift operators successively (see Appendix C). Thus,

$$I^-(uuu) = [I^-(u)](uu) + (u)[I^-(u)](u) + (uu)[I^-(u)]$$
$$= duu + udu + uud.$$

Similarly, successive rows are constructed by replacing u-quarks with s-quarks.

These 10 states are the *only* completely symmetric combinations we can make. Of the remaining 17 of the total of 27, one is completely antisymmetric. The wavefunction

$$(dsu + uds + sud - usd - sdu - dus) \tag{5.4}$$

describes this state, changing sign under interchange of any two quarks, for example $1 \leftrightarrow 2$ or $2 \leftrightarrow 3$. It is easily constructed by adding an s-quark to an antisymmetric u, d combination, i.e., $(ud - du)s$, plus cyclic permutations thereof. This leaves 16 states, which turn out to consist of two octets of mixed symmetry.

5.3. QUARK SPIN AND COLOR

Since the members of the decuplet consist of the spin-$\frac{3}{2}$ baryons of lowest mass, we assume that the quarks sit in the spatially symmetric ground state ($l = 0$). The value $J = \frac{3}{2}$ is then obtained by having the quarks in a symmetric spin state, with spins "parallel", as in $\Delta^{++} = u \uparrow u \uparrow u \uparrow$, for example. Hence the $\frac{3}{2}^{+}$ decuplet is characterized by *symmetry of the three-quark wavefunction in both flavor and spin, as well as space*. This clearly violates the Pauli principle, that two or more fermions may not exist in the same quantum state. This was a problem years ago: subsequently it turned out that another degree of freedom, called *color*, was necessary for other reasons. It is postulated that quarks exist in three colors—say red, green, blue—and that baryons and mesons built from quarks have zero net color, that is, they are *color singlets*. (The hadrons must be colorless; otherwise color would be a necessary and measurable property of hadrons.) So Δ^{++} consists of one red, one green, and one blue u-quark, which makes them nonidentical. "Color" is perhaps an unfortunate name; it is simple a notation for a new property of quarks, quite separate from the flavor quantum number. The three colors specify the "strong charges" of the quarks, in just the same way that the signs $+$ and $-$ specify their electric charges. The evidence for color comes from a number of sources. For example, the predicted rate of decay $\pi^{0} \rightarrow 2\gamma$ is found to be proportional to the square of the number of colors N_{c}, and comparison with experiment gives $N_{c} = 2.98 \pm 0.11$; and the cross-section ratio

$$\sigma(e^{+}e^{-} \rightarrow \text{hadrons})/\sigma(e^{+}e^{-} \rightarrow \mu^{+}\mu^{-})$$

at high energy is proportional to N_{c} and requires a value of 3 within about 10% (see Section 8.6).

5.4. THE BARYON OCTET

We noted that the baryon decuplet members have wavefunctions which are symmetric in both spin and flavor, as well as in space. It is plausible that we can also form three-quark states which are symmetric under *simultaneous* interchange of flavor and spin of any quark pair, although not, as in the decuplet, under each separately. These states are identified with the members of the lowest-lying baryon states of $J^P = \frac{1}{2}^+$, the baryon octet which includes the proton and neutron as members.

To construct the baryon octet wavefunctions we first start with a proton (uud) and put two quarks in a spin singlet state (see (3.40))

$$\frac{1}{\sqrt{2}} (\uparrow\downarrow - \downarrow\uparrow),$$

which is antisymmetric. To make the overall state symmetric, we need a flavor-antisymmetric combination of u and d (since u and u cannot do it), which is the isosinglet (see Section 4.3)

$$\frac{1}{\sqrt{2}} (ud - du).$$

We then add the third quark u, with spin up, to obtain

$$(u\uparrow d\downarrow - u\downarrow d\uparrow - d\uparrow u\downarrow + d\downarrow u\uparrow)u\uparrow.$$

Although the expression in brackets is symmetric under interchange of the first and second quarks (flavor and spin), the whole expression has to be symmetrized by making a cyclic permutation (12 terms in all), giving finally

$$\phi(p, J_z = +\tfrac{1}{2}) = \frac{1}{\sqrt{18}} [2u\uparrow u\uparrow d\downarrow + 2d\downarrow u\uparrow u\uparrow + 2u\uparrow d\downarrow u\uparrow$$

$$-u\downarrow d\uparrow u\uparrow - u\uparrow u\downarrow d\uparrow - u\downarrow u\uparrow d\uparrow$$

$$-d\uparrow u\downarrow u\uparrow - u\uparrow d\uparrow u\downarrow - d\uparrow u\uparrow u\downarrow]. \qquad (5.5)$$

The other members of the baryon octet of $J^P = \frac{1}{2}^+$ can be worked out in similar fashion. This octet is depicted in Fig. 5.3, where the wavefunctions have been indicated as uud, ssu, etc., but it should be understood that these are abbreviations for the properly symmetrized expressions. The eight members consist of the n and $p(939)$ nucleon isospin doublet ($I = \frac{1}{2}, S = 0$), the $\Sigma(1193)$ isotriplet ($I = 1, S = -1$), the $\Xi(1318)$ isodoublet ($I = \frac{1}{2}, S = -2$), and the $\Lambda(1116)$ isosinglet ($I = 0, S = -1$).

The hypothesis that baryon masses differ because of differences in the strange quark content gave for the decuplet

$$\Sigma(1384) - \Delta(1232) = \Xi(1533) - \Sigma(1384) = \Omega^-(1672) - \Xi(1533).$$
$$\qquad 152 \text{ MeV} \qquad\qquad\qquad 149 \text{ MeV} \qquad\qquad\qquad 139 \text{ MeV} \qquad (5.6)$$

In the $\frac{1}{2}^+$ octet, the same hypothesis gives:

$$M_\Sigma = M_\Lambda,$$
$$1193 \text{ MeV} \quad 1116 \text{ MeV}$$
$$\qquad\qquad\qquad\qquad\qquad\qquad (5.7)$$
$$M_\Lambda - M_N = M_\Xi - M_\Lambda.$$
$$177 \text{ MeV} \qquad 203 \text{ MeV}$$

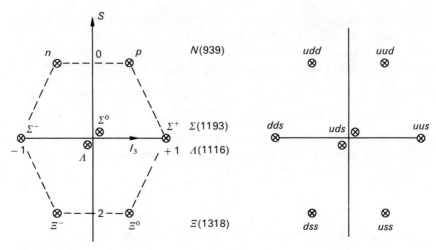

Figure 5.3 The baryon octet of spin-parity $\frac{1}{2}^+$. The observed states are given on the left, and quark flavor assignments on the right.

There is rather poor agreement for each of the three equalities above and a large discrepancy between mass differences in decuplet and octet. Historically, mass relations in multiplets were first obtained by making certain assumptions about the nature of the breaking of unitary symmetry and one prediction was the Gell-Mann–Okubo mass formula for baryons:

$$M = M_0 + M_1 Y + M_2 \left[I(I+1) - \frac{Y^2}{4} \right], \tag{5.8}$$

where the hypercharge $Y = B + S$. In the decuplet of Fig. 5.1, the states are related by $Y = B + S = 2(I - 1)$, so that $M = A + BY$ and we obtain the linear S-dependence of mass as in (5.6). For the baryon octet, (5.8) predicts

$$3M_\Lambda + M_\Sigma = 2M_N - M_\Xi$$
$$ 4541 \text{ MeV} \qquad 4514 \text{ MeV} \tag{5.9}$$

correct to 1% accuracy. However, there are no clear physical reasons for the assumptions on which (5.8) is based, and they certainly do not explain the great difference in the M_0-values for octet and decuplet states. A more quantitative description of baryon and meson masses is obtained in terms of hyperfine-splitting effects of the interquark interactions, and these are discussed in Section 5.10 below.

Baryon multiplets of higher spin are also observed, and these can be accounted for in terms of the u-, d-, s-quark combinations, three at a time, introducing relative orbital angular momentum l of the quarks as necessary to account for the states of high J. This separate quantization of orbital and spin angular momentum of the quarks is of course valid only in the approximation of nonrelativistic motion.

5.5. QUARK-ANTIQUARK COMBINATIONS: THE PSEUDOSCALAR MESONS

As indicated before, the states observed in nature consist of three-quark combinations (the baryons) and quark-antiquark combinations (the mesons). Restricting ourselves to three flavors, we expect families of mesons containing $3^2 = 9$ states, or nonets. Given spin-$\frac{1}{2}$ for quarks and antiquarks, we might expect both spin triplet ($\uparrow \uparrow$) states of $J = 1$ (the vector mesons) and spin singlet ($\downarrow \uparrow$) states of $J = 0$ (the pseudoscalar mesons).

In discussing the baryon multiplets, emphasis was placed on the quark-exchange symmetry. But now we are dealing with quarks and antiquarks—thus the interchange $u \to \bar{u}$, for example. It is necessary therefore to consider the effect of charge conjugation applied to quark wavefunctions. If the baryon number B is conserved, there is no actual physical process $Q \leftrightarrow \bar{Q}$: thus, as a result of the operation of charge conjugation, or particle-antiparticle conjugation, an arbitrary and unobservable phase occurs. We can write $|u\rangle \overset{c}{\to} |\bar{u}\rangle$ or $|u\rangle \overset{c}{\to} |\bar{u}\rangle e^{i\phi}$. The phase ϕ is generally chosen according to the Condon-Shortley convention, which actually introduces a minus sign in some places. In the following table, arrows denote the C-operation:

	Nucleons		Quarks		
I_3	Particle	Antiparticle	Particle	Antiparticle	(5.10)
$+\frac{1}{2}$	$\|p\rangle$	$+\|\bar{n}\rangle$	$\|u\rangle$	$+\|\bar{d}\rangle$	
$-\frac{1}{2}$	$\|n\rangle$	$-\|\bar{p}\rangle$	$\|d\rangle$	$-\|\bar{u}\rangle$	

Assuming that our quark-antiquark combinations are in the $l = 0$ singlet spin state (see (3.40)) and recalling the opposite intrinsic parity of fermion and antifermion (Section 3.4), the quantum numbers will be $B = 0$, $J^P = 0^-$. These correspond to *pseudoscalar mesons*, so called because the wavefunctions have $J = 0$, have odd parity, and change sign under spatial inversion.

With only u- and d-quarks and antiquarks we can make $2^2 = 4$ combinations as follows:

I	I_3	Wavefunction	Q/e	
1	1	$u\bar{d} = \pi^+$	$+1$	
1	-1	$-\bar{u}d = \pi^-$	-1	isospin triplet (5.11)
1	0	$\sqrt{\frac{1}{2}}(d\bar{d} - u\bar{u}) = \pi^0$	0	
0	0	$\sqrt{\frac{1}{2}}(d\bar{d} + u\bar{u}) = \eta$	0	isospin singlet

These assignments are justified by applying the isospin shift operators $I^{\pm}$, in exact analogy with the angular-momentum operators $J^{\pm}$ (see Appendix C):

$$I^{\pm}|\Psi(I, I_3)\rangle = \sqrt{I(I + 1) - I_3(I_3 \pm 1)}|\Psi(I, I_3 \pm 1)\rangle. \qquad (5.12)$$

Thus, applying to single quark states:

$$I^+|d\rangle \to |u\rangle, \qquad I^+|\bar{u}\rangle \to |-\bar{d}\rangle, \qquad I^+|u\rangle = I^+|\bar{d}\rangle = 0. \qquad (5.13)$$

Furthermore,

$$I^-|\Psi(1, 1)\rangle = I^+|\Psi(1, -1)\rangle = \sqrt{2}|\Psi(1, 0)\rangle,$$

$$I^+|\Psi(1, 0)\rangle = \sqrt{2}|\Psi(1, 1)\rangle, \qquad I^+\Psi|1, 1\rangle = I^-\Psi|1, -1\rangle = 0. \qquad (5.14)$$

Applying these results to quark-antiquark combinations (5.11), we obtain for example,

$$I^+|\pi^-\rangle = I^+|-d\bar{u}\rangle = |-u\bar{u} + d\bar{d}\rangle = \sqrt{2}|\pi^0\rangle,$$

$$I^+|\pi^0\rangle = I^+ \frac{|d\bar{d} - u\bar{u}\rangle}{\sqrt{2}} = \frac{|u\bar{d} + 0 - 0 + u\bar{d}\rangle}{\sqrt{2}} = \sqrt{2}|u\bar{d}\rangle = \sqrt{2}|\pi^+\rangle.$$

$$(5.15)$$

The $I = 1$ combinations in (5.11) are thus identified with π^+, π^-, π^0, the lowest-mass pseudoscalar mesons. The fourth combination in (5.11) has the property

$$I^{\pm}|\eta\rangle = I^{\pm} \frac{|d\bar{d} + u\bar{u}\rangle}{\sqrt{2}} = \frac{|u\bar{d} - u\bar{d}\rangle}{\sqrt{2}} = 0. \qquad (5.16)$$

Thus $|\eta\rangle$ is an isospin singlet, which does not transform under an isospin transformation into any other state; it is orthogonal to the $I = 1$ combinations, so that $\langle\eta|\pi^0\rangle = 0$, for example. With the phase convention under C-conjugation which we have adopted, the singlet state is *symmetric* in quark labels ($d \to u$, $\bar{d} \to \bar{u}$), whereas the π^+, π^-, π^0 states all change sign. The singlet is identified with the η-meson, of mass 550 MeV.

Introduction of s-quarks in addition to u and d gives us a total of $3^2 = 9$ states, which are listed in Table 5.2, together with their assignment to the pseudoscalar mesons. In analogy with the singlet state (5.16) obtained with the u, d quark-antiquark combinations, these nine states built from u-, d-, and s-quarks break down into a singlet—the symmetric $Q\bar{Q}$ combination η_0 in the last line of Table 5.2—plus eight states which can be transformed one into another by interchange of the u-, d-, and s-quarks. Note that the eighth member η_8, is $(d\bar{d} + u\bar{u} - 2s\bar{s})/\sqrt{6}$, orthogonal to η_0. The square-root factors are inserted where appropriate so that all states are normalized to unity.

TABLE 5.2 Pseudoscalar-meson states as quark-antiquark combinations

	I	I_3	S	Meson	Quark combination	Decay	Mass, MeV
octet	1	1	0	π^+	$u\bar{d}$	$\pi^\pm \to \mu\nu$	140
	1	-1	0	π^-	$d\bar{u}$		
	1	0	0	π^0	$(d\bar{d} - u\bar{u})/\sqrt{2}$	$\pi^0 \to 2\gamma$	135
	$\frac{1}{2}$	$\frac{1}{2}$	$+1$	K^+	$u\bar{s}$	$K^+ \to \mu\nu$	494
	$\frac{1}{2}$	$-\frac{1}{2}$	$+1$	K^0	$d\bar{s}$	$K^0 \to \pi^+\pi^-$	498
	$\frac{1}{2}$	$-\frac{1}{2}$	-1	K^-	$\bar{u}s$	$K^- \to \mu\nu$	494
	$\frac{1}{2}$	$\frac{1}{2}$	-1	$\bar{K}^0$	$\bar{d}s$	$\bar{K}^0 \to \pi^+\pi^-$	498
	0	0	0	η_8	$(d\bar{d} + u\bar{u} - 2s\bar{s})/\sqrt{6}$	$\eta \to 2\gamma$	549
singlet	0	0	0	η_0	$(d\bar{d} + u\bar{u} + s\bar{s})/\sqrt{3}$	$\eta' \to \eta\pi\pi$ $\to 2\gamma$	958

The linear octet mass formula (5.8) used for the baryons is not satisfied for mesons; it is found that it works if (mass)^2 is used instead of mass:

$$2(M_{K^0}^2 + M_{\bar{K}^0}^2) = \underset{0.988\ \text{GeV}^2}{4M_{K^0}^2} = \underset{0.924\ \text{GeV}^2}{M_{\pi^0}^2 + 3M_\eta^2}. \qquad (5.17)$$

However, the actual states η and η' observed in nature appear to be linear combinations of the wavefunctions η_8 and η_0 written down in the last two lines of Table 5.2. On the basis of mass formulae, the mixing angle is $\theta \simeq 11°$. This matter is discussed more fully below, for the vector-meson nonet.

5.6. THE VECTOR MESONS

The $l = 0$ spin triplet combinations ($\uparrow\uparrow$) of Q and $\bar{Q}$ give us the vector mesons ($J^P = 1^-$) (see Fig. 5.4). In this case the octet-singlet mixing angle is large ($\theta \simeq 35°$). Formally we can write

$$\phi = \phi_0 \sin\theta - \phi_8 \cos\theta,$$
$$\omega = \phi_8 \sin\theta + \phi_0 \cos\theta, \qquad (5.18)$$

where ϕ, ω denote the physical vector-meson states and ϕ_0, ϕ_8 the singlet and octet states, respectively, of $I = S = 0$. Assuming that the matrix element of the Hamiltonian between the states yields the (mass)^2, i.e., $M_\phi^2 = \langle\phi|H|\phi\rangle$, etc., we obtain from (5.18)

$$M_\phi^2 = M_0^2 \sin^2\theta + M_8^2 \cos^2\theta - 2M_{08}^2 \sin\theta\cos\theta, \qquad (5.19a)$$

$$M_\omega^2 = M_8^2 \sin^2\theta + M_0^2 \cos^2\theta + 2M_{08}^2 \sin\theta\cos\theta, \qquad (5.19b)$$

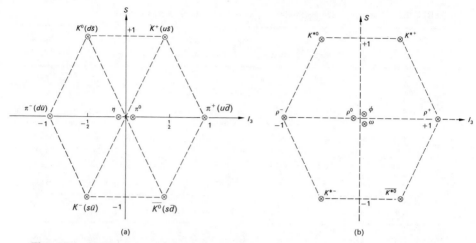

Figure 5.4 (a) The lowest-lying pseudoscalar-meson states ($J^P = 0^-$). Quark flavor assignments are indicated. (b) The vector-meson nonet ($J^P = 1^-$). The quark assignments are the same as in (a).

in an obvious notation. Further, since ϕ and ω are orthogonal, we obtain

$$M_{\phi\omega}^2 = 0 = (M_0^2 - M_8^2) \sin\theta \cos\theta + M_{08}^2(\sin^2\theta - \cos^2\theta). \quad (5.20)$$

Eliminating M_{08} and M_0 between these three equations gives

$$\tan^2\theta = \left(\frac{M_\phi^2 - M_8^2}{M_8^2 - M_\omega^2}\right). \quad (5.21)$$

Using the analogue of (5.17), we have

$$M_8^2 = \tfrac{1}{3}(4M_{K*}^2 - M_\rho^2), \quad (5.22)$$

so that the observed masses (Table 5.3) give $\theta \simeq 40°$. For the particular case $\sin\theta = 1/\sqrt{3}$, $\theta \simeq 35°$, Eq. (5.18) would give

$$\phi = \frac{1}{\sqrt{3}} (\phi_0 - \sqrt{2}\phi_8),$$

$$\omega = \frac{1}{\sqrt{3}} (\phi_8 + \sqrt{2}\phi_0),$$

TABLE 5.3 Vector-meson nonet

State	I	Y	Mass, MeV	Dominant decay mode
ρ	1	0	776	$\rho \to 2\pi$
K^*	$\tfrac{1}{2}$	± 1	892	$K^* \to K\pi$
ω	0	0	783	$\omega \to 3\pi$
ϕ	0	0	1019	$\phi \to K\bar{K}$

where from Table 5.2

$$\phi_0 = (d\bar{d} + u\bar{u} + s\bar{s})/\sqrt{3},$$
$$\phi_8 = (d\bar{d} + u\bar{u} - 2s\bar{s})/\sqrt{6}, \tag{5.23}$$

so that

$$\phi = s\bar{s}$$
$$\omega = (u\bar{u} + d\bar{d})/\sqrt{2}. \tag{5.24}$$

In this case of "ideal mixing"—which is almost true in practice—ϕ is composed entirely of s-quarks, and ω of u and d. We note that these simple expressions predict similar masses for ω and ρ as well as a larger mass for the ϕ—as is observed. Even more importantly, they allow some understanding about the decay modes of ω and ϕ. These are

$$
\left.
\begin{array}{l}
\phi(1020) \to K^+K^- \\
\qquad\ \to K^0\bar{K}^0
\end{array}
\right\} 84\%
\qquad
\left.
\begin{array}{l}
\omega(783) \to \pi^+\pi^-\pi^0 \ \ 90\% \\
\qquad\ \to \pi^+\pi^- \\
\qquad\ \to \pi^0\gamma
\end{array}
\right\} 10\%
\tag{5.25}
$$
$$\phi \to \pi^+\pi^-\pi^0 \ \ 15\%$$

The phase-space factors favor 3π decay of the ϕ, since the Q-value is 600 MeV, compared with $Q = 24$ MeV for $K\bar{K}$ decay. Yet the $K\bar{K}$ decay is dominant. This must be somehow connected with the $s\bar{s}$ composition of the ϕ-meson as in (5.24). To understand this we can draw quark flow diagrams (Fig. 5.5). Diagram (c) is for $\phi \to 3\pi$, involves unconnected quark lines, and is suppressed. This suppression is known as the Zweig rule and is especially

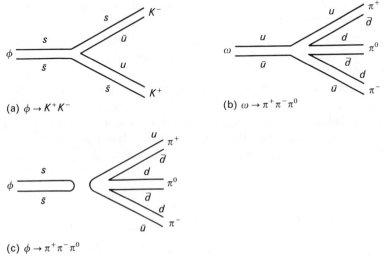

(a) $\phi \to K^+K^-$

(b) $\omega \to \pi^+\pi^-\pi^0$

(c) $\phi \to \pi^+\pi^-\pi^0$

Figure 5.5 Quark flow diagrams for ϕ- and ω-decay.

important in understanding the narrow widths of the more massive mesons (ψ, Υ) built from b and c quarks (see Section 5.13). In the context of a field theory of quark interactions (quantum chromodynamics) the Zweig suppression is viewed in terms of a multigluon intermediate state (Section 5.15)

5.7. LEPTONIC DECAYS OF VECTOR MESONS

As an illustration of the quark-antiquark assignments for the vector mesons mentioned above,

$$\rho^0 = \frac{1}{\sqrt{2}}(u\bar{u} - d\bar{d}),$$

$$\omega^0 = \frac{1}{\sqrt{2}}(u\bar{u} + d\bar{d}), \tag{5.26}$$

$$\phi^0 = s\bar{s},$$

we consider their leptonic decays $V \to l^+l^-$ (where $l = e, \mu$) which are assumed to proceed via exchange of a single virtual photon (see Fig. 5.6). The partial width for decay is given by the Van Royen–Weisskopf formula (1967):

$$\Gamma(V \to l^+l^-) = \frac{16\pi\alpha^2 Q^2}{M_V^2}|\psi(0)|^2, \tag{5.27}$$

where $Q^2 = |\sum a_i Q_i|^2$ is the squared sum of the charges of the quarks, in the meson, $\psi(0)$ is the amplitude of the $Q\bar{Q}$ wavefunction at the origin, and M_V is the meson mass. Apart from numerical factors, the form of this expression follows in straightforward fashion. The square of the propagator of the single photon exchanged introduces a factor q^{-4} (where $|q|^2 = M_V^2$), and the phase-space factor for the two-body final state, a factor q^2. $\sqrt{\alpha}\,Q_i$, where Q_i is the quark charge, measures the coupling amplitude to the photon, and we take a superposition of amplitudes a_i from all quarks in the meson. Including the coupling of the photon to the lepton pair gives a total amplitude $\sqrt{\alpha}\sqrt{\alpha}\sum a_i Q_i$, which has to be squared to get the decay rate. Finally, $|\psi(0)|^2$ is the probability that the quark and antiquark will interact with the photon at a point in space-time (the origin of their relative coordinates). In the expression for the phase space, we have assumed $M_V^2 \gg m_l^2$.

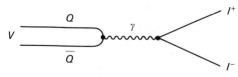

Figure 5.6

Since ρ, ω, ϕ have similar masses, we expect $|\psi(0)|^2/M_V^2$ to be practically constant, and thus $\Gamma_{e^+e^-} \propto Q^2$ (see also Table 5.9). The Q^2-factors are, from (5.26),

$$\rho^0: \quad \left[\frac{1}{\sqrt{2}}\left(\frac{2}{3}-\left(-\frac{1}{3}\right)\right)\right]^2 = \frac{1}{2},$$

$$\omega^0: \quad \left[\frac{1}{\sqrt{2}}\left(\frac{2}{3}-\frac{1}{3}\right)\right]^2 = \frac{1}{18},$$

$$\phi^0: \quad \left(\frac{1}{3}\right)^2 = \frac{1}{9}.$$

The expected and observed leptonic widths are in the ratio

$$\Gamma(\rho^0):\Gamma(\omega^0):\Gamma(\phi^0) = \begin{matrix} 9:1:2 & \text{predicted,} \\ 8.8 \pm 2.6:1:1.70 \pm 0.41 & \text{observed.} \end{matrix} \quad (5.28)$$

This result is a test both of the quark assignments in the vector mesons and of the quark charges.

5.8. DRELL-YAN PRODUCTION OF LEPTON PAIRS BY PIONS ON ISOSCALAR TARGETS

Another test of the quark charge assignments is provided by the process of lepton pair production by pions on nucleons. As shown in Fig. 5.7, we imagine this process to proceed by annihilation of antiquark from the pion with a quark from the nucleon, to form a virtual photon transforming to a muon pair. Again, the cross-section is proportional to the squares of the quark charges. For $\pi^-(=\bar{u}d)$ on an isoscalar ^{12}C nucleus $(=18u + 18d)$ we expect annihilation of $u\bar{u}$; thus

$$\sigma(\pi^-C \to \mu^+\mu^- + \cdots) \propto 18Q_u^2 = 18(\tfrac{4}{9}),$$

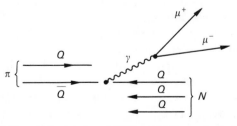

Figure 5.7 The Drell-Yan mechanism for production of lepton pairs is viewed as the fusion of quark and antiquark to a virtual photon, decaying to the pair.

while for incident $\pi^+ (= u\bar{d})$,

$$\sigma(\pi^+ C \rightarrow \mu^+ \mu^- + \cdots) \propto 18Q_d^2 = 18(\tfrac{1}{9}).$$

The cross-section ratio $\sigma(\pi^- C)/\sigma(\pi^+ C)$ is indeed $4:1$ in the region well away from any heavy meson resonances (e.g., $\psi \rightarrow \mu^+ \mu^-$).

5.9. PION-NUCLEON CROSS-SECTIONS

The quark model makes predictions on the relative magnitudes of high-energy hadron-hadron cross-sections, interpreting these as the additive effects of the scattering amplitudes between the individual constituent quarks. The individual collisions between pairs of quarks are treated as quasi-independent of the other constituents. Then the total amplitude $f(0)$ for forward elastic scattering will be the sum of the amplitudes for each quark pair. The optical theorem (4.48) tells us that the total hadron-hadron cross-section $\sigma_T \propto \text{Im} f(0)$. The real part of the amplitude $f(0)$ is assumed to be small—and this is confirmed, for example, by looking at the Coulomb interference in the forward direction in pp elastic scattering. We also need to use the Pomerančuk theorem (Section 4.11) which implies that at high energy, $\sigma(QQ) = \sigma(Q\bar{Q})$, as well as isospin invariance for u-and d-quark cross-sections: $\sigma(uu) = \sigma(dd) = \sigma(ud)$. Putting all these things together, it then follows, from simple quark counting, that we expect a ratio

$$\frac{\sigma(\pi N)}{\sigma(NN)} = \frac{2}{3}.$$

At $E_\pi = 60$ GeV, $\sigma(\pi^+ p)$ and $\sigma(\pi^- p)$ are both equal to 24 mb, while $\sigma(pp) = \sigma(pn) = 38$ mb in an equivalent energy range (see Fig. 4.16). This ratio is indeed approximately $\frac{2}{3}$. For further examples, see Problem 5.4.

5.10. MASS RELATIONS AND HYPERFINE INTERACTIONS

So far, mass differences in hadron multiplets have been simply ascribed to those between the constituent u-, d-, and s-quarks. These alone cannot account for the observed differences, especially those between baryon octet and decuplet members of the same quark constitution. The discrepancy has to be attributed to the effects of quark-quark interactions, for which there is a theory called quantum chromodynamics, discussed in Chapter 8. Here, we shall only sketch out briefly how the most important new contribution, the hyperfine splitting of the hadron energy levels, can be described in terms of the color force between quarks.

First, let us recall that in the energy-level diagram of the hydrogen atom, each level of given n, l, and j quantum numbers is split into two very close hyperfine levels, due to the interactions of the magnetic moments of the constituent proton and electron. In the ground (1S) state, the transition between the two states gives rise to the famous 1420-MHz (21-cm) radio-frequency spin-flip line. Now consider two charged point fermions with magnetic dipole moments μ_i and μ_j separated by distance r_{ij}. The interaction energy is proportional to $\mu_i \cdot \mu_j / r_{ij}^3$. For two particles in a relative S-state, the interaction when averaged over all directions in space is zero except at the origin. The dipole moment has the form expected for a Dirac pointlike fermion:

$$\mu_i = \frac{e_i}{2m_i} \sigma_i \qquad (5.29)$$

in units $\hbar = c = 1$, where e_i and m_i are the electric charge and mass of the particle, σ_i its spin-vector ($\sigma_i^2 = 1$). The interaction energy due to this dipole-dipole interaction is then*

$$\Delta E = \frac{2\pi}{3} \frac{e_i e_j}{m_i m_j} |\psi(0)|^2 \sigma_i \cdot \sigma_j, \qquad (5.30)$$

where $\psi(0)$ is the wavefunction of our two-particle system at the origin ($r_{ij} = 0$). The numerical factor arises from the angular integration for the S-state wavefunction.

Turning now to quarks, the normal magnetic interaction associated with the electric charge and spin of the quarks is small on the scale of hadron masses; it is of order of the electromagnetic mass differences (~ 1 MeV). But quarks carry a strong color charge, and at small interquark separation the color potential is assumed to have the same $(1/r)$ form as the Coulomb potential [see (5.47)]. Associated with the color charges of spinning quarks is a color magnetic interaction, of the same form as (5.30) but with electric charges replaced by color charges. The numerical coefficient in the expression for ΔE depends on whether the interaction is between a quark pair or a quark-antiquark pair, and on the eigenvalues of certain operators associated with the color symmetry. The actual expressions are (see Appendix G)

$$\Delta E(Q\bar{Q}) = \frac{8\pi\alpha_s}{9m_i m_j} |\psi(0)|^2 \sigma_i \cdot \sigma_j, \qquad (5.31)$$

$$\Delta E(QQ) = \frac{4\pi\alpha_s}{9m_i m_j} |\psi(0)|^2 \sigma_i \cdot \sigma_j, \qquad (5.32)$$

where α_s is the strong coupling constant (Section 8.8) and equal to the square of the color charge, in analogy with the fine structure constant of electromagnetism, $\alpha = e^2/4\pi\hbar c$. The product of the Pauli vectors σ_i, σ_j depends in

---

* See for example J. D. Bjorken and S. D. Drell, *Relativistic Quantum Mechanics*, McGraw-Hill, New York, 1964, p. 58.

magnitude and sign on the relative quark spin orientations, just as for two bar magnets the force depends on orientation. Denoting the spin vectors of the quarks by $\mathbf{s}_i$, $\mathbf{s}_j$ (where $s_z = \pm\frac{1}{2}$) and the total spin by $\mathbf{S} = \mathbf{s}_i + \mathbf{s}_j$, one obtains

$$\boldsymbol{\sigma}_i \cdot \boldsymbol{\sigma}_j = 4\mathbf{s}_i \cdot \mathbf{s}_j = 2[S(S+1) - s_i(s_i+1) - s_j(s_j+1)]$$

$$= \begin{cases} +1 & \text{for} \quad S = 1, \\ -3 & \text{for} \quad S = 0. \end{cases} \qquad (5.33)$$

Turning to baryons, consisting of three quarks, we have to sum (5.33) over the quark spins, to obtain, with $\mathbf{S} = \mathbf{s}_i + \mathbf{s}_j + \mathbf{s}_k$,

$$\sum \boldsymbol{\sigma}_i \cdot \boldsymbol{\sigma}_j = 4 \sum \mathbf{s}_i \cdot \mathbf{s}_j = 2[S(S+1) - 3s(s+1)]$$

$$= \begin{cases} +3 & \text{for} \quad S = \frac{3}{2}, \\ -3 & \text{for} \quad S = \frac{1}{2}. \end{cases} \qquad (5.34)$$

This formula will be satisfactory for the nucleon N- or Δ-states, where the three (u and d) quark masses in the denominator of (5.32) are equal. For a $\Sigma^+(1193)$ hyperon (uus) in the octet, the different mass of s and u quarks has to be taken into account in summing (5.33) over the three-quark state. The like pair is $u\!\uparrow\! u\!\uparrow$, in a triplet spin state with $\boldsymbol{\sigma}_u \cdot \boldsymbol{\sigma}_u = 1$, so that from (5.34), $2\boldsymbol{\sigma}_u \cdot \boldsymbol{\sigma}_s = \sum \boldsymbol{\sigma}_i \cdot \boldsymbol{\sigma}_j - \boldsymbol{\sigma}_u \cdot \boldsymbol{\sigma}_u = -4$. Thus we find, for example,

$$(\Delta E)_\Delta = +\frac{3}{m_u^2} K,$$

$$(\Delta E)_N = -\frac{3}{m_u^2} K, \qquad (5.35)$$

$$(\Delta E)_\Sigma = \left(\frac{1}{m_u^2} - \frac{4}{m_u m_s}\right) K,$$

where $K = 4\pi\alpha_s |\psi(0)|^2/9$. From the eight isomultiplets of baryons in the decuplet and octet, we are thus able to fit the four unknown parameters, K, m_u, m_s, and the constant M_0 of Eq. (5.8). The resulting mass values are compared with those observed in Table 5.4, which yields

$$m_n(= m_u = m_d) = 363 \text{ MeV},$$

$$m_s = 538 \text{ MeV}, \qquad (5.36)$$

$$K/m_n^2 = 50 \text{ MeV}.$$

In all cases, agreement is within 1% or better. In particular, the 300-MeV mass difference between Δ and N is successfully accounted for. For the pseudoscalar- and vector-meson states, discussed in Sections 5.5 and 5.6, a similar treatment is possible. In particular, the large mass difference between the $\pi(140)$ and $\rho(776)$, representing singlet and triplet spin combinations of u, d quarks and antiquarks, is ascribed to the hyperfine interaction. We note

TABLE 5.4 Masses of baryons predicted from hyperfine-splitting effects (from Rosner 1980)

Baryon and mass (MeV)	Quark composition (n denotes u or d)	$\Delta E/K$	Predicted mass, MeV
$N(939)$	$3n$	$-3/m_n^2$	939
$\Lambda(1116)$	$2n, 1s$	$-3/m_n^2$	1114
$\Sigma(1193)$	$2n, 1s$	$1/m_n^2 - 4/(m_n m_s)$	1179
$\Xi(1318)$	$1n, 2s$	$1/m_s^2 - 4/(m_n m_s)$	1327
$\Delta(1232)$	$3n$	$3/m_n^2$	1239
$\Sigma(1384)$	$2n, 1s$	$1/m_n^2 + 2/(m_n m_s)$	1381
$\Xi(1533)$	$1n, 2s$	$1/m_s^2 + 2/(m_n m_s)$	1529
$\Omega(1672)$	$3s$	$3/m_s^2$	1682

from (5.31) and (5.32) that the coefficient for mesons is twice that for baryons; this is a crucial prediction of the theory of the interquark color field. Again, the predicted masses agree to within $\sim 1\%$ with the observed values. The fitted m_n, m_s values are slightly less than for the baryons, while K/m_n^2 as defined above is 80 MeV (cf. 50 MeV for baryons). However, it is known that the rms radius of the charge distribution of mesons is smaller ($R_0 \simeq 0.6\,\text{fm}$) than it is for baryons ($R_0 \simeq 0.8\,\text{fm}$), so that $|\psi(0)|^2$, which is proportional to $1/R_0^3$, will be almost a factor 2 larger. So the expectedly larger hyperfine splitting in the mesons, compared with baryons, seems to be verified in the data.

5.11. ELECTROMAGNETIC MASS DIFFERENCES AND ISOSPIN SYMMETRY

The actual mass of a charged hadron can be thought of as made up of two components. First there is a sort of "bare" mass m originating from that of the quark constituents and from their strong mutual interactions, of the type described in the previous section. Secondly, there will be a contribution Δm due to the electric charge of the hadron—basically equal to the work required to put the charge on the previously uncharged particle. If all baryons in the $\frac{1}{2}^+$ octet have similar charge distributions, then we might expect similar values of Δm for similar charges:

$$\Delta m_p = \Delta m_{\Sigma^+},$$

$$\Delta m_{\Sigma^-} = \Delta m_{\Xi^-},$$

$$\Delta m_{\Xi^0} = \Delta m_n.$$

Adding the "bare" masses to each side and summing these equations gives

$$m_p + m_{\Sigma^-} + m_{\Xi^0} = m_{\Sigma^+} + m_{\Xi^-} + m_n$$

or

$$(m_p - m_n) = (m_{\Sigma^+} - m_{\Sigma^-}) + (m_{\Xi^-} - m_{\Xi^0}).$$

$$\underset{-1.3 \text{ MeV}}{} \quad \underbrace{\underset{-8.0 \text{ MeV}}{} \qquad \underset{+6.4 \text{ MeV}}{}}_{-1.6 \text{ MeV}} \tag{5.37}$$

This formula was due originally to Coleman and Glashow and is well verified within the errors. The individual mass differences are associated with isospin symmetry breaking, which has already been discussed in Section 4.5 (see Table 4.3). In the context of the quark model, there are several distinct effects to consider in accounting for the mass differences:

(a) *Difference in masses of the u- and d-quarks.* The sign of each term in (5.37) indicates $m_d > m_u$.

(b) *Coulomb energy difference associated with the electrical energy between pairs of quarks.* This will be of order $e^2/R_0 = (e^2/\hbar c)\,(\hbar c/R_0)$, where R_0 is the size of a baryon. With $\hbar c = 197$ MeV fm, $R_0 \simeq 0.8$ fm, we have $e^2/R_0 \simeq 2$ MeV.

(c) *Magnetic energy difference associated with the magnetic-moment (hyperfine) interaction between quark pairs.* From Eq. (5.30) this will be of order of magnitude $(e\hbar/mc)^2(1/R_0)^3$, where m is the quark mass and $|\psi(0)|^2 \simeq R_0^{-3}$. Since $\hbar/mc$ is of order R_0, the magnetic energy is also in the region of e^2/R_0, i.e., 1 or 2 MeV.

Fitting the exact forms of these terms to the numbers in (5.37), it is found that $m_d - m_u = 2$ MeV. While, given the values of the spins, charges, and radii of baryons, any model must predict Coulomb and magnetic energy differences of the above magnitudes, the closeness in mass of the u- and d-quarks could not have been forseen. Thus the property of approximate isospin invariance of the interactions between hadrons and in atomic nuclei can be associated with the near equality of m_u and m_d, and of no more fundamental significance. At the present time, the origin of quark masses and the above near equality is however not understood.

5.12. BARYON MAGNETIC MOMENTS

From the symmetry properties of the three-quark wavefunction in the baryon octet, it is a fairly straightforward matter to compute the magnetic moments of the various members, by assuming that they are equal to the vector sums of the quark moments. If the quarks behave as pointlike Dirac particles, each will have a magnetic dipole moment as in (5.29). For the proton, *uud*, we have already noted that the two u-quarks will be in a symmetric (triplet) spin state described by a spin function $\chi(J = 1; m = 0, \pm 1)$, while the third ($d$) quark can be denoted by $\phi(J = \frac{1}{2}, m = \pm \frac{1}{2})$. The total-angular-momentum function for a spin-up proton will be $\psi(J = \frac{1}{2}, m = \frac{1}{2})$, and the Clebsch-Gordan coefficients for combining $J = 1$ and $J = \frac{1}{2}$ give us (see Appendix C, and Table III)

$$\psi(\tfrac{1}{2}, \tfrac{1}{2}) = \sqrt{\tfrac{2}{3}}\chi(1, 1)\phi(\tfrac{1}{2}, -\tfrac{1}{2}) - \sqrt{\tfrac{1}{3}}\chi(1, 0)\phi(\tfrac{1}{2}, \tfrac{1}{2}). \tag{5.38}$$

For the first combination, the moment will be $\mu_u + \mu_u - \mu_d$, and for the second, just μ_d. Hence we get for the proton moment

$$\mu_p = \tfrac{2}{3}(2\mu_u - \mu_d) + \tfrac{1}{3}\mu_d = \tfrac{4}{3}\mu_u - \tfrac{1}{3}\mu_d. \tag{5.39}$$

The result for the neutron in the same, with the labels u and d interchanged. For Σ^+ one replaces μ_d by μ_s in (5.39), and for Σ^-, μ_d by μ_s and μ_u by μ_d. For the Λ-hyperon, which is a uds combination with $I = 0$, the pair u and d must be in an $I = 0$ (that is, antisymmetric) isospin state. So they must be in an antisymmetric spin state ($J = 0$). Hence the u and d in the Λ make no contribution to the moment, and $\mu_\Lambda = \mu_s$. Values of μ for Σ^0, Ξ^0, and Ξ^- can be worked out similarly. Table 5.5 shows the values observed and predicted, using the values of quark mass deduced from Table 5.4 to evaluate the quark moments using (5.35). All numbers are given in terms of the nuclear magneton (n.m.) $\mu = e\hbar/2Mc$, where M is the proton mass.

The predicted p, n, and Λ moments agree very well with experiment. In particular, the expected ratio $\mu_n/\mu_p = -\frac{2}{3}$ is in impressive agreement with the observed value -0.685. There are however considerable discrepancies for the Σ- and Ξ-states, and these underline a basic weakness of the static or nonrelativistic quark model. In later chapters, we shall see that a more sophisticated picture of baryons (or mesons) has them consisting of the "valence" quarks of the static model, plus neutral vector gluons (the

TABLE 5.5 Magnetic moments of "stable" baryons in the $J^P = \frac{1}{2}^+$ octet

Baryon	Magnetic moment in quark model	Prediction, n.m.	Observed, n.m.
p	$\tfrac{4}{3}\mu_u - \tfrac{1}{3}\mu_d$	2.79	2.793
n	$\tfrac{4}{3}\mu_d - \tfrac{1}{3}\mu_u$	-1.86	-1.913
Λ	μ_s	-0.58	-0.614 ± 0.005
Σ^+	$\tfrac{4}{3}\mu_u - \tfrac{1}{3}\mu_s$	2.68	2.33 ± 0.13
Σ^0	$\tfrac{2}{3}(\mu_u + \mu_d) - \tfrac{1}{3}\mu_s$	0.82	—
Σ^-	$\tfrac{4}{3}\mu_d - \tfrac{1}{3}\mu_s$	-1.05	-1.00 ± 0.12
Ξ^0	$\tfrac{4}{3}\mu_s - \tfrac{1}{3}\mu_u$	-1.40	-1.25 ± 0.014
Ξ^-	$\tfrac{4}{3}\mu_s - \tfrac{1}{3}\mu_d$	-0.47	-1.85 ± 0.75

mediating vector bosons of the color field) some of which can transform to quark-antiquark pairs. The $Q\bar{Q}$ pair content of a baryon, although small, can contribute extra currents and thus affect the magnetic moment. At the present time, however, no comprehensive dynamical theory of quark interactions, applicable to the static baryon states, exists.

A brief discussion of the measurement of hyperon magnetic moments is appropriate here. For illustration we shall discuss that of the Λ-hyperon. The Λ-hyperons can be produced in hadron-hadron collisions, for example in the reaction

$$\pi^- + p \to \Lambda^0 + K^0,$$

and are found to be polarized with spins normal to the ΛK production plane, i.e., along the vector $(\mathbf{p}_K \times \mathbf{p}_\Lambda)$. The sense and degree of polarization can be found by observing the up-down asymmetry in the subsequent parity-violating weak decay $\Lambda \to p + \pi^-$ (Section 7.7). Suppose that the Λ traverses a magnetic field $\mathbf{B}$ perpendicular to $\mathbf{p}_\Lambda$ and in the production plane, as in Fig. 5.8. The Larmor precession frequency is

$$\nu_L = \mu_\Lambda \frac{B}{h},$$

and the angle of rotation is therefore

$$\phi = \mu_\Lambda \int \frac{B \, dl}{h}$$

proportional to the line integral of the field, and is numerically $18° \cdot (\mu_\Lambda/\mu_N)$ for $\int B \, dl = 1$ T m, where μ_N is the nuclear magneton.

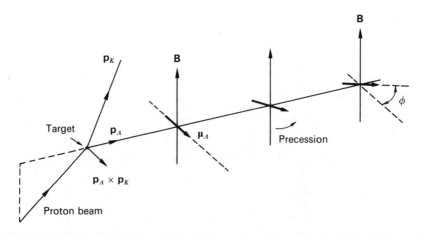

Figure 5.8 Sketch of precession of spin of Λ-hyperons produced in association with neutral kaons, in a magnetic field. Initially, the hyperons are polarized normal to the production plane.

The first experiments on μ_Λ were done with nuclear-emulsion detectors and an incident pion momentum of $\simeq 1.05\ \mathrm{GeV}/c$, where the transverse polarization was very high ($\simeq 100\%$), but the path length in the field was limited by the lifetime to $c\tau_\Lambda \sim 10$ cm. Thus, appreciable precession angles could be obtained only with the use of pulsed fields of order 20 T. The more recent experiments have employed high-energy beams of 300-GeV incident protons, which produce more intense secondary hyperon beams of much higher momentum ($\simeq 100\text{--}250\ \mathrm{GeV}/c$). Thus the useful path length is increased due to relativistic time dilation by two orders of magnitude, and d.c. spectrometer magnets can be employed. The decay products are recorded in multiwire proportional chambers. The greatly reduced polarization ($\simeq 8\%$ instead of $\simeq 100\%$) at high energy is more than compensated by the huge increase in hyperon intensities available—the most recent experiment involving a sample of 3×10^6 Λ-particles.

5.13. HEAVY-MESON SPECTROSCOPY AND THE QUARK MODEL

Over the past decade, our quark Table 5.1 has become enlarged by the discovery of two more flavors, c for charm and b for bottom, associated with much more massive quarks (see Table 5.6). In addition, there is provisional evidence for a top quark, t (see Section 5.16).

It is a characteristic of the strong "color" forces between quarks that these heavier constituents involve somewhat weaker interactions at smaller distances. The series of heavy-meson states ψ, Υ formed from $c\bar{c}$ and $b\bar{b}$, respectively, have narrow widths and present a spectrum of discrete spectroscopic levels similar to (but on a quite different energy scale from) that of positronium (e^+e^-). The level sequence and spacing can be calculated using simple potential models and is in astonishingly close agreement with what is

TABLE 5.6 Quark quantum numbers (as of 1985): $Q/e = I_3 + \frac{1}{2}(B + S + C + B^* + T)^a$

Flavor	I	I_3	S	C	B^*	T	Q/e
u	$\frac{1}{2}$	$\frac{1}{2}$	0	0	0	0	$+\frac{2}{3}$
d	$\frac{1}{2}$	$-\frac{1}{2}$	0	0	0	0	$-\frac{1}{3}$
s	0	0	-1	0	0	0	$-\frac{1}{3}$
c	0	0	0	1	0	0	$+\frac{2}{3}$
b	0	0	0	0	-1	0	$-\frac{1}{3}$
t	0	0	0	0	0	1	$+\frac{2}{3}$

^{a} B denotes baryon number, which is $\frac{1}{3}$ for all quarks.
B^* here denotes the bottom or beauty quantum number.

observed. These heavy-meson spectra indeed present the clearest evidence yet for the existence of basic quark and antiquark constituents of hadrons (see Fig. 5.15).

5.13.1. Charmonium Levels

The ψ series of meson resonances was first observed in 1974 in e^+e^- collisions at SLAC (Stanford), using the e^+e^- collider SPEAR (Augustin et al. 1974); and the lowest-lying state, called ψ or J/ψ, was simultaneously observed in experiments at the Brookhaven AGS in collisions of 28-GeV protons on a beryllium target (Aubert et al. 1974), leading to a massive e^+e^- pair:

SLAC: $e^+e^- \to \psi \to$ hadrons   (5.40)
$$\to e^+e^-, \mu^+\mu^-$$

BNL: $p + Be \to \psi/J +$ anything.   (5.41)
$$\to e^+e^-$$

The original data on the reaction (5.40) are shown in Fig. 5.9 and on (5.41) in Fig. 5.10. In both cases, a sharp resonance ψ is observed, peaking at a mass of 3.1-GeV. In (5.41), massive electron pairs were detected by means of a magnet spectrometer and detectors downstream of the target, electrons and positrons being recorded in coincidence at large angles on either side of the incident proton beam axis. In the e^+e^- experiment, the reaction rate in the beam intersection region was measured as the beam energies were increased in small steps. In addition to the particle ψ, a second resonance ψ' of mass 3.7 GeV was also found in this first SPEAR experiment (Abrams et al. (1974)).

The observed widths of the peaks in Figs. 5.9 and 5.10 are dominated by the experimental resolution, on the secondary-electron momentum in the BNL experiment and on the circulating-beam momentum in the SLAC experiment. The true width of the ψ is much smaller and can be determined from the total reaction rate and the leptonic branching ratio, both of which have been measured. Recalling the Breit-Wigner formula (4.55) for the formation of a resonance of spin J from two particles of spin s_1 and s_2, we can write

$$\sigma(E)_{e^+e^- \to \psi \to e^+e^-} = \frac{4\pi\lambdabar^2(2J + 1)\Gamma^2_{e^+e^-}/4}{(2s_1 + 1)(2s_2 + 1)[(E - E_R)^2 + \Gamma^2/4]}, \quad (5.42)$$

where $\lambdabar$ is the de Broglie wavelength of the e^+ and e^- in the CMS, E is the CMS energy, E_R is the energy at the resonance peak, Γ is the total width of the resonance, and $\Gamma_{e^+e^-}$ is its partial width for $\psi \to e^+e^-$. With $s_1 = s_2 = \frac{1}{2}$ and the assumption $J = 1$, the total integrated cross-section is readily found from (5.42) using the substitution $\tan\theta = 2(E - E_R)/\Gamma$:

$$\int_0^\infty \sigma(E)\,dE = \frac{3\pi^2}{2}\lambdabar^2\left(\frac{\Gamma_{e^+e^-}}{\Gamma}\right)^2\Gamma. \quad (5.43)$$

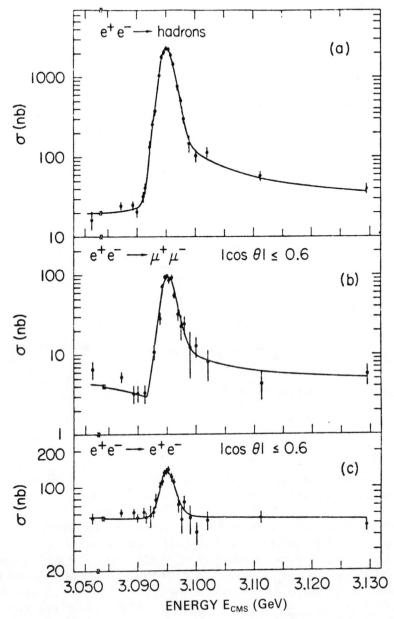

Figure 5.9 Results of Augustin *et al.* (1974) showing the observation of the ψ/J resonance of mass 3.1 GeV, produced in e^+e^- annihilation at the SPEAR storage ring, SLAC.

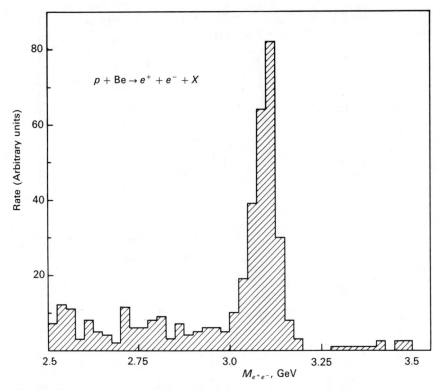

Figure 5.10   Results of Aubert *et al.* (1974) indicating the narrow resonance ψ/J in the invariant-mass distribution of e^+e^- pairs produced in inclusive reactions of protons with a beryllium target. The experiment was carried out with the 28-GeV AGS at Brookhaven National Laboratory.

The integrated cross-section in Fig. 5.9(c) must be equal to $\int \sigma(E)\,dE$, and is numerically 900 nb MeV. The branching ratio $\Gamma_{e^+e^-}/\Gamma = 0.07$, and $\lambdabar = \hbar c/pc$, where $pc = 1500$ MeV and $\hbar c = 200$ MeV fm. Inserting these numbers in (5.43), we obtain $\Gamma = 0.067$ MeV for the true width of the ψ, which is much smaller than the experimental width, of order several MeV. In comparison with other vector mesons such as the $\rho(776\text{ MeV})$ with $\Gamma = 100$ MeV and $\omega(784\text{ MeV})$ with $\Gamma = 11$ MeV, the $\psi(3100\text{ MeV})$ has an extremely small width, and the purely electromagnetic decay $\psi \to e^+e^-$ competes with that into hadrons. Note that the partial width $\Gamma(\psi \to e^+e^-) = 4$ keV and is not so different from that of the other vector mesons. For example, $\Gamma(\omega \to e^+e^-) = 0.8$ keV and $\Gamma(\phi \to e^+e^-) = 1.6$ keV.

The assumption $J^P = 1^-$, i.e., the vector nature of the ψ-particle, is justified by observing the shape of the resonance curve in Fig. 5.9(b). It has the characteristic dispersionlike appearance characteristic of two interfering amplitudes: these are due to the direct channel [Fig. 5.11(a)] and the

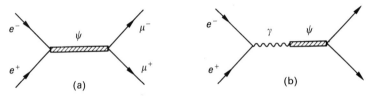

Figure 5.11

production of ψ via an intermediate virtual photon [Fig. 5.11(b)]. The interference between these diagrams is proof that ψ must have the same quantum numbers as the photon. The isospin assignment $I = 0$ is based on the characteristics of hadronic decays. Since the ψ decays predominantly to an odd number of pions, then from G-parity arguments (see Section 4.5.1) it follows that I must be even. The $I = 0$ assignment is confirmed by observations on the decay mode $\psi \to \rho\pi$: the various charge states $\rho^+\pi^-$, $\rho^0\pi^0$, $\rho^-\pi^+$ are found to be equally populated. Reference to the Clebsch-Gordan coefficients (Table III, Appendix) for combining two states of $I = 1$ then shows $I = 0$ is the correct assignment for ψ.

In summary, some properties of the particles ψ and ψ' are listed in Table 5.7. An example of the decay $\psi' \to \psi + \pi^+\pi^-$, $\psi \to e^+e^-$ is shown in Fig. 5.12.

The extreme narrowness of the ψ and ψ' states indicated that there was no possibility of understanding them in terms of u, d, and s (and $\bar{u}$, $\bar{d}$, and $\bar{s}$) quarks. A new type of quark had in fact been postulated some years before by Glashow, Iliopoulos, and Maiani (1970), in connection with the nonexistence of strangeness-changing neutral weak currents (see Section 7.11). This carried a new quantum number, C for *charm*, which, like strangeness, would be conserved in strong and electromagnetic interactions. The large masses of the ψ, ψ' mesons imply that, if they contain such charmed quarks, these in turn must be massive. It is indeed postulated that ψ, ψ' consist of the vector

TABLE 5.7   Charmonium states and decay modes

State	Mass, MeV	J^P, I	Γ, MeV	Branching ratio	
$J/\psi(3100)$	3097 ± 0.1	1^-, 0	0.063	Hadrons [mostly $(2n + 1)\pi$]	86%
				e^+e^-	7%
				$\mu^+\mu^-$	7%
$\psi(3700)$	3686 ± 0.1	1^-, 0	0.215	$\psi + 2\pi$	50%
				$\chi + \gamma$	23%
				e^+e^-	0.9%
				$\mu^+\mu^-$	0.9%

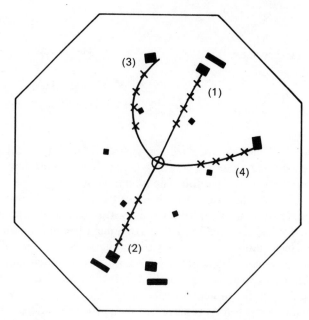

Figure 5.12 Example of the decay $\psi'(3.7) \rightarrow \psi(3.1) + \pi^+ + \pi^-$ observed in a spark chamber detector. The $\psi(3.1)$ decays to $e^+ + e^-$. Tracks (3) and (4) are due to the relatively low-energy (150-MeV) pions, and (1) and (2) to the 1.5-GeV electrons. The magnetic field and the SPEAR beam pipe are normal to the plane of the figure. The trajectory shown for each particle is the best fit through the sparks, indicated by crosses. [From G. S. Abrams *et al.*, *Phys. Rev. Letters* **34**, 1181 (1975).]

combinations of $c\bar{c}$, called *charmonium*, just as the ρ^0 consists of $u\bar{u}$ and $d\bar{d}$. Other combinations with a net charm number, e.g., $c\bar{d}$, form the so-called charmed mesons, which had been observed previously in neutrino experiments (but not clearly identified) and were soon to be catalogued in SLAC experiments. With the exception of the lowest-lying D-meson, of mass 1870 MeV, which decays weakly in a $\Delta C = 1$ transition, all are broad states, just like the ρ and ω. Since $M_\psi < 2M_D$, the decay of the ψ into the meson states with $C = +1$ and -1 is energetically impossible; it follows that ψ must decay into states containing only u-, d-, s-quarks and antiquarks.

We recall that the small branching ratio for the decay $\phi \rightarrow \pi^+\pi^-\pi^0$ in comparison with $\phi \rightarrow K\bar{K}$ (see Fig 5.5) was ascribed to the so-called Zweig rule, where unconnected lines in the quark flow diagram lead to a suppression of the decay amplitude. We can draw similar diagrams for ψ-decay (Fig. 5.13).

According to this rule, the decay $\psi \rightarrow D\bar{D}$ is preferred but is not allowed by energy conservation, leaving the decay into noncharmed mesons as in (b), which is much more strongly suppressed than in ϕ-decay. We shall return to the problem of the ψ width in Section 5.15.

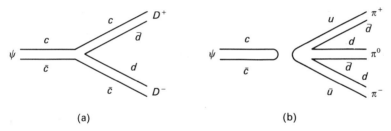

Figure 5.13 Quark diagrams for charmonium decay. The diagram (a) is favored but forbidden by energy conservation for charmonium states $\psi(3.1)$ and $\psi'(3.7)$ with masses below threshold $2M_D = 3.75$ GeV. The "Zweig-forbidden" diagram (b) is therefore the only one allowed for hadronic decay.

Still heavier charmonium states are observed, but these are all above $D\bar{D}$ threshold; successive states rapidly become broader because the decays to $D\bar{D}$ and other charmed mesons are favored by the Zweig rule. Charmed pseudoscalar mesons $D^+ (=c\bar{d})$, $D^0 (=c\bar{u})$, and their antiparticles; charmed vector mesons D^* decaying in the mode $D^* \to \pi D$; and mesons $F(=c\bar{s}$ etc.) with both charm and strangeness, have all been catalogued. The pseudoscalars decay by $\Delta C = 1$ weak interactions into noncharmed states, with the decay into kaons ($D^0 \to K^-\pi^+$ etc.) favored by the Cabibbo suppression factor (see Section 7.9). Figure 2.15 shows a nice example of production and decay of a vector meson D^* in a neutrino reaction in the BEBC hydrogen chamber at CERN.

5.13.2. Upsilon States (Υ)

The discovery of the narrow charmonium ($\psi = c\bar{c}$) states in 1974 was followed in 1977 by the observation of similar narrow resonances in the mass region 9.5–10.5 GeV, attributed to bound states of still heavier "bottom" quarks with charge $\frac{1}{3}$, and generically named $\Upsilon = b\bar{b}$—see Table 5.8.

Figure 5.14 shows the results on the mass spectrum of muon pairs produced in 400-GeV proton-nucleus collisions

$$p + \text{Be, Cu, Pt} \to \mu^+ + \mu^- + \text{anything},$$

as observed in a two-arm spectrometer by Herb *et al.* (1977) and Innes *et al.* (1977) in an experiment at FNAL. A broad peak centered around 10 GeV is

TABLE 5.8 Upsilon states

	$\Upsilon(1\,^3S)$	$\Upsilon'(2\,^3S)$	$\Upsilon''(3\,^3S)$	$\Upsilon'''(4\,^3S)$
Mass, GeV	9.46 ± 0.01	10.02 ± 0.01	10.35 ± 0.01	10.57 ± 0.01
$\Gamma_{e^+e^-}$, keV	1.2 ± 0.2	0.50 ± 0.14	0.40 ± 0.10	0.30 ± 0.10
$\Gamma_{(\text{tot})}$, MeV	0.042 ± 0.015	0.03 ± 0.01	—	14 ± 5

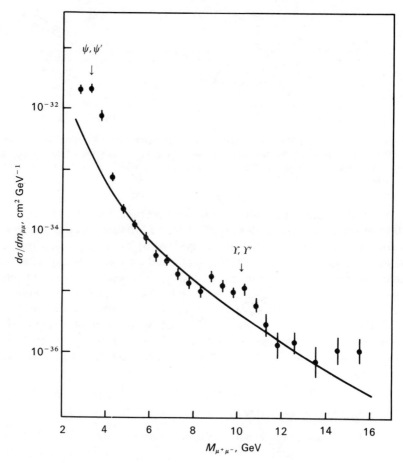

Figure 5.14   First evidence for the upsilon resonances Υ, Υ', obtained by Herb *et al.* (1977) from the spectrum of muon pairs observed in 400-GeV proton-nucleus collisions at Fermilab, near Chicago. The enhancement due to these resonances stands out against the rapidly falling continuum background. The individual states Υ, Υ', etc., are not resolved.

apparent against the falling continuum background. Since the total width ($\simeq 1.2$ GeV) was greater than that arising from the apparatus resolution (0.5 GeV), it was deduced that two or three resonances were present, with masses of 9.4, 10.01, and possibly 10.4 GeV—named Υ, Υ', and Υ'', respectively.

As in the case of charmonium, the states Υ, Υ' were also observed (one year later) in e^+e^- experiments at the DORIS storage ring in Hamburg, where they could be clearly resolved, and at CESR, Cornell, where the narrow state Υ'' and a fourth state Υ''' were identified. As for charmonium, the apparent widths of the three Υ-states are determined by the beam energy resolution. Their masses and leptonic widths are given in Table 5.8.

TABLE 5.9 Leptonic widths of vector mesons

Meson	Quark wavefunction	$\lvert\Sigma a_i Q_i\rvert^2$	$\Gamma_{e^+e^-}$, keV	$\Gamma_{e^+e^-}/\lvert\Sigma a_i Q_i\rvert^2$
ρ	$(u\bar{u} - d\bar{d})/\sqrt{2}$	$\frac{1}{2}$	6.6 ± 0.8	13.2 ± 1.5
ω	$(u\bar{u} + d\bar{d})/\sqrt{2}$	$\frac{1}{18}$	0.71 ± 0.07	12.8 ± 1.3
ϕ	$s\bar{s}$	$\frac{1}{9}$	1.31 ± 0.06	11.8 ± 0.5
ψ	$c\bar{c}$	$\frac{4}{9}$	4.7 ± 1.0	10.5 ± 2.3
Υ	$b\bar{b}$	$\frac{1}{9}$	1.2 ± 0.2	10.6 ± 1.8

In the nonrelativistic quark model, the leptonic decay width of a vector meson coupling to a single virtual photon is expected to be proportional to the square of the quark charges (Rutherford scattering) (see (5.27)). Empirically, as seen in Table 5.9, the ratio $\Gamma_{e^+e^-}/\lvert\Sigma a_i Q_i\rvert^2$ of the leptonic width to the square of the mean quark charge of the ρ, ω, ϕ, and ψ are closely similar, and that of the Υ agrees if one assumes it is built from b, $\bar{b}$ quarks with $Q = \frac{1}{3}$.

It will be noted from Table 5.8 that the $4\,^3S$ level Υ''' has a much larger total width than the lower S-states, indicating that it is above threshold for production of pairs of pseudoscalar B-mesons ($\bar{b}u$, $b\bar{u}$, $b\bar{d}$, $\bar{b}d$) analogous to the D-mesons containing c-quarks. The B-meson mass is 5.27 GeV.

5.14. COMPARISON OF QUARKONIUM AND POSITRONIUM LEVELS

Figures 5.15 and 5.16 show the level scheme of charmonium ($c\bar{c}$), the upsilon series ($b\bar{b}$), and positronium (e^+e^-). The similiarities are striking and compelling evidence for the view that, like positronium, the ψ- and Υ-series can be ascribed to bound states of fermion-antifermion pairs. We want to try to put this comparison on a more quantitative footing.

In the $c\bar{c}$ level scheme, we have already noted the $J^{PC} = 1^{--}$ triplet S-states $\psi(3100)$ and $\psi(3685)$. where $P = (-1)^{l+1}$ and $C = (-1)^{l+s}$ as in Section 3.9. In addition there is a candidate for a singlet ($^1S_0$) state called $\eta_c(2980)$ of $J^{PC} = 0^{-+}$, reached by radiative decay $\psi(3685) \rightarrow \gamma\eta_c$, as well as a candidate for the 2^1S_0 state at a mass of 3590 MeV. The P-states are labeled χ_c (or χ_b in the upsilon system, Fig. 5.15(b)) and are obtained in radiative decays $\psi \rightarrow \gamma\chi_c$. All these states are below $D\bar{D}$ threshold and narrow (Zweig rule). For the Υ-system, there are three sets of narrow S-states and two of P-states below threshold for producing a pair of B-mesons (10.54 GeV).

Coming now to positronium, the level energies using the nonrelativistic Schrödinger equation and a Coulomb potential are

$$E_n = -\frac{\alpha^2 mc^2}{4n^2}, \tag{5.44}$$

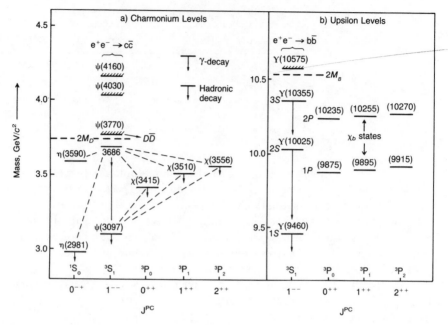

Figure 5.15 Energy levels of (a) the charmonium ($c\bar{c}$) system and (b) the upsilon ($b\bar{b}$) system. Numbers in brackets indicate masses in MeV/c^2.

where n is the radial quantum number, corresponding to $1S, 2S, \ldots$, etc., and m is the electron mass. This is the usual Balmer formula for the hydrogen atom levels, but includes a factor of $\frac{1}{2}$ for the reduced-mass effect in positronium. Relativistically, the levels are split, first by the spin-orbit interaction into $S, P, \ldots$ states of different orbital angular momentum $l(<n)$, and secondly by the spin-spin (magnetic moment) interaction into triplet ($^3S_1$) and singlet ($^1S_0$) states. In atomic spectroscopy these splittings are called fine structure and hyperfine structure, respectively, but in positronium (where both constituents have magnetic moments equal to a Bohr magneton), both are of similar magnitude

$$\Delta E(\text{fine structure}) \sim \frac{\alpha^4 m c^2}{n^3}. \tag{5.45}$$

The triplet and singlet states are called ortho- and para-positronium (see Section 3.9). Equations (5.44) and (5.45) are quoted to leading order in α; in the full quantum electrodynamics (QED) calculation, there are higher-order corrections with relative contributions of magnitude $\alpha, \alpha^2, \ldots$, etc. The measured transition frequencies for $2^3S_1 \to 1^3P_2$ (8628 MHz) and $1^3S_1 \to 1^1S_0$ (203387 MHz) are in exact agreement with the QED prediction, as are the lifetimes (3.46) and (3.47) for decay of the ortho and para ground states.

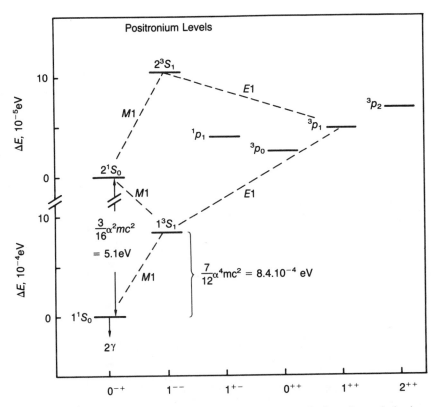

Figure 5.16 Energy levels of positronium. Note the varying scale along the vertical axis.

For positronium, we know the potential has the Coulomb form

$$V_{em} = -\frac{\alpha}{r}, \tag{5.46}$$

where r is the separation of e^+ and e^-. The exact form of the potential in the heavy $Q\bar{Q}$ systems is however unknown. In the field theory of quark-quark interactions, quantum chromodynamics (QCD), the strong color field is mediated by massless vector gluons, just as in QED it is mediated by massless vector photons. Hence the potential might be expected to be of the Coulomb form, and this is verified experimentally from the two-jet cross-section at hadron colliders (see Section 8.9). At large r, the quarks are subject to confining forces, and from the linearity of the plot of J versus (mass)^2 for baryons and mesons it is deduced that the potential at large r is linear (see Section 8.10). Thus a favored QCD potential is

$$V_{QCD} = -\frac{4}{3}\frac{\alpha_s}{r} + kr \tag{5.47}$$

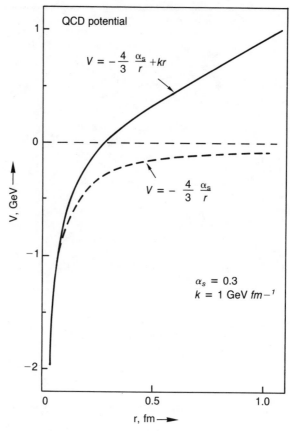

Figure 5.17   Form of the QCD potential $V = \frac{-4}{3}(\alpha_s/r) + kr$, with $\alpha_s = 0.3$ and $k = 1$ GeV fm^{-1}.

where the $\frac{4}{3}$ is a color factor† (see Appendix J), α_s is the quark-gluon coupling analogous to a α in (5.46), and $k \simeq 1$ GeV fm^{-1}. A plot of this potential for $\alpha_s = 0.3$ is shown in Fig. 5.17. The energy levels for this potential have to be calculated numerically. The first term gives a $2S - 1S$ separation proportional to constituent mass as in (5.44), while the second gives one proportional to $m^{-1/3}$. The similarity in the observed separations, 589 MeV for the ψ-states and 565 MeV for Υ (Fig. 5.15) despite the factor 3 difference in constituent mass, is presumably the result of an accidental balance between these two terms.

---

† The factor $\frac{4}{3}$ is justified briefly as follows. The quark has three color degrees of freedom and couples to an octet of color gluon states. Averaging over quark color we obtain a factor $\frac{8}{3}$ as compared with electromagnetism (one "color" of photon and one fermion charge). This has to be divided by 2 because α_s is defined as *twice* the square of the strong color charge.

TABLE 5.10 Positronium and quarkonium level splittings

System	Coupling	Constituent mass	$\Delta E(2^3S_1 - 1^3S_1) \sim \alpha^2 m$	α_s	$\Delta E(2^3S_1 - 1^3P_2) \sim \alpha^4 m$	α_s
e^+e^-	$\alpha = 1/137$	0.51 MeV	5.1 eV	—	$3.6.10^{-5}$ eV	—
$\psi = c\bar{c}$	$4\alpha_s/3$	1870 MeV	589 MeV	~ 1	130 MeV	~ 1
$\Upsilon = b\bar{b}$	$4\alpha_s/3$	5280 MeV	565 MeV	~ 0.6	110 MeV	~ 0.7

In order to get a qualitative feeling for what determines the level splittings, let us consider the analytical solution of (5.47) using just the first, Coulombic, term. From the fact that, for a $1/r$ potential, $\Delta E(2^3S_1 - 1^3S_1) \simeq \alpha^2 m$, and $\Delta E(2^3S_1 - 1^3P_2) \simeq \alpha^4 m$, it is straightforward to deduce values for α_s (in terms of α) assuming masses for c- and b-quarks equal to the D- and B-meson masses, as shown in Table 5.10. The factors of 10^8 in the structure and of 10^{13} in the fine-structure splittings between the quarkonium and positronium levels can be accounted for in terms of the much larger constituent masses and a value for the strong coupling, $\alpha_s \simeq 1$. Since we neglected the linear term in (5.47), it is clear from Fig. 5.17 that these values for α_s will be overestimates. Numerical solutions of the full potential give $\alpha_s \simeq 0.3$ for the ψ- and Υ-systems, which are compatible with values deduced from the frequency of continuum three-jet events in e^+e^- annihilation at high energy, from the two-jet scattering cross-section in hadron-hadron collisions, and from the analysis of deep-inelastic lepton-nucleon scattering, all discussed in Chapter 8.

5.15. INTERPRETATION OF THE ZWEIG RULE

We now discuss the extreme narrowness of the heavy quarkonium states (of order of tens of keV) enshrined in the Zweig rule for unconnected quark lines (Fig. 5.13). These states, as well as the mesons into which they decay, are color singlets, so that the connection between initial and final quark states in QCD must be via a color-singlet gluon combination. Thus, at least two gluons must be exchanged. Since ψ, Υ couple to photons, they have charge-conjugation quantum number $C = -1$, and therefore, by the same argument as was used for positronium, the number of gluons exchanged must be odd. Thus, three-gluon exchange is the simplest possibility. For the ψ, for example, we have from (5.27)

$$\Gamma(\psi \to \gamma \to e^+e^-) = \frac{16\pi\alpha^2}{M^2} |\psi(0)|^2 |Q_i|^2 \simeq 5 \text{ keV}, \qquad (5.48)$$

while for hadronic decay via the three gluons it is found that

$$\Gamma(\psi \to 3G \to \text{hadrons}) = \frac{160(\pi^2 - 9)}{81 M^2} \alpha_s^3 |\psi(0)|^2 \simeq 70 \text{ keV}. \qquad (5.49)$$

TABLE 5.11 Leptonic branching ratios for vector mesons

Meson	$\Gamma_{e^+e^-}/\Gamma_{\text{total}}$	α_s
$\phi(1020)$	$(2.09 \pm 0.07) \times 10^{-3}$	0.44
$\psi(3100)$	$(7.4 \pm 1.2) \times 10^{-2}$	0.21
$\Upsilon(9480)$	$(3.0 \pm 1.0) \times 10^{-2}$	0.18

The last equation is the same as that for $e^+e^- \to 3\gamma$ [Eq. (3.47)] apart from extra numerical factors due to color for $Q\bar{Q} \to 3G$, multiplying the rate by $\frac{5}{18}$ and replacing α by α_s. From (5.48) and (5.49) we obtain values of α_s in Table 5.11. The value from the leptonic branching ratios is in good accord with the previous estimates. Note that the larger values of α_s from the $\phi(1020)$ is consistent with a decrease in α_s as one goes to larger masses and momentum transfers, as discussed in Section 8.12. The value of $\Gamma(\text{total})$ in the case of the ϕ-meson is that for the Zweig-forbidden decay $\phi \to 3\pi$ only and does not include the contribution for $\phi \to K\bar{K}$. The narrowness of the quarkonium states is therefore attributed to the exchange of three "hard" gluons as in Fig. 5.18(a), and this severely limits the hadronic width. In contrast the decay of $\psi(3770) \to D\bar{D}$ can proceed by single "soft" gluon exchange with low-momentum transfer, and this state has a much greater width (25 MeV compared with 0.063MeV).

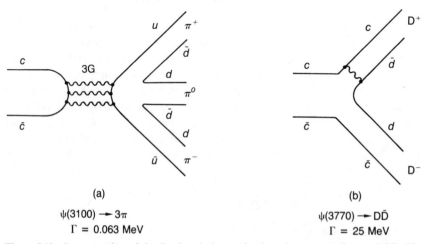

(a) $\psi(3100) \to 3\pi$ $\quad$ $\Gamma = 0.063$ MeV

(b) $\psi(3770) \to D\bar{D}$ $\quad$ $\Gamma = 25$ MeV

Figure 5.18 Interpretation of the Zweig rule in quarkonium decay, according to QCD. The decay (a) of $\psi(3100)$ to noncharmed quark states, via triple gluon exchange, is greatly suppressed in comparison with the decay (b) of $\psi(3770)$ into a $D\bar{D}$ pair containing charmed quarks.

5.16. THE TOP QUARK

It had been anticipated that a t-quark (t for top) of charge $+\frac{2}{3}$ should exist, as a partner of the b-quark, following the discovery of upsilon mesons in 1977. The existence of three pairs of quarks of charges $+\frac{2}{3}$ and $-\frac{1}{3}$ was expected, in parallel with the three pairs of leptons, of charge -1 and 0, on symmetry grounds (see Section 7.12). Experimental searches at the (presently) highest energy e^+e^- collider PETRA have shown that $m_t > 22$ GeV. Preliminary results at the CERN $p\bar{p}$ collider find events consistent with the production and decay of the W-boson according to the scheme $W \to t\bar{b} \to 2$ jets + charged lepton (electron or muon), with $m_t \simeq 40$ GeV (Arnison *et al.* (1984a)).

5.17. THE SEARCH FOR FREE QUARKS

Ever since the quark hypothesis of hadron substructure was formulated in 1964 by Gell-Mann and by Zweig, intensive searches for the existence of free quarks have been carried out. The vast majority of these experiments have failed to find any such evidence, and this has led to the postulate of confinement, that the interaction between quarks is such that they cannot be separated as free particles once they are trapped inside a hadron.

The experimental evidence for or against free quarks has been reviewed extensively, and only a summary will be given here. The experiments have been of two broad types:

(a) Production *in situ* of quarks in high-energy collisions at accelerators, in hadron-hadron, lepton-hadron, and $e^+ e^-$ reactions, and among cosmic-ray secondaries.

(b) Search for quarks already existing in matter, either on earth or elsewhere.

5.17.1. Accelerator Experiments

The searches for quarks produced as secondaries in hadron-hadron (usually proton-nucleus) collisions have used a combination of Čerenkov counters, pulse-height and time-of-flight techniques, and magnetic spectrometers to identify particles of charge $|2e/3|$ and $|e/3|$. They provide limits of $< 10^{-11}$ for the ratio of quarks to pions, and are sensitive to free-quark masses up to about 10 GeV. Generally, such hadronic interactions consist of rather "soft" collisions in which the momentum transfers are small. Lepton-nucleon collisions (see Chapter 8) involve larger momentum transfers and might be more efficient at "kicking out" quarks. The limits obtained are $< 10^{-4}$ quarks per interaction. In e^+e^- experiments the ratio

$\sigma(e^+e^- \to \text{quark} + \text{anything})/\sigma(e^+e^- \to \mu^+\mu^-) < 10^{-2}$. The lepton experiments so far are sensitive only to quark masses up to 20 GeV.

Analyses of secondary cosmic-ray particles have, on more than one occasion, reported a possible quark signal, but the evidence was never convincing, and the only safe statement to be made is that the sea-level quark flux is less than 10^{-10} of the flux of primary nucleons incident on the atmosphere. The level of sophistication and redundancy built into the cosmic-ray experiments is far inferior to that of those at accelerators.

5.17.2. Search for Preexisting Quarks

If quarks exist in ordinary matter, they may be located in "quark atoms" with peculiar physical and chemical properties and with, of course, a net fractional charge. Some experiments have therefore sought to obtain enriched samples by application of electric fields. In this way, the free-quark density in sea water is found to be $< 10^{-24}$ quarks per nucleon. If quarks were present in cosmic rays, for example, and brought to rest in the oceans, this limit implies a cosmic-ray quark flux below 10^{-8} of primary nucleon flux. Searches have also been made in moon rocks, meteorites, deep-ocean sediments, and even oyster shells. The existence of quarks was investigated by optical or mass spectrometry using an arc or ion source. All these searches have proved fruitless and provided limits on quark density not far inferior to that from sea water. Of course, it can be argued that because of the peculiar physical and chemical properties of quark atoms, they are concentrated only in specific materials or even in regions of space outside the solar system.

So far, only one experiment has reported a positive result and actually claimed the existence of free particles of fractional charge (LaRue, Fairbank, and Hebard 1977). The principle of the electrometer method employed is similar to that of Millikan's oil-drop device. It is to levitate small (100-μg) diamagnetic spheres of niobium in a suitably shaped magnetic field, in which they oscillate vertically with a frequency $f \simeq 1$ Hz. The spheres are located between two horizontal capacitor plates, 1 cm apart, to which an oscillatory electric field is applied at the frequency f, but $90°$ out of phase with the free oscillations. The rate of change of the oscillation amplitude is proportional to the electric charge on the sphere. The net charge on a sphere can be changed by $\pm 1e$ by bringing close a radioactive electron or positron source, and this provides the charge calibration. The niobium spheres examined were heat-treated in niobium or tungsten substrates to improve the Q-value of the natural oscillations. In their most recent paper (LaRue et al. 1981) a total of 39 measurements were reported on 13 spheres with radii 100–140 μm. Five of the spheres exhibited fractional charges on different occasions. Of these, 5 measurements gave a negative charge of mean value -0.343 ± 0.011, and 9 gave a positive charge of mean value $+0.328 \pm 0.007$.

The remaining 25 measurements corresponded to integral charges (0.001 ± 0.003). The number of fractional charges observed corresponds to about 10^{-20} quarks per nucleon of niobium if we include the total number of nucleons in the spheres. This is comparable to the upper limits from other quark searches.

A second experiment, by Gallinaro, Marinelli, and Morpurgo (1977), employed (ferromagnetic) iron cylinders (200 μg) suspended in a magnetic field, but found only integral charges (i.e., $\Delta Q < 0.1e$). Their limit on the free-quark concentration in iron is $< 3 \times 10^{-21}$ quarks/nucleon. For a review of magnetic levitation experiments, see Marinelli and Morpurgo (1982).

In view of the above results, no conclusion can be reached about the existence of free quarks in terrestrial conditions, except that, if they do exist at all, it is at the level of $\leq 10^{-20}$ of bound quarks. Quark confinement in hadrons may even be absolute, and any existing particles of fractional charge be "leftovers" from creation in the early universe.

PROBLEMS

5.1. (a) Verify the expressions for the magnetic moments of baryons in Table 5.5. The magnetic moments of the proton and neutron, as well as appropriate combinations of those of the hyperons, depend only on the magnetic moments of the u- and d-quarks. Assuming that the u- and d-quarks each have one-third of the mass of a nucleon, calculate the baryon moments and compare them with the experimental views. (b) The anomalous magnetic moments of the neutron and proton are nearly equal in magnitude but opposite in sign. Show how this result can also be obtained by considering a nucleon to consist part of the time as a Dirac (pointlike) nucleon, and part of the time as a pointlike core with a charged circulating pion in a P-state, contributing to the overall moment as a circulating current.

5.2. Discuss the possible decay modes of the Ω^--hyperon allowed by the conservation laws, and show that weak decay is the only possibility.

5.3. Evaluate $|\psi(0)|^2$ in (5.27) from the typical size of a hadron, and hence estimate the absolute leptonic widths of the vector mesons ρ, ω, ϕ.

5.4. Show that the additive quark model predicts the following cross-section relations:

$$\sigma(\Lambda p) = \sigma(pp) + \sigma(K^- n) - \sigma(\pi^+ p),$$

$$\sigma(\Sigma^- p) = \sigma(pp) + \sigma(K^- p) - \sigma(\pi^- p) + 2[\sigma(K^+ n) - \sigma(K^+ p)],$$

$$\sigma(\Sigma^- n) = \sigma(pp) + \sigma(K^- p) - \sigma(\pi^- p).$$

BIBLIOGRAPHY

Close, F. E., *An Introduction to Quarks and Partons*, Academic Press, 1979.
Berkelman, K., "Upsilon spectroscopy at CESR," *Phys. Rep.* **98**, 145 (1983).
Feldman, G. J., and M. L. Perl, "Recent results in e^+e^- annihilation above 2 GeV", *Phys. Rep.* **33**, 285 (1977).
Franzini, P., and J. Lee-Franzini, "Upsilon resonances", *Ann. Rev. Nucl. Part. Science* **33**, 1 (1983).
Gell-Mann, M., and Y. Ne'eman, *The Eightford Way*, Benjamin, New York, 1964.
Glashow, S. L., "Quarks with color and flavour", *Sci. Am.* **233**, 38 (Oct. 1975).
Greenberg, O. W., "Quarks", *Ann. Rev. Nucl. Science* **28**, 327 (1978).
Jones, L. W., "A review of quark search experiments", *Rev. Mod. Phys.* **49**, 717 (1977).
Kernan, A., and G. Van Dalen, "Charm and beauty production in strong interactions", *Phys. Rep.* **106**, 297 (1984).
Kim, Y. S., "The search for quarks in terrestrial matter", *Contemp. Phys.* **14**, 289 (1973).
Kokkedee, J. J., *The Quark Model*, Benjamin, New York, 1969.
Lederman, L., "The upsilon particle", *Sci. Am.* **239**, 60 (Oct. 1978).
Lipkin, H., *Lie Groups for Pedestrians*, North-Holland, Amsterdam, 1966.
Nambu, Y., "Confinement of quarks", *Sci. Am.* **235**, 48 (Nov. 1976).
Schwitters, R. F. "Fundamental particles with charm", *Sci. Am.* **238**, 56 (Oct. 1977).

CHAPTER 6

Electromagnetic Interactions

6.1. INTRODUCTION

Electromagnetic interactions are familiar as providing the binding forces in atoms and molecules. The energy levels of the hydrogen atom, neglecting spin effects, are given by

$$E = -\frac{\alpha^2 \mu}{2n^2} \left(= -\frac{13.6}{n^2} \text{ eV} \right), \tag{6.1}$$

where $\mu = mM/(m + M) \simeq m$ is the reduced mass and m, M are the masses of electron and proton, and $n = 1, 2, \ldots$ is the principal quantum number. Other elementary systems bound by the Coulomb potential have eigenvalues given by the same formula, such as muonium, $\mu^+ e^-$, and positronium, $e^+ e^-$. (Indeed, we already met the above formula in Section 5.14.) In all these systems, the actual levels are split by spin-orbit (fine structure) and spin-spin (hyperfine structure) interactions, both of magnitude $\sim \alpha^2 E$, and the level energies are affected by higher-order "radiative corrections" (Lamb shift) as predicted by QED.

The point to be emphasized here is that the *bound-state* energy levels—for example, the μ and n-dependence in (6.1)—depend on the form $(1/r)$ of the Coulomb potential and also measure properties of the constituents, such as mass ratio (from μ), magnetic moments, and the coupling, α, to the Coulomb field.

In this chapter, we shall concern ourselves with the continuum system of *unbound states* resulting from the elastic scattering of electrons by protons (or other nuclei). Again, the angular distribution of scattered particles is determined by the form of the potential, the coupling α, and the masses and magnetic moments of the colliding particles (as well, of course, by the incident beam energy). First we discuss the simplest case of spinless scattering and then the modifications when spin effects are included.

6.2. ELASTIC SCATTERING OF SPINLESS ELECTRONS BY NUCLEI

The Rutherford scattering of "spinless" electrons by nuclei can be derived from first-order perturbation theory (and also classically). We employ (4.4) for the transition probability:

$$W = \frac{2\pi}{\hbar} |M_{if}|^2 \rho_f. \tag{6.2}$$

For a perturbing central potential $V(r)$ provided by a stationary nucleus Ze, the matrix element becomes the volume integral

$$M_{if} = \int \psi_f^* V(r) \psi_i \, d\tau, \tag{6.3}$$

where ψ_i, ψ_f are the initial- and final-state wave functions of the scattered electron. The Born approximation assumes the perturbation to be weak, so that only single scattering is considered (this can be shown to be equivalent to the requirement $Z < 137$). We can therefore represent ψ_i and ψ_f as plane waves, before and after scattering. Writing $\mathbf{k}_0$ and $\mathbf{k}$ for the initial and final propagation vectors, we obtain from (6.3)

$$M_{if} = \int e^{i(\mathbf{k}_0 - \mathbf{k}) \cdot \mathbf{r}} V(\mathbf{r}) \, d^3\mathbf{r}. \tag{6.4}$$

As in Eq. (4.6), the differential scattering cross-section is W/v, where v is the velocity of the incident beam relative to the scattering center. Setting the density-of-states factor

$$\rho_f = \frac{p^2 \, d\Omega}{h^3} \frac{dp}{dE_f},$$

where $\mathbf{p} = \hbar \mathbf{k}$ is the momentum of the scattered electron and E_f is the total energy in the final state, gives us

$$\frac{d\sigma}{d\Omega} = \frac{1}{(2\pi)^2 \hbar^4} \frac{p^2}{v} \frac{dp}{dE_f} |M_{if}|^2. \tag{6.5}$$

So far, we have assumed the nucleus to be infinitely massive. In practice we have to consider the nuclear recoil. Let $\mathbf{p}'$, W, and M denote the momentum, total energy, and rest mass of recoiling nucleus, and θ the angular deflection of the electron (see Fig. 6.2). Using units $\hbar = c = 1$, and assuming both incident and scattered electrons are extreme relativistic, we have

$$p_0 = k_0 = E_0, \qquad p = k = E, \qquad v \simeq 1.$$

All quantities refer to the laboratory system. Applying energy-momentum conservation

$$E_i = p_0 + M = E_f = p + W, \qquad \mathbf{p}_0 = \mathbf{p} + \mathbf{p}',$$

one finds

$$E_f = p + \sqrt{p'^2 + M^2} = p + \sqrt{p_0^2 + p^2 - 2pp_0 \cos\theta + M^2},$$

$$\left(\frac{\partial p}{\partial E_f}\right)_\theta = \frac{W}{E_f - p_0 \cos\theta} = \frac{W}{M}\frac{p}{p_0},$$

and

$$\frac{p}{p_0} = \frac{1}{1 + \dfrac{p_0}{M}(1 - \cos\theta)}. \tag{6.6}$$

Equation (6-5) then becomes

$$\frac{d\sigma}{d\Omega} = \frac{1}{4\pi^2} p^2 \frac{W}{M}\frac{p}{p_0} \left| \int e^{i\mathbf{q}\cdot\mathbf{r}} V(\mathbf{r})\, d^3\mathbf{r} \right|^2, \tag{6.7}$$

where the momentum transfer

$$\mathbf{q} = \mathbf{p}_0 - \mathbf{p}.$$

Now let us represent the nucleus by a sphere of charge density $\rho(\mathbf{R})$ as shown in Fig. 6.1 normalized so that

$$\int_0^\infty \rho(\mathbf{R})\, d^3\mathbf{R} = 1.$$

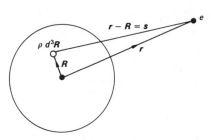

Figure 6.1

Then

$$V(\mathbf{r}) = \frac{Ze^2}{4\pi} \int \frac{\rho(\mathbf{R})\, d^3\mathbf{R}}{|\mathbf{r} - \mathbf{R}|},$$

and

$$
\begin{aligned}
M_{if} &= \frac{Ze^2}{4\pi} \iint \frac{\rho(\mathbf{R})e^{i\mathbf{q}\cdot\mathbf{R}}\, d^3\mathbf{R}\, e^{i\mathbf{q}\cdot(\mathbf{r}-\mathbf{R})}\, d^3\mathbf{r}}{|\mathbf{r} - \mathbf{R}|} \\
&= \frac{Ze^2}{4\pi} \int \rho(\mathbf{R})e^{i\mathbf{q}\cdot\mathbf{R}}\, d^3\mathbf{R} \int \frac{e^{iqs\cos\alpha}2\pi s^2\, ds\, d(\cos\alpha)}{s},
\end{aligned}
\tag{6.8}
$$

where $\mathbf{s} = \mathbf{r} - \mathbf{R}$ and α is the polar angle between $\mathbf{s}$ and $\mathbf{q}$, the momentum-transfer vector. We define the nuclear form factor by

$$F(q^2) = \int \rho(\mathbf{R})e^{i\mathbf{q}\cdot\mathbf{R}}\, d^3\mathbf{R}. \tag{6.9}$$

Then

$$
\begin{aligned}
M_{if} &= \frac{Ze^2}{2} F(q^2) \int s\, ds \int e^{iqs\cos\alpha}\, d(\cos\alpha) \\
&= \frac{Ze^2}{2} F(q^2) \int \frac{s\, ds(e^{iqs} - e^{-iqs})}{iqs}.
\end{aligned}
\tag{6.10}
$$

This integral unfortunately diverges—as indeed it should, since the cross-section for scattering of two particles via an inverse-square-law field is infinite. In fact, the nucleus is of course *screened* at large distances by atomic electrons. The trick is therefore to modify $V(\mathbf{r})$ by a factor $e^{-r/a}$, where a represents a typical atomic radius; afterwards we let $a \to \infty$ if we wish. Since $a \gg R$ (by a factor $\simeq 10^4$), we can set $e^{-r/a} = e^{-s/a}$. Then (6.10) becomes

$$
\begin{aligned}
M_{if} &= \frac{(Ze^2/2)F(q^2)}{iq} \left\{ \int e^{-s(1/a - iq)}\, ds - \int e^{-s(1/a + iq)}\, ds \right\} \\
&= \frac{(Ze^2)F(q^2)}{iq} \left\{ \frac{1}{(1/a) - iq} - \frac{1}{(1/a) + iq} \right\} \\
&= \frac{(Ze^2/2)F(q^2)}{q^2 + (1/a)^2}.
\end{aligned}
\tag{6.11}
$$

Note that if $q < 1/a$ (i.e., very small momentum transfer), $M^2 \propto d\sigma/d\Omega \to$ constant. Now $a \sim 10^{-8}$ cm and thus $1/a \sim 1$ keV only; for the region of interest, $q \gg 1/a$ and we obtain for the differential cross-section

$$\frac{d\sigma}{d\Omega} = \frac{4Z^2(e^2/4\pi)^2}{q^4} p^2 \frac{W}{M} \frac{p}{p_0} [F(q^2)]^2. \tag{6.12}$$

For reasons that will appear later, $(W/M) - 1 = q^2/2M^2 \ll 1$, so we can set $W/M = 1$ in all practical cases—the recoiling nucleus is nonrelativistic. Under the further assumption that the nuclear recoil momentum, $p' = q \ll p_0$, we can set $p = p_0$ and

$$q^2 = 2p_0^2 - 2p_0^2 \cos \theta = 4p_0^2 \sin^2 \tfrac{1}{2}\theta, \tag{6.13}$$

so that

$$\frac{d\sigma}{d\Omega} = \frac{Z^2(e^2/4\pi)^2[F(q^2)]^2}{4p_0^2 \sin^4 \tfrac{1}{2}\theta}. \tag{6.14}$$

Since

$$\alpha = \frac{e^2}{4\pi}$$

and

$$d\Omega = 2\pi \, d(\cos \theta) = \frac{2\pi \, dq^2}{2p_0^2}, \tag{6.15}$$

we can also express this in the form

$$\frac{d\sigma}{dq^2} = \frac{4\pi\alpha^2 Z^2[F(q^2)]^2}{q^4}. \tag{6.16}$$

Expressions (6.14) and (6.16), when applied to an effectively pointlike nucleus—that is, low values of q^2 so that $F(q^2) \simeq 1$—are the famous *Rutherford scattering formula*.

Note that the q^2-dependence (6.16) is another way of expressing the form ($1/r$ in coordinate space) of the Coulomb potential, as was the $1/n^2$ dependence of bound-state energy in (6.1).

6.3. FOUR-MOMENTUM TRANSFER

Equation (6.16) is expressed in terms of the square of the 3-momentum transfer, $q^2 = |\mathbf{q}|^2$ between incident and target particle in the lab system. It is desirable to express it in covariant form, that is, in a way which is independent of the reference frame (lab system, center-of-momentum system), and we do this by redefining q as the *4-momentum transfer*. In Fig. 6.2, P_0, P, and Q represent the four-momenta of the particles involved. Each 4-momentum has three space and one time component.

The square of the 4-momentum vector (of a real particle) is given by

$$P^2 = \mathbf{p}^2 + (iE)^2 = p^2 - E^2 = -m^2$$

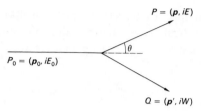

Figure 6.2

and is thus invariant (see Appendix A). Similarly, the 4-momentum transfer squared, between the incident and emergent electron, is invariant, with a value

$$q^2 = (\mathbf{p}_0 - \mathbf{p})^2 - (E_0 - E)^2 = -2m^2 - 2pp_0 \cos \theta + 2EE_0$$
$$= 2pp_0(1 - \cos \theta) = 4pp_0 \sin^2 \tfrac{1}{2}\theta, \qquad (6.17a)$$

if we neglect the electron mass ($m^2 \ll q^2$). Since the scattering angle is a real quantity ($-1 < \cos \theta < 1$), q^2 is positive. For exchange of a real particle, q^2 would be negative, i.e., with the same sign as the energy or time component $(iE)^2$. A value $q^2 > 0$ is sometimes referred to as a *spacelike* momentum transfer, to distinguish it from $q^2 < 0$, which is called *timelike*. All scattering processes necessarily refer to the spacelike region of q^2.

An alternative expression for q^2 is obtained by considering the transfer to the nucleus:

$$q^2 = (-\mathbf{p}')^2 - (M - W)^2 = -2M^2 + 2MW = 2MT, \qquad (6.17b)$$

where the kinetic energy acquired by the nucleus is $T = W - M$. Thus,

$$\frac{W}{M} = 1 + \frac{q^2}{2M^2}. \qquad (6.18)$$

If the nucleus is to recoil coherently, it turns out that $q^2 \ll 2M^2$, so that $W/M \simeq 1$ in a practical case, as assumed in (6.14).

Although the values of q^2 in (6.17) refer to quantities measured in the laboratory frame, its numerical value is the same in all inertial frames. The result of evaluating $|M_{if}|^2$ is as before, except that now, q being a 4-momentum transfer, (6.9) is strictly an integral over space-time, with qR as a scalar product of 4-vectors. However, if $W/M \simeq 1$, the energy transfer (time component) is small, and we can still interpret (6.9) as the integral over a spatial charge distribution.

The expression (6.9) for the nuclear form factor is seen to be simply the Fourier transform of the charge-density distribution $\rho(R)$ in the nucleus. Integrating over angles it is easy to show that

$$F(q^2) = \int \rho(R) \frac{\sin qR}{qR} 4\pi R^2 \, dR. \qquad (6.19)$$

The reader is also referred to the problems (6.1, 6.2) at the end of the chapter. Typically, nuclei have radii of a few fermi. If $R = 4$ fm, for example, $qR = 1$ implies $qc = \hbar c/R = 197$ MeV fm/4 fm $\simeq 50$ MeV. Thus, if $q \ll 50$ MeV/c, the scattering is pointlike, whereas for large values the form factor reduces the cross-section.

Finally, let us note that for $Z = 1$ and $F(q^2) = 1$, that is, for the scattering of a spinless electron by a spinless pointlike proton, the differential cross-section (6.16) assumes the form

$$\frac{d\sigma}{dq^2} = \frac{4\pi\alpha^2}{q^4}.$$ (6.20)

This has a simple interpretation. The matrix element is written as the product of two vertex factors and a photon propagator term (see Section 1.6)

$$M = \sqrt{\alpha} \cdot \sqrt{\alpha} \cdot \frac{1}{q^2},$$

whereas, from (6.6) and (6.15), the element of phase-space density is, for $p \simeq p_0$,

$$p^2 \frac{dp}{dE_f} d\Omega = \pi \, dq^2.$$

Hence,

$$\frac{d\sigma}{dq^2} \simeq |M|^2 \simeq \frac{\alpha^2}{q^4}.$$

6.4. SCATTERING OF ELECTRONS WITH SPIN BY SPINLESS NUCLEI

The relativistic wave equation of Dirac describes pointlike electrons with spin $\frac{1}{2}\hbar$. It is discussed in detail in Appendix D, where it is shown that in a reference frame in which the electron is relativistic, the spin vector σ (where $\sigma^2 = 1$) is *aligned* with the momentum vector $\mathbf{p}$. If $\mathbf{p}$ defines the z-direction, the expectation value $\langle\sigma_z\rangle = \pm 1$, while $\langle\sigma_x\rangle = \langle\sigma_y\rangle = 0$. The degree of longitudinal polarization is termed the *helicity*, H, so that

$$H = \frac{\sigma \cdot \mathbf{p}}{|\mathbf{p}|} = \pm 1.$$ (6.21)

Particles of $H = +1$ are called right-handed, those of $H = -1$ left-handed. These states are denoted by ψ_R and ψ_L, respectively.

In an electromagnetic interaction (or indeed in the coupling of any spin $\frac{1}{2}$ fermion with any vector or axial vector field), it is found (Appendix D)

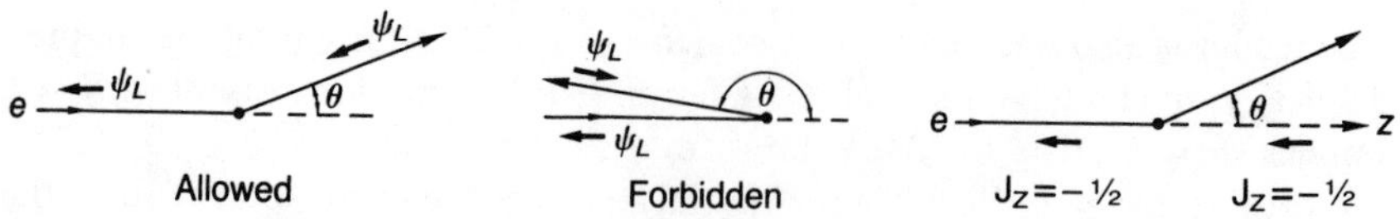

Figure 6.3

that the *helicity is conserved*. The state ψ_L is preserved as ψ_L after the scattering, and similarly with ψ_R. Thus, for the electric or non-spin-flip part of the scattering of the electron by another charged particle, 180° scattering is forbidden by angular-momentum conservation. In other words, the photon exchanged is transverse, with no net angular momentum component along the z-axis (see Fig. 6.3).

An electron, say with *LH* helicity, scattered at arbitrary angle θ will be described by $\psi_L(\theta)$, which is a superposition of $J_z = \frac{1}{2}$ and $J_z = -\frac{1}{2}$ amplitudes. The fraction of the amplitude which describes an electron, originally prepared in a state $J_z = -\frac{1}{2}$, emerging at angle θ also with $J_z = -\frac{1}{2}$ is given by the d-functions (rotation matrices, see Appendix C)

$$d^j_{m,\,m'}(\theta) = d^{1/2}_{-1/2,\,-1/2} = d^{1/2}_{1/2,\,1/2} = \cos\left(\frac{\theta}{2}\right), \tag{6.22}$$

whereas

$$d^{1/2}_{1/2,\,-1/2}(\theta) = \sin\left(\frac{\theta}{2}\right).$$

Hence, the effect of the electron spin on the scattering by a spinless nucleus is to introduce a factor $\cos^2(\theta/2)$ in the cross-section. In a relativistic calculation, it also turns out that the term W/M drops out (see Appendix A on relativistic normalization of wavefunctions), so that writing q^2 as the square of the 4-momentum transfer as in (6.17), (6.16) becomes

$$\left(\frac{d\sigma}{d\Omega}\right)_{\text{Mott}} = \frac{Z^2\alpha^2 \cos^2 \frac{1}{2}\theta}{4p_0^2 \sin^4 \frac{1}{2}\theta[1 + (2p_0/M)\sin^2 \frac{1}{2}\theta]}, \tag{6.23}$$

called the *Mott formula*. The final term in the denominator allows for recoil of the nucleus. Obviously, the square of the elastic form factor $|F(q^2)|^2$ should multiply this expression at large q^2 when the nucleus is not pointlike.

6.5. ELECTRON SCATTERING BY NUCLEONS

The final step is to include in our formula the effect of the spin of the target as well as the projectile (electron). Let us consider scattering of electrons by (hypothetical) pointlike protons. This gives rise to a magnetic interaction (which flips over the spins of the particles) as well as the electrical (Coulomb) interaction. Since the magnetic field due to the proton moment varies as r^{-3}, compared with r^{-2} for the electric field, close collisions and high q^2 will be more important for the magnetic interaction. The actual form of the *Dirac cross-section* is (see Appendix G):

$$\left(\frac{d\sigma}{d\Omega}\right)_{\text{Dirac}} = \left(\frac{d\sigma}{d\Omega}\right)_{\text{Rutherford}} \left(\cos^2 \tfrac{1}{2}\theta + \frac{q^2}{2M^2} \sin^2 \tfrac{1}{2}\theta\right), \qquad (6.24)$$

where

$$\left(\frac{d\sigma}{d\Omega}\right)_{\text{Rutherford}} = \frac{\alpha^2}{(4p_0^2 \sin^4 \tfrac{1}{2}\theta)[1 + (2p_0/M) \sin^2 \tfrac{1}{2}\theta]},$$

and where the first term in the right-hand bracket is for the electric (Mott) scattering, while the magnetic, spin-flip scattering in the second term contains a factor $\sin^2 \tfrac{1}{2}\theta$ from (6.22). The cross-section (6.24) would apply if both electron and proton were pointlike, Dirac particles with magnetic moments $\mu = e\hbar/2mc$, where m is the particle mass.

Protons and neutrons are not pointlike, and their magnetic moments are anomalous, in the sense that they differ from the predictions for pointlike Dirac particles:

	μ (Dirac)	μ (observed)
Proton	$e\hbar/2Mc = 1$ n.m.	$+2.79$ n.m.
Neutron	0	-1.91 n.m.

The structure of protons and neutrons, like that of nuclei, is described empirically by suitable form factors: two form factors (one electric, one magnetic) are needed for each. In the way in which they are defined, the cross-section assumes the form

$$\frac{d\sigma}{d\Omega} = \left(\frac{d\sigma}{d\Omega}\right)_{\text{Mott}} \left\{\left(\frac{G_E^2 + (q^2/4M^2)G_M^2}{1 + (q^2/4M^2)}\right) + \frac{q^2}{4M^2} \cdot 2G_M^2 \tan^2 \frac{\theta}{2}\right\}, \qquad (6.25)$$

where $G_E = G_E(q^2)$, $G_M = G_M(q^2)$ with the normalization $G_E^P(0) = 1$, $G_E^N(0) = 0$, $G_M^P(0) = 2.79$, $G_M^N(0) = -1.91$. This is called the *Rosenbluth formula*. The essential feature is that

$$\left(\frac{d\sigma}{d\Omega}\right)\Big/\left(\frac{d\sigma}{d\Omega}\right)_{\text{Mott}} = A(q^2) + B(q^2) \tan^2 \frac{\theta}{2}, \qquad (6.26)$$

so if one plots the cross-section for different incident momenta and different scattering angles, such that q^2 remains fixed, a linear dependence on $\tan^2(\theta/2)$ should be obtained. The basic assumption on which (6.25) is derived is that of the Born approximation—that only a single collision, mediated by *single virtual photon exchange*, is involved. In principle, double photon exchange may also occur. The equality of the cross-sections for e^+p and e^-p scattering

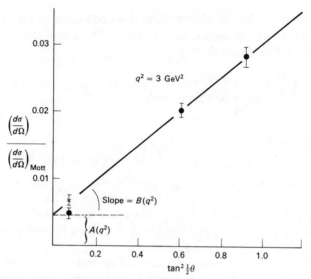

Figure 6.4 The electron-proton scattering cross-section plotted for fixed q^2 and different scattering angles θ (Rosenbluth plot). (After Weber (1967).)

(for which the two-photon exchange terms should have opposite signs) indicates that two-photon exchange is unimportant. This is reinforced by the observed linearity of the Rosenbluth plot (Fig. 6.4).

The experimental determination of the proton form factor has been carried out by directing high-energy (400-MeV to 16-GeV) electron beams at a hydrogen target, and making precision measurements of the momentum and angle of scattered electrons by means of magnetic spectrometers. Elastic events can be selected using the kinematic relation (6.6) between the energy and angle of the scattered electron (with M as the proton mass). For the neutron, scattering is observed with deuterium targets, and a subtraction procedure employed:

$$\frac{d\sigma}{d\Omega}\,(en) = \frac{d\sigma}{d\Omega}\,(ed) - \frac{d\sigma}{d\Omega}\,(ep) + \text{correction factor},$$

where the correction factor involves the nuclear physics of the deuteron. Because of this, the neutron data are less precise. The first experiments, demonstrating the deviation of the scattering from that expected for a point particle—and thus measuring the form factors—were carried out by Hofstadter and his collaborators in 1961, at Stanford, and have since been extended in numerous laboratories. The end result of the present experiments is that the form factors obey the simple *scaling law*

$$G_E^p(q^2) = \frac{G_M^p(q^2)}{|\mu_p|} = \frac{G_M^n(q^2)}{|\mu_n|} = G(q^2), \qquad G_E^n(q^2) = 0, \qquad (6.27)$$

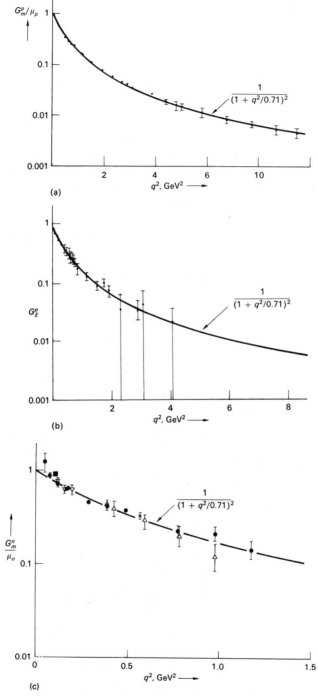

Figure 6.5 Comparison of the magnetic and electric form factors of neutron and proton. They are consistent with the scaling law (6.27). (a) Proton magnetic form factor; (b) proton electric form factor; (c) neutron magnetic form factor. (After Weber (1967).)

and the empirical *dipole formula*

$$G(q^2) = \left(1 + \frac{q^2}{M_V^2}\right)^{-2}, \quad \text{with} \quad M_V^2 = (0.84 \text{ GeV})^2. \quad (6.28)$$

The relations (6.27) hold over the experimental range so far investigated, $q^2 = 0$–2 GeV^2 (see Fig. 6.5). At higher q^2, only electron-proton scattering data exist, and, as (6.25) indicates, these refer essentially to the magnetic form factor $G_M^p(q^2)$, the electric contribution being small and unmeasurable. For this scattering, the dipole formula (6.28) fits to within 10% accuracy up to $q^2 = 25$ GeV^2, as shown in Fig. 6.6. Over this range, the form factor squared falls by a factor of 10^6. Using the inverse of (6.19), one finds that (6.28) can be interpreted in terms of an exponential charge–magnetic-moment distribution of the proton of density

$$\rho(R) = \rho_0 \exp(-M_V R), \quad (6.29)$$

with a root-mean-square radius

$$R_{rms} = \frac{\sqrt{12}}{M_V} = 0.80 \text{ fm}. \quad (6.30)$$

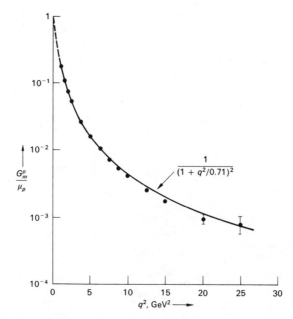

Figure 6.6 The electromagnetic form factor of the proton, in the region of high q^2. The observations are fairly well fitted by the dipole formula. The data are principally from the SLAC laboratory. (After Panofsky (1968).)

The elastic form factors reflect the spatial distribution of charge and magnetic moment in the nucleon. If we had a complete theory of hadron structure, these form factors could be calculated. At the present time no such theory exists and the subject must be treated on a largely empirical basis.

6.6. THE PROCESS $e^+e^- \rightarrow \mu^+\mu^-$

The cross-section for this electromagnetic process has a magnitude

$$\sigma(e^+e^- \rightarrow \mu^+\mu^-) = \frac{4\pi\alpha^2}{3s}, \qquad (6.31)$$

with an angular distribution of the simple form

$$\frac{d\sigma}{d\Omega} \propto (1 + \cos^2\theta), \qquad (6.32)$$

where θ is the angle of emission of the muons with respect to the incident beam direction in the CMS, s is the square of the CMS energy ($s = 4E_1E_2$, where E_1 and E_2 are the energies of the electron and positron colliding head-on), and all lepton masses have been neglected in comparison with the CMS energy.

The exact form of the above equations are obtained using the Dirac equation and making a proper trace calculation (see Appendix G). However, their general form in the relativistic limit follows from simple arguments. First, the cross-section must be proportional to α^2, from the coupling of the exchanged photon at each vertex (Fig. 6.7(b)). Second, σ has dimensions of (length)^2 or (energy)$^{-2}$ and therefore can only vary as $1/s$, since this is the only energy scale in the problem, if $s \gg m_e^2, m_\mu^2$. (The numerical factor $4\pi/3$ comes from integration over solid angle and averaging over spins).

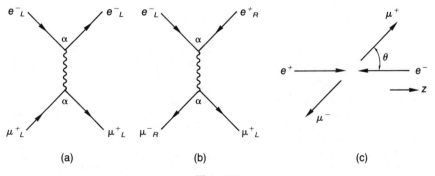

(a)   (b)   (c)

Figure 6.7

Finally, the angular distribution follows directly from helicity conservation for relativistic fermions in a vector interaction. Consider first the scattering process $e^-\mu^+ \rightarrow e^-\mu^+$ (see Fig. 6.7(a)). Since helicity is conserved, $e_L^- \rightarrow e_L^-$ and $e_R^- \rightarrow e_R^-$ (and similarly for the muon). This diagram is related to that for $e^+e^- \rightarrow \mu^+\mu^-$ (Fig. 6.7(b)) by replacing an incoming (outgoing) particle by an outgoing (incoming) antiparticle, which will therefore have the opposite helicity. It is called the *crossed diagram*. It follows that e_L^- and e_R^+ (or e_R^- and e_L^+) are coupled, but not e_L^+ and e_L^- (or e_R^+ and e_R^-). Hence, the intermediate photon must have $J = 1$, $J_z = \pm 1$ only, not $J_z = 0$ (as can be seen by viewing in the CMS frame, Fig. 6.7(c)). Because electromagnetic interactions conserve parity, both $J_z = +1$ and -1 states occur with equal probability. If the initial state has $J_z = +1$, say, so must the final state, and the amplitude for emitting the μ^+ at angle θ to the e^+ in the frame of Fig. 6.7(c) is given by the d-function (see Appendix C)

$$d_{m,m'}^j = d_{1,1}^1 = \tfrac{1}{2}(1 + \cos\theta). \tag{6.33}$$

If the initial state has $J_z = -1$, we have to replace θ (defined relative to the positive z-direction) by $\pi - \theta$, that is, the amplitude becomes $\tfrac{1}{2}(1 - \cos\theta)$. Squaring and adding the amplitudes of these two orthogonal states, we obtain (6.32)

$$\frac{d\sigma}{d\Omega} \propto (1 + \cos^2\theta).$$

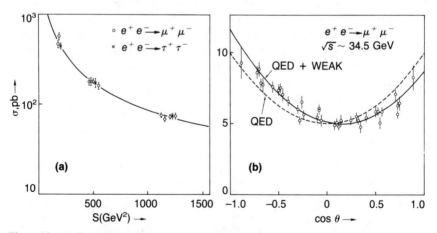

Figure 6.8 (a) Compilation of results (Wu, (1984)) on total cross-sections for $e^+e^- \rightarrow \mu^+\mu^-$ and $e^+e^- \rightarrow \tau^+\tau^-$ from the e^+e^- collider PETRA at DESY. The QED prediction (6.31) is shown by the curve. For the total cross-section, neutral-current effects (Z^0-exchange) are small and unmeasurable. (b) The CMS angular distribution for the process $e^+e^- \rightarrow \mu^+\mu^-$ as measured at PETRA. The pure QED prediction (6.32) is shown by the dashed line. The full-line curve shows the small forward-backward asymmetry expected from combination of both γ- and Z^0-exchange. For more details, see Section 9.7.2.

If the muon mass cannot be neglected, that is, the muons are not pure helicity ± 1 states, one should modify this expression to $(1 + \cos^2 \theta + 2(m_\mu^2/s)\sin^2 \theta)$. The actual derivation of these results is given in Appendix G.

The experimental data on the total and differential cross-sections at high energy for the process $e^+e^- \to \mu^+\mu^-$ and $e^+e^- \to \tau^+\tau^-$ are shown in Fig. 6.8. The total cross-section is in good accord with the prediction. The angular distribution has the general $(1 + \cos^2 \theta)$ form, but with a marked backward-forward asymmetry. This arises because, in addition to the pure photon-exchange diagram (Fig. 6.7(b)), there is a parity-violating (backward-forward asymmetric) weak amplitude arising from Z^0-exchange. Since the weak amplitude is of order G, compared with $4\pi\alpha/s$ for the electromagnetic, the γ-Z^0 interference term is expected to make a relative contribution of magnitude $f = a_{\text{weak}} \cdot a_{em}/a_{em}^2$, that is, $f \sim Gs/4\pi\alpha \sim 10^{-4}s$ where s is in GeV^2. Thus, the expected asymmetry $(B - F)/(B + F) = f \simeq 10\%$ at $s = 1000$ GeV^2, as observed. Such asymmetries are important tests of the electroweak theory, and are discussed quantitatively in Chapter 9.

6.7. BHABHA SCATTERING $e^+e^- \to e^+e^-$

The dimensional arguments used in the previous section apply equally to Bhabha scattering, to predict a $1/s$ dependence of the total cross-section. However, the angular distribution is more complex, because (see Fig. 6.9) two diagrams contribute. Figure 6.9(a) is analogous to ep scattering and Fig. 6.9(b) to $e^+e^- \to \mu^+\mu^-$. The first will dominate at small angles. We do not

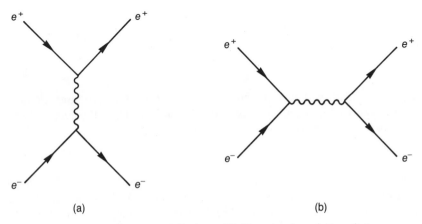

(a)   (b)

Figure 6.9   Diagrams contributing to Bhabha scattering, $e^+e^- \to e^+e^-$.

quote the formula, but remark that at small angles the cross-section is large and is used to monitor the intensity in e^+e^- colliders. The angular distribution, overall, is in excellent agreement with QED.

6.8. QUANTUM ELECTRODYNAMICS AND HIGHER-ORDER PROCESSES

The previous discussion of ep scattering and e^+e^- annihilation was made on the assumption of single photon exchange. Naturally, exchange of 2, 3,... virtual photons is also possible, but they are suppressed relative to single photon exchange by the coupling factors α^2, α^4, ..., etc. With the attainable precision of scattering experiments, it is hardly possible to measure their effects.

These so-called higher-order processes are all calculable in quantum electrodynamics. Very precise experiments are needed to check the predictions, and necessarily involve stationary states, which can be measured with great accuracy. The main classes of measurement are those on lepton magnetic moments, and those on the bound-state energy levels of the hydrogen and other atoms, of muonium μ^+e^-, and positronium e^+e^-. There is not space here to describe the experiments in the detail required, but we can outline the motivations for them and their results.

6.8.1. Lepton Magnetic Moments

According to the Dirac theory, an electron or muon is pointlike and possesses a magnetic moment equal to the Bohr magneton

$$\mu_B = \frac{eh}{2mc}, \tag{6.34}$$

where m is the lepton mass. Generally, the magnetic moment $\boldsymbol{\mu}$ is related to the spin vector $\mathbf{s}$ by

$$\boldsymbol{\mu} = g\mu_B\mathbf{s}, \tag{6.35}$$

where g is called the Landé g-factor, and $g\mu_B = \mu/s$ is the gyromagnetic ratio, i.e., the ratio of magnetic to mechanical moment. Thus the Dirac theory predicts for particles of spin-$\frac{1}{2}$

$$g = 2. \tag{6.36}$$

The actual g-values of the electron and muon have been determined with great precision and differ by a small amount (0.2%) from the value 2. Thus the Dirac picture of a structureless, point particle is not exact. Similar departures are encountered in atomic levels: for example, the famous Lamb shift is a 1 % correction to the $2P_{1/2}$-$2S_{1/2}$ level separation in the fine structure of hydrogen.

The magnetic moment of a charged particle depends on the ratio e/m and thus, classically, for a rotating structure, on the spatial distributions of charge and mass. (If the two distributions are the same, a value $g = 1$ is obtained on classical arguments.) Deviation from the g-value of 2 for a spin-$\frac{1}{2}$ particle therefore argues that processes are taking place which distort the relative charge and mass distributions. For example, the proton has a g-value of 5.59, which must arise from its composite structure and different values of e/m for the components (quarks).

At any instant of time, an electron has a certain probability of consisting of a "bare" or pointlike object, surrounded by a cloud of one or more virtual photons, which are continually being emitted and reabsorbed. Qualitatively, we can see that although the charge e resides on the electron, part of the mass energy is carried by the photon cloud, and hence the value of e/m for the electron itself will be slightly increased. This increased value would obtain if one measures the magnetic moment of the electron with an external magnetic field B. The amount of the correction will be proportional to the photon emission probability, determined by the coupling constant $\alpha = e^2/4\pi\hbar c$. It is obvious that correction terms of order $\alpha, \alpha^2, \ldots$ correspond to emission of 1, 2, $\ldots$ virtual photons at the moment the field B is applied.

Figure 6.10 gives examples of diagrams for the various virtual processes to be considered. Fig. 6.10(a) corresponds to the Dirac-type interaction with a "bare" electron, Fig. 6.10(b) to emission of one virtual photon and therefore of order α, whereas Fig. 6.10(c) shows α^2 processes, one of two-photon emission and one of virtual pair formation. Since the pair will be polarized by, and tend to shield, the electron charge from external fields, this last effect is called "vacuum polarization".

The spectrum of virtual photons of 4-momentum k in such processes is found to be of the form $f(k)\ dk \simeq dk/k$. Since there is no limit on the momenta in virtual states, the total correction, that is the integral over the spectrum, will be logarithmically divergent. For example the mass or "self-energy" of an electron will apparently become infinite. However, it is found

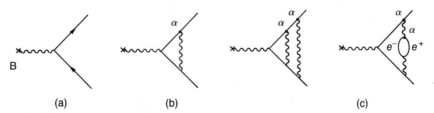

Figure 6.10 Graphs indicating some of the QED radiative corrections to the magnetic moments of leptons.

that the theory has the remarkable property of *renormalizability*; the divergences can always be absorbed into a redefinition of the "bare" lepton charges and masses, which are in any case arbitrary, as being equal to the physically measured values. All other predicted physical quantities, such as level shifts in atoms, magnetic moments of leptons, or collision cross-sections, are then finite.

The predictions for the lepton magnetic moments appear as power series in α, with coefficients which have been calculated by summing all the appropriate Feynman graphs. For the electron the predicted anomaly is

$$\left(\frac{g-2}{2}\right)_e^{\text{QED}} = 0.5\frac{\alpha}{\pi} - 0.32848\left(\frac{\alpha}{\pi}\right)^2 + 1.19\left(\frac{\alpha}{\pi}\right)^3 + \cdots$$

$$= (1\,159\,652.4 \pm 0.4) \times 10^{-9}, \tag{6.37}$$

whereas that for the muon is

$$\left(\frac{g-2}{2}\right)_\mu^{\text{QED}} = 0.5\frac{\alpha}{\pi} + 0.76578\left(\frac{\alpha}{\pi}\right)^2 + 24.45\left(\frac{\alpha}{\pi}\right)^3 + \cdots$$

$$= (1\,165\,851.7 \pm 2.3) \times 10^{-9}. \tag{6.38}$$

Note that the leading correction $0.5(\alpha/\pi)$ is obviously the same for electron and muon, while the α^2 term differs in both magnitude and sign in the two cases. This arises because the momenta of the particles in the intermediate virtual states scale in proportion to the parent particle mass. The vacuum-polarization term for the muon is much larger.

The QED result (6.38) differs significantly from the observed value by some 9 standard deviations (Table 6.1). There are however additional effects which arise because a virtual photon can transform not only into a lepton pair but also into hadrons. This contribution can be computed directly from the results of electron-positron colliding-beam experiments, $e^+e^- \to \gamma \to$ hadrons. It is negligible for the $g-2$ of the electron, but for the muon the correction is

$$\Delta a_{\text{S.I.}} = \frac{m_\mu^2}{12\pi^3} \int \frac{\sigma(e^+e^- \to \text{hadrons})\, ds}{s} \simeq 7 \times 10^{-8}, \tag{6.39}$$

TABLE 6.1 $(g-2)/2$ anomaly for leptons

	$10^9(g-2)/2$	
	Electron	Muon
QED prediction	1 159 652.4 $\pm$ 0.4	1 165 851.7 $\pm$ 2.3
With strong-interaction correction	—	1 165 918 $\pm$ 10
Observed value	1 159 652.4 $\pm$ 0.2	1 165 924 $\pm$ 9

revising (6.38) to

$$\left(\frac{g-2}{2}\right)_{\mu}^{\text{theory}} = (1\,165\,918 \pm 10) \times 10^{-9}, \qquad (6.40)$$

in perfect agreement with experiment. Contributions from other effects (e.g., weak interactions, QED production of pairs of heavy $\tau^{\pm}$ leptons) can be neglected at the present level of accuracy.

Experimental determination of the g-factors of electrons and muons are some of the most precise and beautiful experiments in physics, and are being continually improved. The reader is referred to the bibliography for details. Table 6.1 shows the comparison of experiment with theory.

6.8.2. Hyperfine Structure

As an example of the testing of QED predictions for bound states, we quote comparison with experiment for the ground state $1^3S_1 \to 1^1S_0$ spin-flip transitions in muonium (μ^+e^-) and positronium (e^+e^-), see Table 6.2. Such discrepancies as occur between experiment and theory are thought to be due to neglect of higher-order corrections. The frequencies above are related, to leading order, by simple numerical factors (see Problem 6.6).

On a historical note, we recall that in the Dirac theory, the hydrogen-atom levels are determined by n (the principal quantum number) and j, the total angular momentum of the electron. In the more refined theory (QED), the energy levels depend also on the orbital angular momentum l, so that the $2p_{1/2}$ and $2s_{1/2}$ electronic levels are not degenerate. The difference is the famous Lamb shift (1057-MHz transition frequency). The correct prediction of this shift was one of the first successes of QED.

6.8.3. High-Energy Tests of QED

Finally, we may ask what the possible limits to the validity of QED at very high energies are by supposing that the theory could fail at very high momentum transfers. These are discussed in terms of a "cutoff" parameter, Λ, introduced via an arbitrary structure factor of the form $F = 1 + s/\Lambda^2$, where s

TABLE 6.2 Hyperfine structure interval ($^3S_1 - {}^1S_0$) in muonium and positronium (after Bodwin and Yennie (1978) and Berko and Pendleton (1980)).

	μ^+e^-	e^+e^-	Hydrogen (e^-p)
Theory	4463.304 (± 6) MHz	203400 (± 10) MHz	
Experiment	4463.302 (± 5) MHz	203387 (± 2) MHz	1420.4057 MHz ($\lambda = 21$ cm)

is the square of the CMS energy, or $F = 1 + q^2/\Lambda^2$. If one introduces such a factor into the QED cross-section for $e^+e^- \to \mu^+\mu^-$, then the experimental data provide the limit $\Lambda > 100$ GeV. Other elementary processes provide similar limits.

PROBLEMS

6.1. Show that an exponential charge-density distribution for the proton of the form (6.29) leads to a form factor corresponding to the dipole formula (6.28), and verify that for $M_V = 0.84$ GeV the rms radius is 0.8 fm.

6.2. The process $e^+e^- \to \pi^+\pi^-$ effectively measures the electromagnetic form factor of the pion in the timelike region of q^2. If this process is dominated by the ρ intermediate state, show that the rms radius of the pion is 0.64 fm (for $M_\rho = 765$ MeV).

6.3. In an e^+e^- colliding-beam experiment, the ring radius is 10 m and each beam is of 10 mA, with a cross-sectional area of 0.1 cm². Assuming the electrons and positrons are bunched and the two bunches meet head-on twice per revolution, calculate the luminosity in $cm^{-2} s^{-1}$ (a luminosity L provides a reaction rate of σL per second for a process of cross-section σ). From the Breit-Wigner formula (4.55) calculate the cross-section for the reaction $e^+e^- \to \pi^+\pi^-\pi^0$ at the peak of the ω-resonance, assuming the branching ratio for $\omega \to e^+e^-$ is simply α^2. Hence, deduce the event rate per hour for this process with the above luminosity.

6.4. A 10-GeV electron collides with a proton and emerges from the collision with a 10° deflection and an energy of 7 GeV. Calculate the rest mass W of the recoiling hadronic state.

6.5. A pencil electron beam of energy 15 GeV and intensity 10^{14} particles s^{-1} impinges on a liquid-hydrogen target of length 1 m parallel to the beam and of cross-section sufficient to cover the beam. Estimate the number of electrons per second scattered elastically through 0.1 rad and into a solid angle of 10^{-4} sr for (a) pointlike spinless protons, (b) protons with the form factors of Eq. (6.25). (Hydrogen density = 0.06.)

6.6. The $1^3S_1 - 1^1S_0$ level separation in the ground state of the hydrogen atom, positronium, and muonium is proportional to the product of the magnetic moments of the fermions involved (see (5.30)). From the transition frequency (1420 MHz) in hydrogen, calculate that in muonium and compare with the experimentally observed value in Table 6.2. For positronium, an extra factor of $\frac{7}{16}$ is involved because of the contribution from an annihilation diagram not present in the other systems. Remembering to allow for reduced-mass effects (see (6.1)) calculate the value of the splitting in positronium and compare with the results in the table. (You should obtain agreement with experimental values within $\simeq \frac{1}{2}\%$; what is the reason for the discrepancy?).

BIBLIOGRAPHY

Combley, F., F. J. M. Farley, and E. Picasso, "The CERN muon $(g - 2)$ experiment", *Phys. Rep.* **68**, 93 (1981).

Combley, F., "$(g - 2)$ factors for the muon and electron and the consequences for QED", *Rep. Prog. Phys.* **42**, 1889 (1979).

Farley, F. J. M., and E. Picasso, "The muon $(g - 2)$ experiment", *Ann. Rev. Nucl. Science* **29**, 243 (1979).

Perez-y-Yorba, Y., and F. Renard, "The physics of e^+e^- colliding beams", *Phys. Rep.* **31**, 1 (1977).

Wu, S. L., "e^+e^- physics at PETRA—the first five years", *Phys. Rep.* **107**, 59 (1984).

CHAPTER 7

Weak Interactions

7.1 CLASSIFICATION OF WEAK INTERACTIONS

Weak interactions were first observed in the slow process of nuclear β-decay. This takes place, and can be observed, in circumstances where the much faster strong or electromagnetic decays are forbidden by the conservation laws. Otherwise, although all hadrons and leptons take part in weak interactions, the effects are usually swamped by the stronger electromagnetic or strong couplings.

Weak interactions frequently involve leptons among the interaction products. In particular, the neutral leptons—neutrinos—are unique in that they can only take part in the weak interactions. Experiments with beams of neutrinos produced by the decay in flight of pions and kaons, carried out 20 years ago, together with recent $e^+ e^-$ experiments, showed that charged and neutral leptons (see Table 1.2) appear in doublets,

$$
\begin{array}{c c c c}
Q & L_e = 1 & L_\mu = 1 & L_\tau = 1 \\
\hline
0 & \begin{pmatrix} \nu_e \\ e^- \end{pmatrix} & \begin{pmatrix} \nu_\mu \\ \mu^- \end{pmatrix} & \begin{pmatrix} \nu_\tau \\ \tau^- \end{pmatrix}
\end{array}
\tag{7.1}
$$

to each of which one has to assign a conserved lepton number (L_e, L_μ, L_τ). Antileptons have opposite charge and lepton number to those of leptons.

Thus, muon neutrinos v_μ, produced by the decay $\pi^+ \to \mu^+ v_\mu$ in flight, were shown to result in muons in their subsequent interactions (e.g., $v_\mu n \to p\mu^-$) and never electrons. By the same token, the decay $\mu^+ \to e^+\gamma$ is forbidden.

Apart from conservation of lepton number (if leptons are involved), weak reactions may or may not conserve isospin I and strangeness S. Some typical weak processes—with the reasons why electromagnetic or strong transitions are forbidden—are as follows:

Process	Mean lifetime, s		
Neutron decay:	10^3	E.m. decay forbidden	
$n \to pe^-\bar{v}_e$		by charge conservation	(7.2)
Inverse neutron decay:		Neutrinos have only	
$\bar{v}_e p \to ne^+$		weak interactions	(7.3)
Lambda decay:	10^{-10}	$\Delta S = 1$: strong/e.m.	
$\Lambda \to p\pi^-$		decay forbidden	(7.4)
Pion decay:	10^{-8}	Leptons are the only	
$\pi^+ \to \mu^+ v_\mu$		lighter particles	(7.5)

The lifetimes for weak decays depend on phase-space factors as well as the weak coupling constant G but are long compared with typical lifetimes for electromagnetic decays ($\simeq 10^{-19}$ s) or strong decays ($\simeq 10^{-23}$ s). Weak cross-sections are correspondingly small; for example, the cross-section for $v_\mu + N \to N + \pi + \mu^-$ at 1 GeV is 10^{-38} cm^2, a factor 10^{12} times smaller than for $\pi + N \to \pi + N$ at the same energy.

The weak interactions are sometimes classified as to whether they involve leptons only, leptons and hadrons, or hadrons only, as in Table 7.1.

On a cosmic scale, weak interactions are of great importance. They control the thermonuclear reaction rate in the main sequence stars. The first stage of energy production in the hydrogen of the solar core is believed to be the weak interaction

$$pp \to de^+ v_e. \tag{7.6}$$

TABLE 7.1 Classification and examples of weak interactions

Leptonic	$\mu^+ \to e^+ v_e \bar{v}_\mu$	$v_e e^- \to e^- v_e$
Semileptonic	$\Delta S = 0$: $n \to pe^-\bar{v}_e,$ $\bar{v}_e p \to ne^+$	$\Delta S = 1$: $K^+ \to \pi^0 + e^+ + v_e,$ $K^+ \to \mu^+ v_\mu$
Nonleptonic	$N + N \to N + N,$ parity violation in nuclei	$\Lambda \to \pi^- p,$ $K^+ \to \pi^+ \pi^0,$ $\to \pi^+ \pi^- \pi^+$

The smallness of the cross-section is evidenced by the fact that, even at the enormous density in the core (~ 100 times that of water), any one proton survives on average for $\sim 10^{10}$ years before undergoing this reaction. Hence the long life of the solar system and the impossibility of observing this collision process directly in the laboratory. Weak effects between nucleons can, however, be detected in suitable circumstances; they give rise to parity-violating effects in nuclear transitions (see Section 3.5).

7.2. NUCLEAR β-DECAY: FERMI THEORY

Historically, the prototype weak interaction was nuclear β-decay,

$$n \rightarrow p + e^- + \bar{\nu}_e,$$

or, in terms of quark constituents,

$$d \rightarrow u + e^- + \bar{\nu}_e. \tag{7.7}$$

This process is described by virtual W-exchange as in Fig. 7.1(a), but at the low momentum transfers involved, $q^2 \ll M_W^2$, the interaction is effectively pointlike and described by an apparent four-fermion coupling $G = g^2/M_W^2$, the Fermi constant, as in (1.17) and Fig. 7.1(b).

In the Fermi theory, the transition probability, or decay rate per unit time is given by (4.4)

$$W = \frac{2\pi}{\hbar} G^2 |M|^2 \frac{dN}{dE_0}, \tag{7.8}$$

where E_0 is the energy in the final state, dN/dE_0 is the density of final states and $|M|^2$ is the square of the matrix element, involving integration over angles and spin directions of the particles concerned, and is a constant of order unity. In fact if the total angular momentum J (leptons) $= 0, |M|^2 \simeq 1$, whereas if J (leptons) $= 1, |M|^2 \simeq 3$ (the spin-multiplicity factor). These two types of transition are referred to as Fermi and Gamow-Teller transitions,

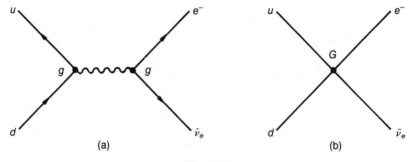

Figure 7.1

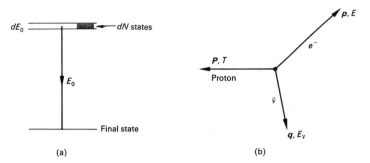

Figure 7.2 (a) An initial state with a spread in energy dE_0 decays to a final (stable) state of unique energy, with energy release E_0. (b) The momentum vectors in neutron β-decay.

respectively. The density-of-states factor is determined by the number of ways it is possible to share out the available energy $E_0 \rightarrow E_0 + dE_0$ between, for example, p, e^-, and $\bar{\nu}$ in the neutron decay (7.7). The quantity dE_0 arises because of the spread in energy of the final state corresponding to the finite lifetime of the initial state. In Fig. 7.2, $\mathbf{p}$, $\mathbf{q}$, and $\mathbf{P}$ are the momenta of the electron, neutrino, and proton; E, E_ν, and T are their kinetic energies.

Then in the rest frame of the initial state (neutron),

$$\mathbf{P} + \mathbf{q} + \mathbf{p} = 0.$$

$$T + E_\nu + E = E_0.$$

Assume $m_\nu = 0$, so $E_\nu = qc$. In order of magnitude, $E_0 \simeq 1$ MeV, so $Pc \simeq 1$ MeV. Thus, if the recoiling nucleon mass is M, its kinetic energy $T = P^2/2M \simeq 10^{-3}$ MeV only and can be neglected. The nucleon serves to conserve momentum, but we can regard E_0 as shared entirely between electron and neutrino. Thus, $qc = E_0 - E$. The number of states in phase space available to an electron, confined to a volume V with momentum $p \rightarrow p + dp$ defined inside the element of solid angle $d\Omega$, is

$$\frac{V \, d\Omega}{h^3} p^2 \, dp.$$

If the fermion wavefunctions are normalized to unit volume ($V = 1$) and we integrate over all space angles, the electron phase-space factor is

$$\frac{4\pi p^2 \, dp}{h^3}.$$

Similarly, for the neutrino it is

$$\frac{4\pi q^2 \, dq}{h^3}.$$

We disregard any possible correlation in angle between $\mathbf{p}$ and $\mathbf{q}$ and treat these two factors as independent, since the proton will take up the resultant momentum. There is no phase-space factor for the proton, since its momentum is now fixed: $\mathbf{P} = -(\mathbf{p} + \mathbf{q})$. So the number of final states is

$$d^2 N = \frac{16\pi^2}{h^6} p^2 q^2 \, dp \, dq.$$

Also, for given values of p and E the neutrino momentum is fixed,

$$q = (E_0 - E)/c,$$

within the range $dq = dE_0/c$. Hence

$$\frac{d^2 N}{dE_0} = \frac{16\pi^2}{h^6 c^3} p^2 (E_0 - E)^2 \, dp.$$

If $|M|^2$ is regarded as a constant, this last expression gives the electron spectrum

$$N(p) \, dp \propto p^2 (E_0 - E)^2 \, dp, \qquad (7.9)$$

and thus if we plot $[N(p)/p^2]^{1/2}$ against E, a straight line cutting the x-axis at $E = E_0$ should result. This is called a Kurie plot. For many β-transitions, the Kurie plot is linear, as shown for the decay $^3\text{H} \to {}^3\text{He} + e^- + \bar{\nu}$. It is necessary to include a Coulomb correction factor $F(Z, p)$, which can be calculated exactly, to take account of the energy lost (e^-) or gained (e^+) from the nuclear Coulomb field. This is important only for low-energy electrons (positrons) and nuclei of large Z.

For a nonzero neutrino mass m_ν, it is straightforward to show that the effect is to modify (7.9) to

$$N(p) \, dp \, F(Z, p) \propto p^2 (E_0 - E)^2 \sqrt{1 - \left(\frac{m_\nu c^2}{E_0 - E} \right)^2} \, dp. \qquad (7.10)$$

It is left as an exercise to show that the Kurie plot turns over to cut the axis vertically at $E = E_0 - m_\nu c^2$, as in Fig. 7.3(a). Thus, the shape of the plot near the endpoint allows one in principal to determine the neutrino mass. The situation is complicated in practice by the spectrometer resolution (see Fig. 7.3(b)) as well as the fact that the daughter atom may be left in an excited state. Table 7.2 lists limits or estimates on the neutrino mass. At the present time, the question of a finite neutrino mass from β-decay measurements is open.

The total decay rate is obtained by integrating (7.9) over the electron spectrum. This can be done exactly, but as a crude approximation we can consider the electrons as extreme relativistic, $E \simeq pc$ (not for tritium!), whence we obtain

$$N \simeq \int_0^{E_0} E^2 (E_0 - E)^2 \, dE = \frac{E_0^5}{30}. \qquad (7.11)$$

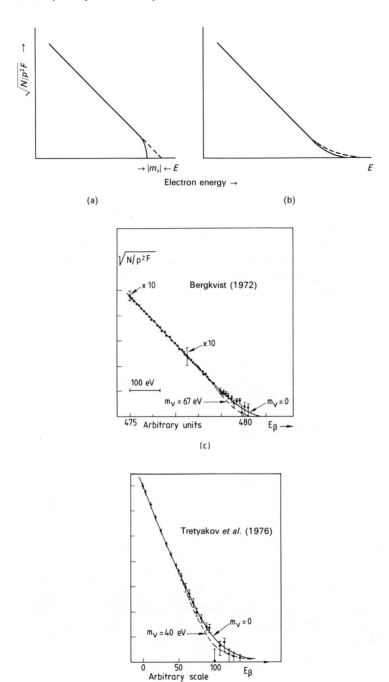

Figure 7.3 Kurie plot of allowed transition for zero and finite neutrino mass: (a) perfect resolution, (b) finite resolution. (c) and (d), plots of measurements of tritium β-decay.

TABLE 7.2 Some neutrino $(\bar{\nu}_e)$ mass estimates in eV/c^2 from Kurie plot of tritium β-decay ($Q = 18.6$ keV)

Langer and Moffatt (1952)	$< 10{,}000$
Bergkvist (1972)	< 65
Tretyakov et al. (1976)	< 35
Lyubimov et al. (1980)	$\simeq 30$
Fritschi et al. (1986)	< 18

Under these conditions, the disintegration constant (i.e., the decay rate) varies as the fifth power of the disintegration energy—the *Sargent rule.*

The value of G can be found from the observed decay rate and (7.8), after integrating over phase space. For example, the decay $O^{14} \rightarrow N^{14*} + e^+ + \nu_e$ is a Fermi transition ($J^P = 0^+ \rightarrow J^P = 0^+$); thus $|M|^2 = 1$ and has lifetime $\tau = 3100$ s. (Actually, we need to double the value of τ, since ^{14}O is a ^{12}C core plus *two* protons, either of which can decay.) Inserting the numerical constants, we get

$$G = 1.02 \times 10^{-5} \hbar c \left(\frac{\hbar}{M_p c} \right)^2$$

$$= \frac{1.02 \times 10^{-5}}{M_p^2} \text{ in units } \hbar = c = 1$$

$$= 1.16 \times 10^{-5} \text{ GeV}^{-2}. \tag{7.12}$$

7.3. INTERACTION OF FREE NEUTRINOS: INVERSE β-DECAY

The cross-section for the inverse reaction (7.3) of free antineutrinos on protons can be calculated from (7.8). In this case, there are only two particles in the final state, so that using (4.6) we obtain (in units $\hbar = c = 1$)

$$\sigma(\bar{\nu}_e p \rightarrow n e^+) = \frac{W}{v_i} = \frac{G^2}{\pi} |M|^2 \frac{p^2}{v_i v_f}, \tag{7.13}$$

where v_i, v_f are the relative velocities of the particles in the initial and final states ($v_i = v_f \simeq c$) and p is the numerical value of the CMS momentum of the neutron and positron. We are dealing with a mixed transition, with $M_F^2 = 1$ for the Fermi contribution ($\Delta J = 0$) and $M_{GT}^2 \simeq 3$ for the spin-multiplicity factor for the Gamow-Teller contribution ($\Delta J = 1$). Thus,

$$\sigma = \frac{M_F^2 + M_{GT}^2}{\pi} G^2 p^2 \simeq \frac{4G^2 p^2}{\pi}. \tag{7.14}$$

For neutrinos in the MeV energy range, incident on a fixed nucleon target, the CMS momentum and laboratory neutrino energy above threshold ($Q = 1.8$ MeV) are related by $p \simeq (E_\nu - Q)/c$. For $pc \simeq 1$ MeV and G from

(7.12) we obtain therefore

$$\sigma = \frac{4}{\pi} \times 10^{-10} \left(\frac{\hbar}{M_p c}\right)^2 \left(\frac{p}{M_p c}\right)^2 \simeq 10^{-43} \text{ cm}^2. \qquad (7.15)$$

This corresponds to a mean free path for antineutrino absorption in water of 10^{20} cm or 100 light years. The first observation of such interactions was made by Reines and Cowan in 1959. They employed a reactor as the source. The uranium fission fragments are neutron-rich and undergo β-decay, emitting electrons and antineutrinos (on average, six per fission, with a spectrum centered at 1 MeV). For a 1000-MW reactor, the useful flux of antineutrinos is of order 10^{13} cm^{-2} s^{-1}. The reaction

$$\bar{\nu}_e p \rightarrow n e^+$$

was observed, using a target of cadmium chloride ($CdCl_2$) and water. The positron produced in this reaction rapidly comes to rest by ionization loss and forms positronium, which annihilates to γ-rays, in turn producing fast electrons by the Compton effect. The electrons are recorded in a liquid scintillation counter. The time scale for this process is of order 10^{-9} s, so the positron gives a so-called prompt pulse. The function of the cadmium is to capture the neutron after it has been moderated (i.e., reduced to thermal energy by successive elastic collisions with protons) in water—a process which delays by several microseconds the γ-rays coming from eventual radiative capture of the neutron in cadmium. Thus, the signature of an event consists of two pulses microseconds apart. Figure 7.4 shows schematically the experimental arrangement employed.

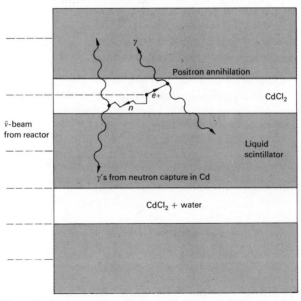

Figure 7.4 Schematic diagram of the experiment by Reines and Cowan (1959), detecting the interactions of free antineutrinos from a reactor.

7.4. PARITY NONCONSERVATION IN β-DECAY

In 1956, following a critical review of the experimental data then available, Lee and Yang came to the conclusion that the weak interactions were not invariant under spatial inversion, i.e., did not conserve parity—largely on the basis of the fact that the K^+ could decay in the two decay modes $K^+ \to 2\pi$, $K^+ \to 3\pi$, in which the final states have opposite parities (so-called τ-θ paradox).

To test parity conservation, an experiment was carried out by Wu *et al.* (1957), who employed a sample of ^{60}Co at $0.01°$K inside a solenoid. At this temperature a high proportion of ^{60}Co nuclei are aligned. ^{60}Co $(J = 5)$ decays to ^{60}Ni*$(J = 4)$—a pure Gamow-Teller transition. The relative electron intensities along and against the field direction were measured (see Fig. 7.5). The degree of ^{60}Co alignment could be determined from observations of the angular distribution of γ-rays from ^{60}Ni*. The results for electron intensity were consistent with a distribution of the form

$$I(\theta) = 1 + \alpha\left(\frac{\boldsymbol{\sigma} \cdot \mathbf{p}}{E}\right)$$

$$= 1 + \alpha \frac{v}{c} \cos \theta, \qquad (7.16)$$

where $\alpha = -1$; $\boldsymbol{\sigma}$ is a unit spin vector in the direction $\mathbf{J}$, $\mathbf{p}$ and E are the electron momentum and total energy, and θ is the angle of emission of the electron with respect to $\mathbf{J}$. The variation with electron velocity was checked over the range $0.4 < v/c < 0.8$.

The fore-aft asymmetry of the intensity in (7.16) implies that the interaction as a whole violates parity conservation. Imagine the whole system reflected in a mirror normal to the z-axis. The first term (unity) does not change sign under reflection—it is *scalar* (even parity). $\boldsymbol{\sigma}$, being an axial vector, does not change sign, while the polar vector $\mathbf{p} \to -\mathbf{p}$. Thus, the product $\boldsymbol{\sigma} \cdot \mathbf{p} \to -\boldsymbol{\sigma} \cdot \mathbf{p}$ under reflection; it is called a *pseudoscalar* (odd parity). Conservation of the z-component of angular momentum in the above transition implies that the electron spin must also point in the direction $\mathbf{J}$, so that if $\boldsymbol{\sigma}$ is now the electron spin vector (normalized so that $\sigma_z = \pm 1$), the intensity is

$$I = 1 + \alpha \frac{\boldsymbol{\sigma} \cdot \mathbf{p}}{E}. \qquad (7.17)$$

The average longitudinal polarization or net *helicity* H can be defined as

$$H = \frac{I_+ - I_-}{I_+ + I_-} = \alpha \frac{v}{c},$$

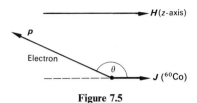

Figure 7.5

where I_+ and I_- represent the intensities for σ parallel and antiparallel to **p**. Experimentally,

$$\alpha = \begin{cases} +1 & \text{for } e^+, H = +v/c, \\ -1 & \text{for } e^-, H = -v/c. \end{cases} \tag{7.18}$$

The property of helicity is not limited to leptons. We have already seen that massless photons possess two possible spin states, $J_z = \pm 1$ (where the z-axis defines the propagation vector). Right-handed ($J_z = +1$) and left-handed ($J_z = -1$) photons, have helicities $+1$ and -1, respectively. Parity is conserved in electromagnetic processes involving photons because the two types of photon are always emitted with equal amplitude, and one does not therefore observe a net circular polarization. On the contrary, in weak interactions, β-processes consist of the emission of *either* electrons, with a net left-handed spin polarization, *or* positrons, which are predominantly right-handed.

7.5. HELICITY OF THE NEUTRINO

The result (7.18), if applied to a neutrino ($m = 0$), implies that such a particle must be fully polarized, $H = +1$ or -1. In order to determine the type of operators occurring in the matrix element, the sign of the neutrino helicity turns out to be crucial. The neutrino is here defined as the neutral particle emitted together with the positron in β^+-decay, or following K-capture. The antineutrino then accompanies negative electrons in β^--decay. The neutrino helicity was determined in a classic and beautiful experiment by Goldhaber *et al.* in 1958. The steps in this experiment are indicated in Fig. 7.6:

(i) ^{152}Eu undergoes K-capture to an excited state of ^{152}Sm with $J = 1$ (Fig. 7.6(a)). To conserve angular momentum, **J** must be parallel to the spin of the electron but opposite to that of the neutrino, so the recoiling ^{152}Sm* has the same polarization sense as the neutrino (Fig. 7.6(b)).

(ii) Now, in the transition ^{152}Sm* $\rightarrow$ ^{152}Sm $+ \gamma$, γ-rays emitted in the forward (backward) direction with respect to the line of flight of ^{152}Sm* will be polarized in the same (opposite) sense to the neutrino, as in Fig. 7.6(c). Thus, the polarization of the "forward" γ-rays is the same as that of the neutrino.

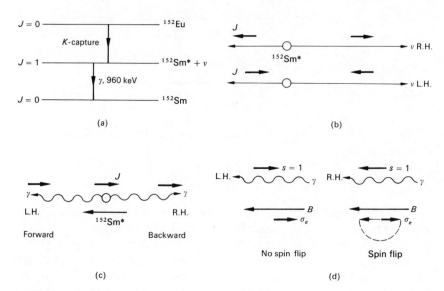

Figure 7.6 Principal steps in the experiment of Goldhaber *et al.* to determine the neutrino helicity.

(iii) The next step is to observe resonance scattering of the γ-rays in a ^{152}Sm target. Resonance scattering is possible with γ-rays of just the right frequency to "hit" the excited state:

$$\gamma + {}^{152}\text{Sm} \rightarrow {}^{152}\text{Sm*} \rightarrow \gamma + {}^{152}\text{Sm}. \qquad (7.19)$$

To produce resonance scattering, the γ-ray energy must slightly exceed the 960 keV to allow for the nuclear recoil. It is precisely the "forward" γ-rays, carrying with them a part of the neutrino-recoil momentum, which are able to do this, and which are therefore automatically selected by the resonance scattering.

(iv) The last step is to determine the polarization sense of the γ-rays. To do this, they were made to pass through magnetized iron before impinging on the ^{152}Sm absorber. An electron in the iron with spin σ_e opposite to that of the photon can absorb the unit of angular momentum by spin-flip; if the spin is parallel it cannot. This is indicated in Fig. 7.6(d). If the γ-ray beam is in the same direction as the field **B**, the transmission of the iron is greater for left-handed γ-rays than for right-handed.

A schematic diagram of the apparatus is shown in Fig. 7.7. By reversing **B** the sense of polarization could be determined from the change in counting rate. When allowance was made for various depolarizing effects, it was concluded that neutrinos were left-handedly spin polarized.

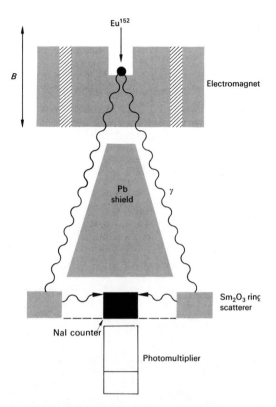

Figure 7.7 Schematic diagram of apparatus used by Goldhaber *et al.* in which γ-rays from the decay of ^{152}Sm*, produced following K-capture in ^{152}Eu, undergo resonance scattering in Sm_2O_3, and are recorded by a sodium iodide scintillator and photomultiplier. The transmission of photons through the iron surrounding the ^{152}Eu source depends on their helicity and the direction of the magnetic field **B**.

In conclusion, the helicity assignments for the leptons emitted in nuclear β-decay are therefore as follows:

$$
\begin{array}{ccccc}
\text{Particle} & e^+ & e^- & v & \bar{v} \\
\text{Helicity} & +v/c & -v/c & -1 & +1
\end{array}
\qquad (7.20)
$$

7.6. THE *V-A* INTERACTION

The formulae (7.16), (7.18), (7.20) have been presented from a purely empirical viewpoint. Their actual form follows from the so-called *V-A* theory of weak interactions. This involves the Dirac description of relativistic fermions; a brief account of this theory, the definition of the γ-matrices and the derivation of the results given below can be found in Appendix D.

Fermi developed his theory of β-decay in analogy with electromagnetic interactions. Consider first the process of electromagnetic scattering of an electron and a proton:

$$e^- + p \rightarrow e^- + p. \tag{7.21}$$

Since electrons and baryons are conserved in this process, we may describe it as the interaction of two *currents*, via single virtual-photon exchange (see Fig. 1.5(c)), and the matrix element is proportional to

$$M \propto \frac{e^2}{q^2} J_{\text{baryon}} \cdot J_{\text{lepton}}, \tag{7.22}$$

where q is the momentum transfer.

Relativistic fermions in the Dirac theory are described by 4-component, or spinor, wavefunctions, and they are operated on by 4×4 matrix operators O. Electromagnetic currents involve the *vector* operator $O_{\text{e.m.}} = \gamma_4 \gamma_\mu$ (where $\mu = 1, \ldots, 4$ indicates space-time components) and the currents have the form

$$J_{\text{lepton}} = \psi_e^* \gamma_4 \gamma_\mu \psi_e \equiv \bar{\psi}_e \gamma_\mu \psi_e,$$
$$J_{\text{baryon}} = \bar{\psi}_p \gamma_\mu \psi_p, \tag{7.23}$$

where $\bar{\psi} = \psi^* \gamma_4$.

By analogy, Fermi assumed that for the weak process of neutron decay, one could write

$$M = G J_{\text{baryon}}^{\text{weak}} \cdot J_{\text{lepton}}^{\text{weak}} = G(\bar{\psi}_p O \psi_n)(\bar{\psi}_e O \psi_\nu), \tag{7.24}$$

where the grouping of the wavefunctions is made more plausible if we write the β-decay process

$$n \rightarrow p + e^- + \bar{\nu}$$

in the equivalent form

$$\nu + n \rightarrow p + e^-.$$

Fermi assumed that the matrix operator O in (7.24) would again be a vector operator as in (7.23). The only differences are that in β-decay one has a constant G instead of e^2, that the weak currents are assumed to interact at a point i.e., there is no $1/q^2$ factor as in (7.22), and that the electric charges of the lepton and baryon change by one unit in the interaction. Hence these β-decay reactions are referred to as charge-changing or simply charged-current weak interactions (see Fig 1.5(a) and (b)).

The vector interaction was satisfactory (prior to the discovery of parity violation in 1956) in describing Fermi transitions. It did not account for Gamow-Teller transitions, since it could not produce a flipover of the nucleon spin. Within certain requirements of relativistic invariance, one can in fact have five possible independent forms for the operator O in (7.24). They are called scalar (S), vector (V), tensor (T), axial vector (A), and pseudoscalar (P). These names are associated with the transformation properties of the weak currents under space inversions. The S- and V-interactions produce Fermi transitions, while T and A produce Gamow-Teller transitions. The pseudoscalar interaction P is unimportant in β-decay, since it couples spinor components proportional to particle velocity and thus introduces a factor in M^2 of v^2/c^2, where v is the nucleon velocity ($v^2/c^2 \simeq 10^{-6}$ in β-decay).

As shown in Appendix D, V- and A-interactions result in lepton and antilepton (e^+ and v_e, for example) of opposite helicities, while S, T-, and P-interactions produce lepton and antileptons with the same helicity. The result (7.20) shows that V- and A-interactions are involved, and not S, T, or P. Thus, we can write in place of (7.24) for a general β-interaction

$$M = G \sum_{i=V,A} C_i(\bar{\psi}_p O_i \psi_n)(\bar{\psi}_e O_i \psi_v), \qquad (7.25)$$

where C_V and C_A are appropriate coefficients. For example, for a pure Fermi transition, $C_A = 0$. The matrix element (7.25) is, however, a scalar quantity. It implies that parity is conserved. This is clearly wrong. We need to add another term to (7.25) to give us a pseudoscalar quantity so that the matrix element contains both scalars and pseudoscalars and thus has no well-defined parity. We can do this by adding a similar expression, with one bracket multiplied by a matrix γ_5 ($\gamma_5 = \gamma_1\gamma_2\gamma_3\gamma_4$). Thus, if we insert a $1/\sqrt{2}$ factor to retain the definition of G,

$$M = \frac{G}{\sqrt{2}} \sum_{i=V,A} [\bar{\psi}_p O_i \psi_n][\bar{\psi}_e O_i (C_i + C_i'\gamma_5)\psi_v]. \qquad (7.26)$$

If, as all the evidence suggests, the interaction is invariant under time reversal, it can be shown that C_i and C_i' must be real coefficients; furthermore, if a neutrino (or antineutrino) is to be completely spin-polarized ($H = \pm 1$), one must have $C_i' = \pm C_i$. Since scalar and pseudoscalar terms then occur with equal magnitude, this is called the principle of *maximum parity violation*. For this case, the operator $(1 \pm \gamma_5)$ acting on the neutrino wavefunction in (7.26) projects out one sign of helicity:

$(1 + \gamma_5) \rightarrow$ L.H. neutrino state (R.H. antineutrino),

$(1 - \gamma_5) \rightarrow$ R.H. neutrino state (L. H. antineutrino).

Since the experiment in Section 7.5 shows the neutrino to be left-handed, we need to take the first expression, corresponding to $C_i' = C_i$. This type of weak interaction is called the *two-component neutrino theory*, and was proposed independently by Lee and Yang, by Landau, and By Salam. It also correctly predicts the helicities $\pm v/c$ for the charged leptons in (7.18). Hence, (7.26) becomes

$$M = G/\sqrt{2} \sum_{i=V,A} C_i[\bar{\psi}_p O_i \psi_n][\bar{\psi}_e O_i (1 + \gamma_5)\psi_v].$$

Inserting the Dirac matrix expressions for the operators $O_V = \gamma_\mu$ and $O_A = i\gamma_\mu\gamma_5$, and setting $C_A/C_V = \lambda$, and $C_V = 1$ (defining the original Fermi coupling G), we obtain

$$M = [G/\sqrt{2}][\bar{\psi}_p\gamma_\mu(1 - \lambda\gamma_5)\psi_n][\bar{\psi}_e\gamma_\mu(1 + \gamma_5)\psi_\nu]. \tag{7.27}$$

The ratio of the A- and V-couplings $\lambda = -1$ in muon decay, which involves leptons only. In hadronic processes, the effects of the strong (quark-quark) interaction can modify C_A (but not C_V—see Problem 7.1), and λ departs from unity. For example, $\lambda = -1.25$ in neutron decay, $n \to p + e^- + \bar{\nu}_e$, and $\lambda = -0.69$ for lambda-hyperon β-decay, $\Lambda \to p + e^- + \bar{\nu}_e$. We note that had the weak nucleon current been pure V-A (with $\lambda = -1$), the operators in nucleon and lepton brackets would have been identical, so that *all* fermions in β-decay would have helicity $H = -v/c$, all antifermions $H = +v/c$; and the result would differ from the pure vector interaction of Fermi only by the inclusion of the projection operator $(1 + \gamma_5)$ in both terms.

7.7. PARITY VIOLATION IN Λ-DECAY

Parity nonconservation is a general property of weak interactions not solely associated with leptonic processes. The "τ-θ paradox", associated with the nonleptonic decay of kaons into pions, has already been mentioned. Another nonleptonic decay is that of the Λ-hyperon, for which the dominant decay modes are

$$\Lambda \to \pi^- p, \qquad \Lambda \to \pi^0 n. \tag{7.28}$$

The Λ-hyperon does in fact also undergo leptonic β-decay, $\Lambda \to p + e^- + \bar{\nu}$, with a small branching ratio.

Parity violation can be demonstrated by considering the decay of Λ-hyperons produced in the associated production process

$$\pi^- p \to \Lambda K^0 \tag{7.29}$$

In this process, the Λ can be (and in general is) spin-polarized. Parity conservation in such a strong interaction implies that the Λ must be polarized with spin $\boldsymbol{\sigma}$ *transverse* to the production plane, i.e., of the form $\boldsymbol{\sigma} \propto (\mathbf{p}_\Lambda \times \mathbf{p}_K)$, which does not change sign under inversion. Note that spin polarization *in* the production plane in general does change sign, and is not allowed.

In practice, the mean transverse polarization

$$P_\Lambda = (N\uparrow - N\downarrow)/(N\uparrow + N\downarrow) \simeq 0.7$$

for an incident pion momentum just above 1 GeV/c. In the decay process (7.28) let us define the direction of $\boldsymbol{\sigma}$ as the z-axis of the Λ rest frame (Fig. 7.8). In this frame, the distribution in angle of emission (θ, ϕ) of the pion or proton will depend on their orbital angular momentum l.

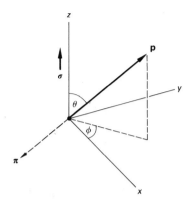

Figure 7.8 Definition of axes and directions in Λ-decay.

Since $J_\Lambda = \frac{1}{2}$, $J_z = \pm\frac{1}{2}$, we can have either $l = 0$ (proton and Λ-spins parallel) or $l = 1$ (spins antiparallel). Thus, we generally expect a combination of s-and p-waves. Denote by m_1 the z-component of the proton spin vector, and by m_2 the z-component of l. In the s-wave case, $m_2 = 0$ and the angular-momentum wavefunction is $Y_l^m = Y_0^0$. Thus, for $J_z = +\frac{1}{2}$, the total wavefunction is the product

$$\psi_s = a_s Y_0^0 \chi^+, \tag{7.30}$$

where a_s denotes the s-wave amplitude and χ^+ the proton spin-up state of $m_1 = +\frac{1}{2}$. For the p-wave, $m_1 + m_2 = J_z = \frac{1}{2}$, with either $m_1 = +\frac{1}{2}$ and $m_2 = 0$, or $m_1 = -\frac{1}{2}$ and $m_2 = +1$.

From Table III in the appendix for the Clebsch-Gordan coefficients for adding $J = 1$ and $J = \frac{1}{2}$; the entry for $J_{\text{total}} = m = +\frac{1}{2}$ gives

$$\psi_p = a_p[\sqrt{\tfrac{2}{3}} Y_1^1 \chi^- - \sqrt{\tfrac{1}{3}} Y_1^0 \chi^+]. \tag{7.31}$$

Here, a_s and a_p are, in general, complex amplitudes. Thus, if *both* s- and p-waves are present, the total amplitude will be

$$\psi = \psi_s + \psi_p = [a_s Y_0^0 - a_p \sqrt{\tfrac{1}{3}} Y_1^0]\chi^+ + [a_p \sqrt{\tfrac{2}{3}} Y_1^1]\chi^-.$$

Recalling the orthogonality of the spin states χ^+ and χ^-, we see that the angular distribution becomes

$$\psi\psi^* = (a_s Y_0^0 - a_p \sqrt{\tfrac{1}{3}} Y_1^0)(a_s Y_0^0 - a_p^* \sqrt{\tfrac{1}{3}} Y_1^0) + a_p^2 (\sqrt{\tfrac{2}{3}} Y_1^1)^2,$$

where one of the phases must be arbitrary and we take a_s to be real. Also, $Y_0^0 = 1$, $\sqrt{\tfrac{1}{3}} Y_1^0 = \cos\theta$, $\sqrt{\tfrac{2}{3}} Y_1^1 = -\sin\theta$, so that

$$\psi\psi^* = |a_s|^2 + |a_p|^2 \cos^2\theta + |a_p|^2 \sin^2\theta - a_s \cos\theta[a_p + a_p^*]$$
$$= |a_s|^2 + |a_p|^2 - 2a_s \operatorname{Re} a_p^* \cos\theta. \tag{7.32}$$

If we set

$$\alpha = \frac{2a_s \, \mathrm{Re} \, a_p^*}{|a_s|^2 + |a_p|^2},$$

the angular distribution has the form

$$I(\theta) = 1 - \alpha \cos \theta. \tag{7.33}$$

The angle θ is defined relative to $\boldsymbol{\sigma}$; physically, one can only measure relative to the normal to the production plane, so if we redefine θ in this way, the above result becomes

$$I(\theta)| = 1 - \alpha P \cos \theta,$$

where P is the average polarization; experiment shows that $\alpha P \simeq -0.7$. Thus, the parity violation in the Λ-decay is manifested as an up-down asymmetry of the decay pion (or proton) relative to the production plane.

      Note that (7.33) has the same form as (7.16), that the parity-violation parameter α is finite only if *both* a_s and a_p are finite, and thus that the parity violation arises from interference of the s (even) and p (odd) waves.

7.8. PION AND MUON DECAY

The lepton helicities first observed in 1957 in nuclear β-decay were detected simultaneously in the decay of pions and muons. We recall that the pion and muon decay schemes are

$$\pi^+ \rightarrow \mu^+ v, \tag{7.34}$$

$$\mu^+ \rightarrow e^+ v \bar{v}. \tag{7.35}$$

      Since the pion has spin zero, the neutrino and muon must have antiparallel spin vectors, as shown in Fig. 7.9. If the neutrino has helicity $H = -1$, as in β-decay, the μ^+ must have negative helicity. In the subsequent muon decay, the positron spectrum is peaked in the region of the maximum energy, so the most likely configuration is that shown—the positron having positive helicity. In fact, the positron spectrum has the shape indicated in Fig. 7.11. In the experiments, positive pions decayed in flight, and those decay muons projected in the forward direction—thus with negative helicity—were selected. These μ^+ were stopped in a carbon absorber, and the e^+ angular

Figure 7.9 Sketches indicating sense of spin polarization in pion and muon decay.

distribution relative to the original muon momentum $\mathbf{p}_\mu$ was observed. The muon spin $\boldsymbol{\sigma}$ should be opposite to $\mathbf{p}_\mu$ if there is no depolarization of the muons in coming to rest (true in carbon). The angular distribution observed was of the form

$$\frac{dN}{d\Omega} = 1 - \frac{\alpha}{3} \cos\theta,$$

where θ is the angle between $\mathbf{p}_\mu$, the initial muon momentum vector, and $\mathbf{p}_e$, the electron momentum vector, and $\alpha = 1$ within the errors of measurement. The same value of α was found for μ^+ and μ^-. This is exactly the form predicted by the $V\text{-}A$ theory. The helicity of the electrons (positrons) was also measured and shown to be $\mp v/c$.

The muon lifetime in the $V\text{-}A$ theory is given by

$$\tau^{-1} = \frac{G^2 m_\mu^5}{192\pi^3},$$

where the dependence on mass to the fifth power follows the Sargent rule for three-body decay. From the measured values $m_\mu c^2 = 105.6593$ MeV, $\tau = 2.19709\ \mu\text{s}$, the value of the Fermi constant is

$$G = 1.16637 \times 10^{-5}\ \text{GeV}^{-2}, \tag{7.36}$$

where a small radiative correction (0.5 %) has been taken into account.

7.8.1. The $\pi \to \mu$ and $\pi \to e$ Branching Ratios

Pion decay provides an important test of the $V\text{-}A$ theory, via the decay branching ratio

$$\frac{\pi \to e\nu}{\pi \to \mu\nu}. \tag{7.37}$$

The pion decay is an example of the four-fermion interaction, if the pion is represented as a quark-antiquark pair (see Fig. 7.10). In nuclear β-decay

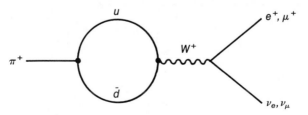

Figure 7.10

terminology, the transition is from a hadronic state of $J^P = 0^-$ (pion) to $J^P = 0^+$ (vacuum), so of the five operators mentioned above, only the A (axial vector) and P (pseudoscalar) operators could be involved. In the pion rest frame, the two leptons must be emitted in opposite directions, but with the same helicities, in order to conserve angular momentum. From our previous discussion, we know that the P- and A-interactions favor leptons with the same and opposite helicities, respectively. In any case the neutrino must have helicity -1. From (7.17) it is apparent that the probability for the A-interaction to produce an e^+ (or μ^+) of velocity v and unfavored helicity $-v/c$ will be proportional to $(1 - v/c)$; whereas for the P coupling, favoring the same helicity for e^+ and neutrino, it will be proportional to $(1 + v/c)$.

Prediction of the branching ratio (7.37) then follows if one includes the phase-space factor. This is

$$\frac{dN}{dE_0} = \text{const } p^2 \frac{dp}{dE_0}.$$

Here, $\mathbf{p}$ is the momentum of the charged lepton in the pion rest frame, v its velocity, and m its rest mass. The neutrino momentum is then $-\mathbf{p}$:

In units $c = 1$, the total energy is

$$E_0 = m_\pi = p + \sqrt{p^2 + m^2}.$$

Hence,

$$p^2 \frac{dp}{dE_0} = \frac{(m_\pi^2 + m^2)(m_\pi^2 - m^2)^2}{4m_\pi^4},$$

$$1 + \frac{v}{c} = \frac{2m_\pi^2}{m_\pi^2 + m^2},$$

$$1 - \frac{v}{c} = \frac{2m^2}{m_\pi^2 + m^2}.$$

Thus, for A-coupling the decay rate is given by

$$p^2 \frac{dp}{dE_0}\left(1 - \frac{v}{c}\right) = \frac{m^2}{2}\left(1 - \frac{m^2}{m_\pi^2}\right)^2,$$

and for P-coupling it is

$$p^2 \frac{dp}{dE_0}\left(1 + \frac{v}{c}\right) = \frac{m_\pi^2}{2}\left(1 - \frac{m^2}{m_\pi^2}\right)^2.$$

The predicted branching ratios become, with the approximation $m_e^2/m_\pi^2 \ll 1$,

$$A\text{-coupling:} \quad R = \frac{\pi \to ev}{\pi \to \mu v} = \frac{m_e^2}{m_\mu^2} \frac{1}{(1 - m_\mu^2/m_\pi^2)^2} = 1.275 \times 10^{-4}, \quad (7.38)$$

$$P\text{-coupling:} \quad R = \frac{\pi \to ev}{\pi \to \mu v} = \frac{1}{(1 - m_\mu^2/m_\pi^2)^2} = 5.5. \quad (7.39)$$

The dramatic difference in the branching ratio for the two types of coupling just stems from the fact that angular-momentum conservation compels the electron or muon to have the "wrong" helicity for A-coupling. The phase-space factor for the electron decay is greater than for muon decay, but the factor $(1 - v/c)$ strongly inhibits decay to the lighter lepton.

The current measured value for the ratio is

$$R_{\text{exp}} = (1.267 \pm 0.023) \times 10^{-4}. \quad (7.40)$$

This result was a major triumph for the V-A theory, and proves that the pseudoscalar coupling is zero or extremely small. Figure 7.11 shows a typical positron spectrum observed from positive pions stopping in an absorber in one experiment. The rare $\pi \to ev$ process yields positrons of unique energy, about 70 MeV. They are accompanied by the much more numerous positrons from the decay sequences $\pi \to \mu v$, $\mu \to ev\bar{v}$. The spectrum from muon decay extends to 50 MeV. Rejection of electrons from $\pi \to \mu \to e$ decays is based on

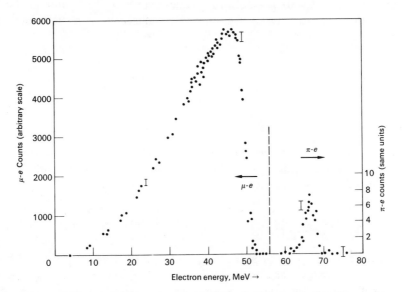

Figure 7.11 Electron spectrum from stopping positive pions, as measured in the experiment of Anderson *et al.* (1960). The broad distribution extending up to 53 MeV is from $\mu^+ \to e^+ + v + \bar{v}$. The narrow peak around 70 MeV is from $\pi^+ \to e^+ + v$ decay. Note the change in vertical scale for these very rare events.

momentum, timing (the mean life of the pion is 25 ns, that of the muon 2200 ns), as well as the absence of a muon pulse in the counters.

The above formulae may be applied also to the kaon-decay branching ratio, for which the predicted value for A-coupling is

$$R_K = \frac{K \to ev}{K \to \mu v} = 2.5 \times 10^{-5},$$

compared with an experimentally measured value of $(2.43 \pm 0.14) \times 10^{-5}$. Here the electron is even more relativistic, and therefore this decay is more strongly suppressed than in pion decay.

In this analysis, it will be noted that we have tacitly assumed the same coupling G for $\pi \to \mu v$ and $\pi \to ev$, and the data support this. This hypothesis of universality in the coupling of electrons and muons (and τ leptons) has to be modified for quarks, as we discuss next.

7.9. WEAK DECAYS OF STRANGE PARTICLES: CABIBBO THEORY

The nonleptonic decays of strange particles appear to be characterized by the selection rule $\Delta S = 1$, $\Delta I = \frac{1}{2}$, as we might expect if one replaces a strange quark $s(S = -1, I = 0)$ by a nonstrange quark $d(S = 0, I = \frac{1}{2})$, and the effects of the other quarks may be neglected. As evidence we cite the branching ratios for Λ decay:

Decay	Quark description	
$\Lambda \to p\pi^-$	$uds \underset{\text{WI}}{\to} udd \underset{\text{SI}}{\to} uud + \bar{u}d$	(7.41)
$\to n\pi^0$	$\to udd + \bar{u}u$	

Since $I = 0$, the $\Delta I = \frac{1}{2}$ rule states that the nucleon and pion must be in a state of $I = \frac{1}{2}$. Referring to the table of Clebsch-Gordan coefficients (Table III, Appendix), the rule predicts

$$\frac{\Gamma(\Lambda \to n\pi^0)}{\Gamma(\Lambda \to n\pi^0) + \Gamma(\Lambda \to p\pi^-)} = \frac{1}{3} \times 1.036 = 0.345, \qquad (7.42)$$

where the correction factor accounts for the slightly different phase space in $n\pi^0$- and $p\pi^-$-states. The observed value is 0.358 ± 0.005, in agreement with this prediction, when account is taken of small electromagnetic radiative corrections.

In semileptonic decays of strange particles, the isospin of the final state cannot be specified, but they obey the selection rule $\Delta Q = \Delta S$, where ΔQ

and ΔS are the changes in charge and strangeness of the hadrons. $\Delta Q = \Delta S = 1$ is expected if $\Delta I_3 = \frac{1}{2}$, from the relation

$$Q = I_3 + \tfrac{1}{2}(B + S).$$

As examples of the rule, the decay

Decay	Quark description
$\Sigma^- \rightarrow n + e^- + \bar{\nu}_e$	$dds \rightarrow ddu + e^- + \bar{\nu}_e (\Delta S = \Delta Q = 1)$

(7.43)

is observed, with branching ratio 1.08×10^{-3}, but not

Decay	Quark description
$\Sigma^+ \rightarrow n + e^+ + \nu$	$uus \rightarrow udd + e^+ + \nu_e (\Delta S = -\Delta Q = 1)$

(7.44)

for which the branching ratio is less than 5×10^{-6}.

Deviations from the $\Delta I = \frac{1}{2}$ rule are certainly observed in cases where the final-state hadrons can have both $I = \frac{1}{2}$ and $I = \frac{3}{2}$. For example, the branching ratio

$$\frac{\Gamma(\Omega^- \rightarrow \Xi^0 \pi^-)}{\Gamma(\Omega^- \rightarrow \Xi^- \pi^0)} = 2.94 \pm 0.35,$$

while the $\Delta I = \frac{1}{2}$ rule predicts 2.00. Thus, the $I = \frac{3}{2}$ amplitude is also present, although heavily suppressed.

Of more interest are the absolute decay rates for $\Delta S = 1$ transitions, as compared with $\Delta S = 0$ transitions. $\Delta S = 1$ transitions are suppressed by a factor of about 20 relative to ones with $\Delta S = 0$. The reduced decay rate for $\Delta S = 1$ processes is accounted for by the Cabibbo theory (1963). In this model, the d- and s-quark states participating in the weak interactions are "rotated" by a mixing angle θ_C, called the Cabibbo angle. In analogy with the lepton doublets involved in charge-changing weak lepton currents, the u, d, s quarks are organized in a doublet:

Leptons   Quarks

$$\binom{\mu}{\nu_\mu}, \quad \binom{e}{\nu_e}, \qquad \binom{u}{d_C} = \binom{u}{d \cos \theta_C + s \sin \theta_C}$$

(7.45)

For either of these sets of doublets, the weak coupling is specified by the Fermi constant G. For a $\Delta S = 0$ semileptonic decay, involving u- and d-quarks, the coupling will then be $G \cos \theta_C$, and for a $\Delta S = 1$ decay, involving s-quarks, it will be $G \sin \theta_C$. We illustrate this in Table 7.3 by

TABLE 7.3

	Hadronic spin-parity	Quark description	Rate	Equation
$p \to n e^+ v_e$ (^{14}O)	$0^+ \to 0^+$	$u \to d e^+ v_e$	$G^2 \cos^2 \theta_C$	(7.46)
$\pi^- \to \pi^0 e^- \bar{v}_e$	$0^- \to 0^-$	$d \to u e^- \bar{v}_e$	$G^2 \cos^2 \theta_C$	(7.47)
$K^- \to \pi^0 e^- \bar{v}_e$	$0^- \to 0^-$	$s \to u e^- \bar{v}_e$	$G^2 \sin^2 \theta_C$	(7.48)
$\mu^+ \to e^+ v_e \bar{v}_\mu$	—	—	G^2	(7.49)

writing out muon decay, ^{14}O β-decay, pion β-decay ($\pi^- \to \pi^0 e^- \bar{v}_e$), and leptonic K-decay ($K^- \to \pi^0 e^- \bar{v}_e$) in terms of the quarks involved.

We note that (7.46) and (7.47) are allowed Fermi transitions ($J^P = 0^+ \to 0^+$ or $0^- \to 0^-$) and should have the same value of $|M|^2$—as indeed they do (see Problem 7.1). From the various processes of semileptonic decay of hadrons we can obtain estimates of θ_C via the relations

$$\Delta S = 0, \qquad G_\pi^2 = G_n^2 = G_\mu^2 \cos^2 \theta_C,$$
$$\Delta S = 1, \qquad G_K^2 = G_\mu^2 \sin^2 \theta_C. \tag{7.50}$$

Thus, comparison of the ^{14}O decay rate with that for μ-decay ((7.46) and (7.49)) gives $\cos^2 \theta_C$ and hence,

$$\theta_C(\text{vector}) = 0.210 \pm 0.025, \tag{7.51}$$

while comparison of reactions (7.48) and (7.47) for pure vector $\Delta S = 1$ and $\Delta S = 0$ transitions gives

$$\theta_C(\text{vector}) = 0.25 \pm 0.01. \tag{7.52}$$

An estimate of θ_C can also be obtained by comparing the pure axial-vector transition rates

$$\frac{K^+ \to \mu^+ v}{\pi^+ \to \mu^+ v} \to \theta_C(\text{axial}) = 0.269 \pm 0.001. \tag{7.53}$$

The results (7.51), (7.52), and (7.53) do not give precisely equal values, but the reasons for the small discrepancies can be understood. This would take us on to subjects like nonconservation of the axial-vector currents and quark mass effects. It is sufficient to state that when these are taken into consideration, the relative rates of strangeness-changing and -nonchanging decays are consistent with a unique value of the Cabibbo angle. However, this theory was formulated at a time when only three flavors of quark (u, d, s) were known. Modifications to be expected in a six-flavor model are discussed in Section 7.12.

7.10. WEAK NEUTRAL CURRENTS

The production at accelerators of intense beams of high-energy neutrinos and antineutrinos, from the early 1960s onward, led to dramatic developments in our understanding of the weak interactions. The layout of a neutrino beam is sketched out in Fig. 7.12. The basic principle is to first produce secondary pions and kaons from high-energy proton collisions in a target *T*. Secondaries of one sign of charge and in a small band of momentum can be selected

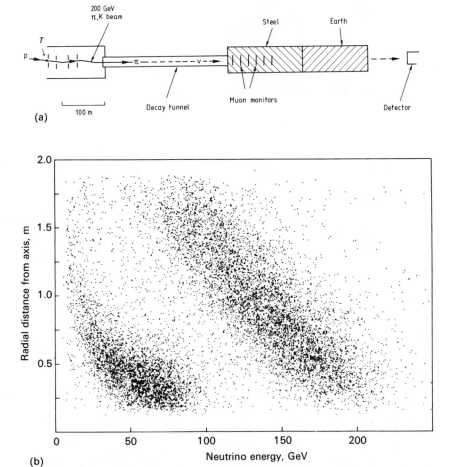

Figure 7.12 (a) Layout of CERN 200-GeV narrow-band neutrino beam. (b) There is a kinematic relation between the energy and direction of neutrinos from decay of a monochromatic beam of pions or kaons (see Problem 7.2). Hence the energies of neutrino events are correlated with the distance from the beam axis. The high-energy band is due to neutrinos from kaon decay, the low-energy events to those from pion decay. Data from the calorimeter detector of Fig. 2.24 (de Groot *et al.* (1979)).

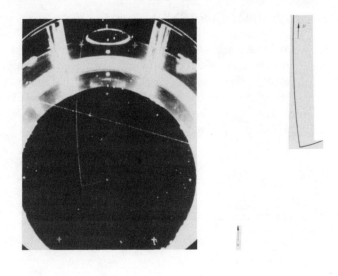

(a)

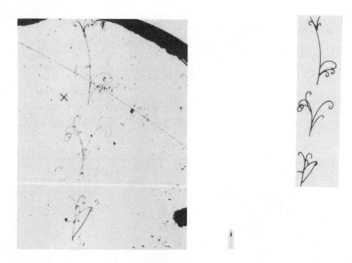

(b)

Figure 7.13 Early examples of "elastic" neutrino interactions observed in a heavy-liquid bubble chamber at CERN in 1963. (a) shows an event attributed to the reaction $\nu_\mu + n \rightarrow p + \mu^-$. The noninteracting muon passes out of the chamber, and the recoil proton comes to rest inside the volume. (b) Event produced by interaction of an electron-neutrino ν_e: $\nu_e + n \rightarrow p + e^-$. The incident beam consists mostly of muon-neutrinos ν_μ, with a very small admixture of ν_e ($\sim\frac{1}{2}\%$) from the three-body decays in flight, $K^+ \rightarrow \pi^0 + e^+ + \nu_e$. The high-energy electron secondary is recognized by the characteristic shower it produces by the processes of bremsstrahlung and pair production. (Courtesy CERN.)

by means of dipole bending magnets, quadrupole focusing magnets, and collimating slits. The pions and kaons enter a decay tunnel, where a fraction decay to muons and neutrinos ($\pi^+ \to \mu^+ \nu_\mu$, $\pi^- \to \mu^- \bar{\nu}_\mu$). Muons are ranged out by a thick iron shield, and the interactions of the neutrinos are observed in the detector. Figures 2.24 and 2.14 show examples of large electronic and bubble-chamber neutrino detectors.

As mentioned before, such experiments gave the first evidence, in 1961, for the separate identity of electron and muon neutrinos, ν_e and ν_μ (see Fig. 7.13 for examples of events). In 1973, the existence of ν_μ interactions without a charged lepton in the final state were first demonstrated in a bubble-chamber experiment at CERN (Hasert *et al.* 1973, 1974). These were termed *neutral-current* events and ascribed to reactions of the form

$$\nu_\mu N \to \nu_\mu X,$$
$$\bar{\nu}_\mu N \to \bar{\nu}_\mu X, \tag{7.54}$$

where X is any hadronic state. They occurred at a fraction of the rate of the more usual *charged-current* events

$$\nu_\mu N \to \mu^- X,$$
$$\bar{\nu}_\mu N \to \mu^+ X. \tag{7.55}$$

As discussed in Chapter 1, the neutral- and charged-current reactions are viewed in terms of exchange of mediating bosons $W^\pm$ and Z^0, as in Fig. 7.14(a) and (b). If we denote the coupling of the leptons or quarks by g, the matrix element contains a propagator term of the form (1.10), which at low q^2 (relevant to all the processes discussed so far) is related to the four-fermion coupling G by (see (1.17)):

$$G = \underset{q^2 \to 0}{Lt} \frac{g^2}{(q^2 + M_W^2)} = \frac{g^2}{M_W^2} \tag{7.56}$$

In the electroweak theory, discussed in Chapter 9, the coupling g of the $W^\pm$ and Z^0 is related to the coupling of the photon to the charged leptons and quarks. Crudely, $g \simeq e$ within numerical factors, so that predictions follow for the weak boson masses. In 1983, the production and decay of these particles

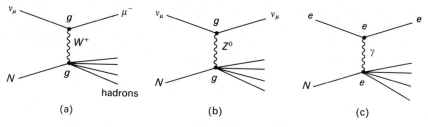

Figure 7.14

$(W \to ev, \mu v,$ and $Z^0 \to e^+e^-, \mu^+\mu^-)$ was observed at the CERN $(270 + 270$ GeV) $p\bar{p}$ collider (see Section 7.13).

The first experiments gave for the ratio of the cross-sections of neutral- and charged-current reactions (7.54) and (7.55)

$$\frac{\sigma(v_\mu N \to v_\mu X)}{\sigma(v_\mu N \to \mu^- X)} \simeq 0.25,$$

$$\frac{\sigma(\bar{v}_\mu N \to \bar{v}_\mu X)}{\sigma(\bar{v}_\mu N \to \mu^+ X)} \simeq 0.45, \qquad (7.57)$$

so that, on the basis of a hundred or so events, the rates for neutral- and charged-current reactions were comparable.

In addition to these semileptonic reactions, purely leptonic neutral-current events of the type

$$\bar{v}_\mu + e^- \to e^- + \bar{v}_\mu \qquad (7.58)$$

were also observed in the same experiment; the first such event, shown in Fig. 1.6, was obtained before the reactions (7.54) were established. The results of the more recent data on semileptonic and leptonic neutral-current reactions are given in Section 9.7. The main point to be made here is that neutral-current couplings exist, with strength comparable to that of charged currents.

7.11. ABSENCE OF $\Delta S = 1$ NEUTRAL CURRENTS: THE GIM MODEL AND CHARM

All neutral current processes observed are characterized by the selection rule $\Delta S = 0$. Indeed, one of the reasons that the early theories of weak interactions, incorporating neutral currents, were discounted was that they had never been observed in decay processes. For example, the ratio of neutral- to charged-current rates in kaon decay is

$$\frac{K^+ \to \pi^+ v\bar{v}}{K^+ \to \pi^0 \mu^+ v_\mu} < 10^{-5} \qquad (\Delta S = 1). \qquad (7.59)$$

In Chapter 9, we shall discuss interpretation of the particles W^+, W^-, Z^0 introduced in Fig. 7.14 as members of a "weak isospin" triplet of bosons interacting with quarks (and leptons). Leaving aside the matrix operators of (7.27), the charged (or charge-raising) weak current, for example in neutron decay, will be $J^+ = \bar{u}d \cos \theta_C$. We then expect from (7.45) that the neutral-current coupling (Fig. 7.15) will be of the form

$$\underbrace{u\bar{u} + (d\bar{d} \cos^2 \theta_C + s\bar{s} \sin^2 \theta_C)}_{\Delta S = 0} + \underbrace{(s\bar{d} + \bar{s}d) \sin \theta_C \cos \theta_C,}_{\Delta S = 1} \qquad (7.60)$$

so that $\Delta S = 1$ neutral currents should be possible, since $\sin \theta_C \neq 0$. In a classic paper in 1970, Glashow, Iliopoulos, and Maiani (GIM) proposed the introduction of a new quark with flavor labeled c for "charm", with charge $+\frac{2}{3}$. They proposed for the quark states in weak interactions a further doublet, consisting of the c-quark and the s, d combination orthogonal to d_c in (7.45). Thus, the two quark doublets were

$$\begin{pmatrix} u \\ d_c \end{pmatrix} = \begin{pmatrix} u \\ d \cos \theta_C + s \sin \theta_C \end{pmatrix}, \qquad \begin{pmatrix} c \\ s_c \end{pmatrix} = \begin{pmatrix} c \\ s \cos \theta_C - d \sin \theta_C \end{pmatrix}. \quad (7.61)$$

In this way, extra terms have to be added to Fig. 7.15, as shown in Fig. 7.16, and when these are included we obtain for the weak-interaction neutral-current matrix element

$$\underbrace{u\bar{u} + c\bar{c} + (d\bar{d} + s\bar{s}) \cos^2 \theta_C + (s\bar{s} + d\bar{d}) \sin^2 \theta_C}_{\Delta S = 0}$$

$$+ \underbrace{(s\bar{d} + \bar{s}d - \bar{s}d - s\bar{d}) \sin \theta_C \cos \theta_C}_{\Delta S = 1}. \quad (7.62)$$

Hence, at the price of a new quark and a second quark doublet, the unwanted $\Delta S = 1$ neutral currents have been canceled.

It was therefore a tremendous triumph for theory when the heavy c-quark states ($\psi = c\bar{c}$, $D = c\bar{u}$, etc.) were observed in 1974–1976, as described in Section 5.13. We note that the assignment (7.61) also predicts that in the decays ($\Delta C = 1$) of charmed mesons to noncharmed mesons, we expect $c \to s$ transitions (coupling $\cos^2 \theta_C$) to dominate $c \to d$ transitions (coupling $\sin^2 \theta_C$). Thus, the $D^0(1893)$ meson decays predominantly in the mode $D^0 \to K^- + \pi's$, and $D^0 \to \pi's$ is strongly suppressed (see Fig. 2.15 for an example of such a decay).

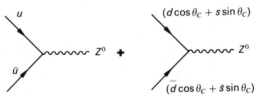

Figure 7.15

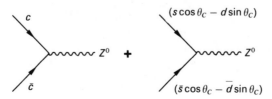

Figure 7.16

By the same mechanism that cancels $\Delta S = 1$ neutral currents, we also expect $\Delta C = 1$ neutral currents to be absent. Limits on these rates can be set from neutrino experiments, looking for (single) charmed-particle production in neutral-current reactions. They are currently known to be less than 3% of the $\Delta C = 0$ neutral-current cross-sections.

7.12. WEAK MIXING ANGLES WITH SIX QUARKS

In Sections 7.9 and 7.11 we have restricted ourselves to u-, d-, s-, and c-quarks and the Cabibbo-GIM formalism for the mixing matrix, which specifies the quark states involved in the weak interactions. The weak charged current is then expressed in matrix form by

$$J_{\text{weak}}^{+} = ([\bar{u}, \bar{c}]) \begin{pmatrix} \cos \theta_C & \sin \theta_C \\ -\sin \theta_C & \cos \theta_C \end{pmatrix} \begin{pmatrix} d \\ s \end{pmatrix}, \tag{7.63}$$

where the coupling constant and the space-time structure operator $\gamma_\mu (1 + \gamma_5)$ have been omitted. The mixing is specified here by a single parameter, the Cabibbo angle θ_C. All data on weak decays involving u-, d-, s-, c-quarks seem to be consistent with a unique value of θ_C, as described previously.

The treatment must be extended if we intend to incorporate further quarks—the b-quark of $Q = -\frac{1}{3}$, which had to be postulated following the discovery of the Υ-resonances (Section 5.13), and its proposed partner, the t-quark of $Q = +\frac{2}{3}$ (see Section 5.16). The existence of three pairs of quarks, in parallel with three pairs of leptons is esthetically attractive, and even receives some justification on the basis of the absence of certain "triangle anomalies." We do not discuss these here, but remark that the condition for canceling such anomalies is that the net charge of all fermions should be zero. The three lepton doublets $(e, v_e; \mu, v_\mu; \tau, v_\tau)$ contribute $-3|e|$; and the three quarks (u, c, t) of $Q = +\frac{2}{3}$, plus three (d, s, b) of charge $Q = -\frac{1}{3}$, will contribute $+3|e|$, if allowance is made for the factor 3 for color.

With six quark flavors, the weak currents will be described by unitary transformations among three quark doublets, characterized by three Euler angles and six phases (Kobayashi and Maskawa (1972)). Of the latter, five are arbitrary and unobservable, leaving one nontrivial phase, δ. Then

$$J_{\text{weak}}^{+} = (\bar{u}, \bar{c}, \bar{t}) M \begin{pmatrix} d \\ s \\ b \end{pmatrix}, \tag{7.64}$$

where the Kobayashi-Maskawa 3×3 matrix M is given by

$$M = \begin{vmatrix} c_1 & c_3 s_1 & s_1 s_3 \\ -c_2 s_1 & c_1 c_2 c_3 - s_2 s_3 e^{i\delta} & c_1 c_2 s_3 + c_3 s_2 e^{i\delta} \\ s_1 s_2 & -c_1 c_3 s_2 - c_2 s_3 e^{i\delta} & -c_1 s_2 s_3 + c_2 c_3 e^{i\delta} \end{vmatrix}, \tag{7.65}$$

with $c_i = \cos \theta_i$ and $s_i = \sin \theta_i$. Here θ_1, θ_2, θ_3 are the three mixing angles, replacing the single angle θ_C of the 2×2 matrix (7.63). Since under time reversal $e^{i\delta} \to e^{-i\delta}$, the phase angle δ introduces the possibility of a T- or CP-violating amplitude (see Section 7.14.3).

Comparing (7.63) and (7.65), we see that the $u \to d$ mixing, specified by a coefficient $\cos \theta_C$ in the Cabibbo theory, is given by $\cos \theta_1$ in the more general model. This coefficient, it will be recalled, is equal to the ratio of coupling G in μ-decay and $G \cos \theta_C$ in nuclear β-decay. The semileptonic decays of hyperons and kaons, previously used to determine $\sin \theta_C$ (Eqs. (7.52) and (7.53)) now determine the product $\sin \theta_1 \cos \theta_3$. Since θ_C from β-decay and from kaon decay are equal within errors, $\cos \theta_3 \simeq 1$ and thus $\sin \theta_3$ must be small.

Information on θ_2 can be obtained from observations on single charmed-particle production in high-energy neutrino collisions; written in terms of quarks, this reaction is expressed by

$$\nu_\mu + d \to \mu^- + c$$
$$\qquad \hookrightarrow s + \mu^+ + \nu_\mu. \qquad (7.66)$$

In this equation, the charmed quark produced in the $\Delta C = 1$ weak reaction is indicated as decaying semileptonically into an s-quark and a lepton pair. Such events are very characteristic—they contain a pair of opposite-sign muons ($\mu^+\mu^-$) of high energy. The rate of the $d \to c$ transitions, compared with the more common $\Delta C = 0$ $d \to u$ transitions, can be obtained from the observed dilepton event rate and the known branching ratios for semileptonic decay of charmed mesons (D^+, D^0, etc.). According to (7.65), it determines the product $\sin \theta_1 \cos \theta_2$; experiments indicate that $\cos \theta_2 \simeq 1$ and hence $\sin \theta_2$ is small.

Although the precision of available data is not very high, it seems therefore that s_1, s_2, and s_3 are all small. If this is so, the model makes predictions about the weak decay sequence of mesons containing b-quarks. The decay chain

$$b \to c \to s \to u,$$

amplitude: $c_1 c_2 s_3 + c_3 s_2 e^{i\delta},$

should be favored over

$$b \to u,$$

amplitude: $s_1 s_3,$

and hence, in the decay of such bottom states, strange particles (kaons) should be frequent among the hadron secondaries. This is indeed the case.

7.13. OBSERVATION OF THE $W^{\pm}$- AND Z^0-BOSONS

7.13.1. *W*- and *Z*-production in $p\bar{p}$ collisions

The masses and decay characteristics of the vector bosons $W^{\pm}$ and Z^0, mediating the charged- and neutral-current weak interactions are predicted in the electroweak theory, discussed in detail in Chapter 9. The result is $M_W \simeq 82$ GeV, $M_{z^0} \simeq 93$ GeV. As is clear from Fig. 7.14, among the decay modes should be the leptonic decays $W \rightarrow l + v_l$, $Z^0 \rightarrow l\bar{l}$ or $v_e \bar{v}_e$ (where $l = e$, μ, τ), as well as the more numerous decays to hadrons via quark pairs: W, $Z \rightarrow Q\bar{Q} \rightarrow$ hadrons.

Such bosons were first observed at the CERN $p\bar{p}$ collider (Arnison *et al.* 1983, Banner *et al.* 1983, Bagnaia *et al.* 1983) via the elementary production and decay processes

$$u + \bar{d} \rightarrow W^+ \rightarrow e^+ + v_e, \mu^+ + v_\mu, \qquad (7.67(a))$$

$$\bar{u} + d \rightarrow W^- \rightarrow e^- + \bar{v}_e, \mu^- + \bar{v}_\mu, \qquad (7.67(b))$$

$$\left.\begin{array}{r} u + \bar{u} \\ d + \bar{d} \end{array}\right\} \rightarrow Z^0 \rightarrow e^+ e^-, \mu^+ \mu^-. \qquad (7.67(c))$$

The cross-section for 7.67(a), producing a W-boson at rest will be given by the usual Breit-Wigner resonance formula (4.55)

$$\sigma(u\bar{d} \rightarrow W^+ \rightarrow e^+ v_e) = \frac{1}{N_c} \cdot \frac{4\pi \lambda^2 \Gamma_{u\bar{d}} \Gamma_{ev}/4}{(2s_d + 1)(2s_u + 1)[(E - M_W)^2 + \Gamma^2/4]} \cdot \frac{(2J + 1)}{3}$$
$$(7.68)$$

where E is the invariant mass of the $u\bar{d}$ system, $\lambda = 2/E$ is the de Broglie CMS wavelength of the colliding particles, and Γ, $\Gamma_{u\bar{d}}$, Γ_{ev} are the total and partial widths for decay. $s_d = s_u = \frac{1}{2}$ are the quark spins. In the (V-A) theory, only L.H. (R.H.) helicity states of the $u(\bar{d})$ are involved, so the spin-multiplicity factor for the W is $(2J + 1)/3 = 1$ (see Fig. 7.17). The color factor $1/N_c = \frac{1}{3}$ is the probability of matching, say, a red quark with an antired antiquark. The peak cross-section is then

$$\sigma_{max}(u\bar{d} \rightarrow W^+ \rightarrow e^+ v_e) = \frac{4\pi}{3 M_W^2} \cdot \frac{\Gamma_{u\bar{d}} \Gamma_{ev}}{\Gamma^2} = \frac{\pi}{36 M_W^2}$$
$$= 5.2 \text{ nb}, \qquad (7.69)$$

where we have assumed $M_W = 83$ GeV, and each of the decays $W \rightarrow u\bar{d}$, $c\bar{s}$, $t\bar{b}$ carries a color factor 3, while $W \rightarrow ev_e$, μv_μ, τv_τ have weight 1, so that $\Gamma_{u\bar{d}}/\Gamma = \frac{1}{4}$ and $\Gamma_{ev}/\Gamma = \frac{1}{12}$.* The cross-section for (7.67(c)) involves the

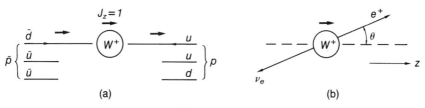

Figure 7.17

neutral-current couplings to quarks and leptons, and depends on the weak mixing angle ($\sin^2 \theta_w$). It turns out to be a factor 10 smaller (see Appendix H).

The cross-sections for the reactions $p\bar{p} \to W^+ + \cdots$, $W^+ \to e^+ \nu$ and $p\bar{p} \to Z^0 + \cdots$, $Z^0 \to e^+ e^-$ result from integrating the elementary quark cross-section (7.68) over the boson width and the momentum distributions of the quarks inside the nucleon. For 270-Gev protons on 270-GeV antiprotons, $\sigma(p\bar{p} \to W \to e\nu)$ is of order 1 nb (10^{-33} cm^2) and $\sigma(p\bar{p} \to Z^0 \to e^+ e^-)$ of order 0.1 nb, to be compared with a value of $\sigma(tot) = 40$ mb $= 4.10^7$ nb for the total $p\bar{p}$ cross-section. The extraction of such rare events, at the level of 10^{-8}–10^{-9} of the total, is possible because the electrons and muons and neutrinos from the decays have very high transverse momenta; $p_T \leq M_W/2 \simeq 40$ GeV/c. Other sources of leptons, for example from hadron decay, give much smaller p_T values (see Fig. 4.20).

One of the calorimeter detectors employed in the experiments is shown in Fig. 7.18. It consists essentially of central tracking chambers to detect individual secondary particles, surrounded by an electromagnetic calorimeter (to detect showers due to electrons and photons), a much larger calorimeter used in the measurement of hadron jets (see Section 8.9), and an outside muon detector. The calorimeters are segmented in intervals of polar and azimuthal angles, θ and ϕ. Apart from the difference in scale, the principals are however much the same as those in the JADE detector (Fig. 2.22).

Figure 1.7 shows a reconstruction of a typical $W \to e\nu$ event. The signature is:

(i)   a high p_T electron, appearing as a single isolated track of high momentum in the central tracking detector, pointing to

(ii) a shower in the electromagnetic calorimeter, with no significant energy deposition in the nearby hadronic calorimeter segments, and

(iii) missing p_T in the event overall, when a summation is made over all secondaries produced (clearly, the other quarks involved in the collision generate secondary hadrons, mostly of low p_T). The missing p_T is associated with the neutrino from the W-decay.

---

* Fermion masses are neglected and all decay modes assigned equal phase-space.

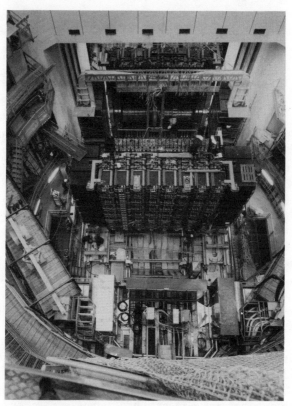

Figure 7.18 UA1 calorimeter detector, viewed from above. The two halves of the magnet/ hadron calorimeter are drawn apart, revealing the elements of the electromagnetic calorimeter that surround the central, cylindrical tracking detector (top of picture). The hadron calorimeter consists of the iron of the magnet yoke instrumented with plastic scintillation counters. The central detector surrounds the vacuum pipe containing the p, $\bar{p}$ intersecting beams. Total detector mass is 2000 tons. (Courtesy CERN.)

Figure 7.19(a) is of a plot of missing p_T in the electron-beam plane against the p_T of the electron, showing that the neutrino and electron have approximately equal and opposite p_T values, as expected (the W-bosons themselves are produced with values of p_T relative to the beam of order 5 GeV/c only).

For the $Z^0 \to e^+e^-$ events (of which Fig. 7.20 shows an example) the selection is based on two isolated tracks from the vertex, pointing to localized track clusters in the electromagnetic calorimeter, and involving large p_T values and invariant mass $M_{e^+e^-} > 50$ GeV. The figure includes a plot of energy deposition in the calorimeter segments, against θ and ϕ, and the concentration in two clusters is very clear.

Decays to muons $W^{\pm} \to \mu\nu$, $Z^0 \to \mu\mu$, have also been observed. In this case, the detection requirement is of one (or two) particles of high p_T which penetrate through the hadron calorimeter (of order six interaction lengths) and are observed in the external muon chambers.

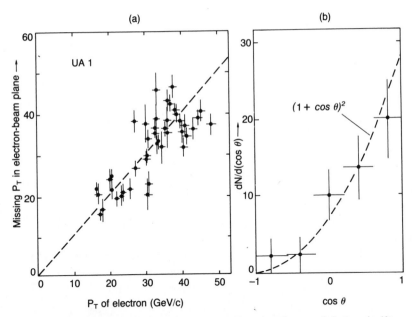

Figure 7.19 (a) Missing p_T in electron/beam plane, plotted against p_T of electron in 43 events attributed to $W \rightarrow e + \nu_e$, from the detector in Fig. 7.18. The p_T of the W-boson and the net p_T of the other particles produced in the event is small, so the missing p_T has to be ascribed to the neutrino. (b) Angular distribution of decay electrons in rest frame of W-boson. The angle θ is that between the e^- and the proton beam direction, or between the e^+ and the antiproton direction.

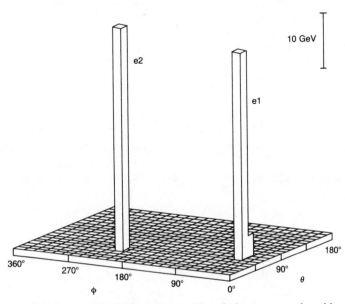

Figure 7.20 $Z^0 \rightarrow e^+ e^-$ event in UA2 experiment. The calorimeter energy deposition is plotted as a function of polar and azimuthal angles.

7.13.2. Boson masses, widths, and decay characteristics

From a sample of W- and Z-decays observed in the UA1 and UA2 calorimeter detectors, the measured masses and production cross-sections are

	Observed	Predicted
M_W (GeV)	81.2 ± 1.3	83.0 ± 2.7
M_{Z^0} (GeV)	92.8 ± 1.5	93.8 ± 2.2
$\sigma(p\bar{p} \to W \to l\nu)$ (nb)	0.55	~ 0.5
$\sigma(p\bar{p} \to Z \to l^+l^-)$ (nb)	0.06	~ 0.05

The error in the predicted masses arises from uncertainty in the Weinberg angle, θ_W. There is excellent agreement between the observed and expected mass values (see Section 9.8). The widths of these boson resonances are calculated in Appendix H. On dimensional grounds, we expect that

$$\Gamma_{W(Z)} \sim GM_{W(Z)}^3; \tag{7.70}$$

since G has dimensions of mass^{-2}. In fact $\Gamma_W = (\sqrt{2}/\pi)GM_W^3 = 2.9$ GeV, if we insert the value of the Fermi constant G from (7.12), while $\Gamma_Z = 2.8$ GeV. The observed values are consistent with these expectations. Figure 7.17 shows that in the V-A theory, the W's are polarized, so we expect a parity-violating asymmetry in the decay, with the e^+ from W^+-decay pointing preferentially in the same direction as the incident antiproton, while for reaction (7.67(b)), the e^- (or μ^-) from the W^--decay points preferentially in the direction of the incident proton. The angular distribution in the W rest frame is easily obtained. If the antiproton beam in Fig. 7.17 defines the z-axis, then the W^+ is produced in the state $J_z = +1$ only. The probability that the e^+ and ν_e, with helicities $+1$ and -1, respectively, emerge at angles θ and $(\pi - \theta)$ and with $J_z = +1$ is clearly given by the square of the appropriate d-function (Appendix C).

$$\frac{dN}{d(\cos\theta)} = |d_{1,1}^1|^2 = (1 + \cos\theta)^2 \tag{7.71}$$

Here θ is the angle between the positron and antiproton directions. The same distribution applies for W^--decays if θ is the angle between electron and proton directions. This angular distribution is well verified experimentally (Fig. 7.19(b)).

7.14. K^0 DECAY

The Gell-Mann–Nishijima formula

$$\frac{Q}{e} = I_3 + \frac{B + S}{2}$$

indicates, in addition to the charged kaons $K^\pm$, of $S = \pm 1$, two neutral kaons, K^0 and $\bar{K}^0$ to complete $I = \frac{1}{2}$ doublets:

	I_3	
S	$+\frac{1}{2}$	$-\frac{1}{2}$
$+1$	K^+	K^0
-1	$\bar{K}^0$	K^-

$$(7.72)$$

The K^0 can be produced by nonstrange particles in association with a hyperon:

$$\pi^- + p \to \Lambda + K^0.$$
$$S \quad 0 \qquad 0 \quad -1 \quad +1$$

$$(7.73)$$

However, a $\bar{K}^0$ can be produced only in association with a kaon or antihyperon, of $S = +1$:

$$\pi^+ + p \to K^+ + \bar{K}^0 + p,$$
$$S \quad 0 \qquad 0 \quad +1 \quad -1 \quad 0$$

$$(7.74)$$

$$\pi^- + p \to \bar{\Lambda} + \bar{K}^0 + n + n.$$
$$S \quad 0 \qquad 0 \quad +1 \quad -1 \quad 0 \quad 0$$

$$(7.75)$$

The threshold pion energy for (7.73) is 0.91 GeV, while for (7.74) and (7.75) it is much higher: 1.50 and 6.0 GeV, respectively. Thus, it is possible to produce a pure K^0-beam by choosing incident pions of suitable energy.

K^0 and $\bar{K}^0$ are particle and antiparticle, and are connected by the process of charge conjugation, which involves a reversal of the value of I_3 and a change of strangeness, $\Delta S = 2$. Strong interactions conserve I_3 and S, so that as far as *production* is concerned, the separate neutral-kaon eigenstates are K^0 and $\bar{K}^0$.

Now suppose K^0- and $\bar{K}^0$-particles propagate through empty space. Since both are neutral, both can decay to pions by the weak interaction, with $|\Delta S| = 1$. Thus, *mixing* can occur via (virtual) intermediate pion states:

$$K^0 \underset{3\pi}{\overset{2\pi}{\rightleftharpoons}} \bar{K}^0$$

These transitions are $\Delta S = 2$ and thus second-order weak interactions. Although extremely weak therefore, this implies that if one has a pure K^0-state at $t = 0$, at any later time t one will have a superposition of both K^0 and $\bar{K}^0$, so that the state can be written

$$|K(t)\rangle = \alpha(t)|K^0\rangle + \beta(t)|\bar{K}^0\rangle. \qquad (7.76)$$

We know that objects which decay by weak interactions are eigen-states of CP, not of strangeness S. The operation of CP on the states K^0 and $\bar{K}^0$ gives

$$CP|K^0\rangle \rightarrow \eta|\bar{K}^0\rangle,$$
$$CP|\bar{K}^0\rangle \rightarrow \eta'|K^0\rangle, \tag{7.77}$$

where η, η' are arbitrary phase factors, which we can define as $\eta = \eta' = 1$. Clearly $|K^0\rangle$ and $|\bar{K}^0\rangle$ are not CP-eigenstates, but we can form the linear combinations

$$|K_1\rangle = \frac{1}{\sqrt{2}}(|K^0\rangle + |\bar{K}^0\rangle), \qquad CP = +1,$$
$$|K_2\rangle = \frac{1}{\sqrt{2}}(|K^0\rangle - |\bar{K}^0\rangle), \qquad CP = -1, \tag{7.78}$$

where

$$CP|K_1\rangle \rightarrow |K_1\rangle,$$
$$CP|K_2\rangle \rightarrow -|K_2\rangle. \tag{7.79}$$

Unlike K^0 and $\bar{K}^0$, distinguished by their mode of *production*, K_1 and K_2 are distinguished by their mode of *decay*. Consider 2π- and 3π-decay modes:

(a) $\pi^0\pi^0$, $\pi^+\pi^-$. By Bose symmetry, the total wavefunction in either case must be symmetric under interchange of the two particles. Since no spin is involved, this is equivalent to the operation C followed by P, so that $CP = +1$.

(b) $\pi^+\pi^-\pi^0$. The small Q-value (70 MeV) of the decay suggests $l = 0$, that is, the three pions are in a relative S-state. By the previous argument, the CP-parity of $\pi^+\pi^-$ is $+1$. The π^0 has $C = +1$ (since it decays to two γ's) and $P = -1$, and therefore $CP = -1$. So, combining the π^0 with the $\pi^+\pi^-$ system, we obtain one of $CP = -1$. For $l > 0$, both positive and negative CP-eigenvalues can result, but such decays are strongly suppressed by angular-momentum barrier effects.

(c) $\pi^0\pi^0\pi^0$. Any orbital angular momentum, l, between any two pions must be even, by Bose symmetry. Hence the l-value of the remaining pion about the dipion is also even, since $J_K = 0$. So the overall parity is the product of the intrinsic pion parities, that is, $P = -1$. Since the neutral pion has $C = +1$, it follows that $CP = -1$, irrespective of the l-values involved.

TABLE 7.4 Neutral-kaon eigenstates

Production	Decay	Lifetime, s
$\lvert K^0 \rangle$ $(S = +1)$	$\lvert K_1 \rangle = (1/\sqrt{2})(\lvert K^0 \rangle + \lvert \bar{K}^0 \rangle) \to 2\pi \;(CP = +1)$	$\tau_1 = 0.9 \times 10^{-10}$
$\lvert \bar{K}_0 \rangle$ $(S = -1)$	$\lvert K_2 \rangle = (1/\sqrt{2})(\lvert K^0 \rangle - \lvert \bar{K}^0 \rangle) \to 3\pi \;(CP = -1)$	$\tau_2 = 0.5 \times 10^{-7}$

To summarize, the 2π state must have $CP = +1$, whereas the 3π-state can have $CP = +1$ or -1, with $CP = -1$ heavily favored kinematically. The two- and three-pion decay modes have different Q-values, and thus different phase-space factors and disintegration rates, the two-pion decay being much faster. In summary, the production and decay states of neutral kaons are listed in Table 7.4.

7.14.1. Strangeness Oscillations

So far, we have discussed K_1- and K_2-amplitudes without regard to their precise time dependence. To be more specific we should include phase factors with the amplitudes. The relative phase of K_1- and K_2-states of a given momentum will only be constant with time if these particles have identical masses. K_1 and K_2 are not charge-conjugate states, having quite different decay modes and lifetimes, so that, in the same sense that the mass difference between neutron and proton can be attributed to differences in their electromagnetic coupling, a $K_1 - K_2$ mass difference—but a very much smaller one—is to be expected because of their different weak couplings.

The amplitude of the state K_1 at time t can be written as

$$a_1(t) = a_1(0)e^{-(iE_1/\hbar)t}e^{-\Gamma_1 t/2\hbar}, \qquad (7.80)$$

where E_1 is the total energy of the particle; so $E_1/\hbar$ is the circular frequency ω_1, and $\Gamma_1 = \hbar/\tau_1$ is the width of the state, τ_1 being the mean lifetime in the frame in which the energy E_1 is defined. The second term must have the form shown, in accord with the law of radioactive decay of the intensity:

$$I(t) = a_1(t)a_1^*(t) = a_1(0)a_1^*(0)e^{-\Gamma_1 t/\hbar}$$
$$= I(0)e^{-t/\tau_1}.$$

Set $\hbar = c = 1$ and measure all times in the rest frame, so that τ_1 is the proper lifetime and $E_1 = m_1$, the particle rest mass. Then the K_1-amplitude is

$$a_1(t) = a_1(0)e^{-(\Gamma_1/2 + im_1)t}. \qquad (7.81)$$

Similarly, for K_2,

$$a_2(t) = a_2(0)e^{-(\Gamma_2/2 + im_2)t}. \qquad (7.82)$$

Now suppose that at $t = 0$, a beam of unit intensity consists of pure K^0. Then from (7.78), $a_1(0) = a_2(0) = 1/\sqrt{2}$. After a time t for free decay *in vacuo*, the K^0-intensity will be

$$I(K^0) = \frac{(a_1(t) + a_2(t))}{\sqrt{2}} \frac{(a_1^*(t) + a_2^*(t))}{\sqrt{2}}$$

$$= \tfrac{1}{4}[e^{-\Gamma_1 t} + e^{-\Gamma_2 t} + 2e^{-[(\Gamma_1 + \Gamma_2)/2]t} \cos \Delta m t], \qquad (7.83)$$

where $\Delta m = |m_2 - m_1|$. Similarly, the $\bar{K}^0$-intensity will be, writing the amplitude as $[a_1(t) - a_2(t)]/\sqrt{2}$,

$$I(\bar{K}^0) = \tfrac{1}{4}[e^{-\Gamma_1 t} + e^{-\Gamma_2 t} - 2e^{-[(\Gamma_1 + \Gamma_2)/2]t} \cos \Delta m t]. \qquad (7.84)$$

Thus, the K^0- and $\bar{K}^0$-intensities *oscillate* with the frequency Δm. Figure 7.21 shows the variation to be expected for $\Delta m = 0.5/\tau_1$. If one measures the number of $\bar{K}^0$ interaction events (i.e., the hyperon yield) as a function of position from the K^0-source, one can therefore deduce $|\Delta m|$. The present value is given by

$$\Delta m \tau_1 = 0.477 \pm 0.002, \qquad (7.85)$$

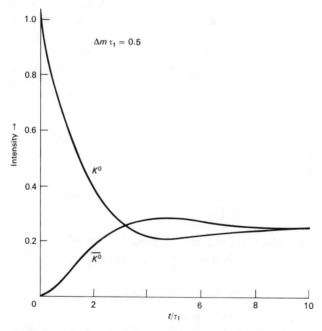

Figure 7.21 Oscillations of K^0- and $\bar{K}^0$-intensities, for an initially pure K^0-beam, as calculated from Eqs. (7.83) and (7.84). A value $\Delta m \tau_1 = 0.5$ has been assumed.

where $\Delta m = m_2 - m_1$, and in separate regeneration experiments, it is found that $m_2 > m_1$.

The actual value of the mass difference is tiny:

$$\Delta m = 3.52 \times 10^{-6} \text{ eV},$$

or a fractional mass difference

$$\frac{\Delta m}{m} = 0.7 \times 10^{-14}. \tag{7.86}$$

The magnitude of the self-energy (i.e., contribution to the mass) of K_1 and K_2 can be estimated crudely by inspection of Fig. 7.22. In these diagrams the "weak" mass generated is clearly proportional to G^2 (all higher-order terms being negligible in comparison), and thus to an effect of the second order in the weak coupling constant—the only one so far observed experimentally. Since $G = 10^{-5}/M_p^2$, we must introduce a mass in order to get the dimensions right. It is reasonable to choose this as the kaon mass. Since we have two strangeness-changing transitions, we should in fact replace G by $G \sin \theta_C$, as in Eq. (7.50). Hence we may estimate

$$\Delta m(\text{for } \Delta S = 1 \text{ coupling}) \simeq G^2 m_K^5 \sin^2 \theta_C \simeq 10^{-4} \text{ eV}, \tag{7.87}$$

within an order of magnitude of the observed value (7.86). By a similar argument, one can set very stringent limits on the strength of a possible direct $\Delta S = 2$ transition between K^0- and $\bar{K}$-states. If fG denotes the coupling strength, then

$$\Delta m_K(\text{for } \Delta S = 2 \text{ coupling}) \simeq fGm_K^3 = 10^3 f \text{ eV}, \tag{7.88}$$

giving $f < 10^{-8}$.

An exact calculation of the mass difference, based on the quark exchanges and the GIM mechanism gives the result (Gaillard *et al.* (1975)):

$$\Delta m = \frac{G^2}{4\pi^2} f_K^2 m_K m_C^2 \cos^2 \theta_c \sin^2 \theta_c,$$

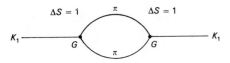

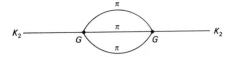

Figure 7.22

where $f_K \simeq 1.2\, m_\pi$ is the kaon decay constant and m_C is the mass of the charmed quark. This expression agrees with experiment for $m_C \simeq 1.5$ GeV. Indeed, this formula predicted the approximate mass of the c-quark before it was discovered.

7.14.2. The K^0 Regeneration Phenomenon

The K_2-decay mode was not observed until some time after its prediction by Gell-Mann and Pais. However, Pais and Piccioni (1955) had observed that the $K_1 K_2$ phenomenon would lead to the process of *regeneration*. Suppose we start with a pure K^0-beam, and let it coast *in vacuo* for the order of $100\, K_1$ mean lives, so that all the K_1-component has decayed and we are left with K_2 only. Now let the K_2-beam traverse a slab of material and interact. Immediately, the strong interactions will pick out the strangeness $+1$ and -1 components of the beam, i.e.,

$$|K_2\rangle = \frac{1}{\sqrt{2}}(|K^0\rangle - |\bar{K}^0\rangle). \tag{7.89}$$

Thus, of the original K^0-beam intensity, 50 % has disappeared by K_1-decay. The remainder, called K_2, upon traversing a slab where its nuclear interactions can be observed, should consist of 50 % K^0 and 50 % $\bar{K}^0$. The existence of $\bar{K}^0$ ($S = -1$) a long way from the source of an originally pure K^0-beam ($S = +1$) was confirmed by the observation of production of hyperons in 1956. Thus, $\bar{K}^0 + p \to \Lambda + \pi^+$, for example.

The K^0- and $\bar{K}^0$-components in (7.89) must be absorbed differently; K^0-particles can only undergo elastic and charge-exchange scattering, while $\bar{K}^0$-particles can also undergo strangeness exchange giving hyperons. With more strong channels open, the $\bar{K}^0$ is therefore absorbed more strongly than K^0. After emerging from the slab, we shall therefore have a K^0-amplitude $f|K^0\rangle$ and a $\bar{K}^0$-amplitude $\bar{f}|\bar{K}^0\rangle$, say, where $\bar{f} < f < 1$. If we then ask what are the characteristics of the emergent beam with respect to decay, we must write, in place of (7.89):

$$\frac{1}{\sqrt{2}}(f|K^0\rangle - \bar{f}|\bar{K}^0\rangle) = \frac{f+\bar{f}}{2\sqrt{2}}(|K^0\rangle - |\bar{K}^0\rangle) + \frac{f-\bar{f}}{2\sqrt{2}}(|K^0\rangle + |\bar{K}^0\rangle)$$

$$= \tfrac{1}{2}(f+\bar{f})|K_2\rangle + \tfrac{1}{2}(f-\bar{f})|K_1\rangle. \tag{7.90}$$

Since $f \neq \bar{f}$, it follows that some of the K_1-state has been regenerated (Fig. 7.23). This regeneration of short-lived K_1's in a long-lived K_2-beam was confirmed by experiment.

The K^0-regeneration phenomenon is simply a consequence of the concepts of superposition and quantization in quantum mechanics. The

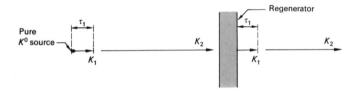

Figure 7.23 Regeneration of short-lived K_1-mesons when a pure K_2-beam traverses a regenerator.

behavior of an atomic beam in an inhomogeneous magnetic field—the classical Stern-Gerlach experiment—provides a fairly close analogy. Referring to Fig. 7.24, suppose we have an initially unpolarized atomic beam of spin-$\frac{1}{2}$ moving along the z-axis, which then traverses an inhomogeneous field H_y directed along the y-axis (representing the strong interaction in the K^0 case). The atoms are then quantized in two spin eigenstates, $\sigma_y = +\frac{1}{2}$ and $\sigma_y = -\frac{1}{2}$, particles in the first state being deflected upward and those in the second downward. (These are analogous to the K^0 and $\bar{K}^0$ S-eigenstates.) We select the part of the beam deflected upward, and pass it through an inhomogeneous field along the x-axis, H_x (the analogue of the weak interactions). The beam again splits into two components, 50% having $\sigma_x = -\frac{1}{2}$ and being deflected to the left, and 50% having $\sigma_x = +\frac{1}{2}$, deflected to the right. Note that *all* information about quantization along the y-axis has necessarily been lost. In turn, if we take the $\sigma_x = \frac{1}{2}$ component (corresponding to K_2), we may again pass it through a field along the y-axis, recover the components $\sigma_y = \pm\frac{1}{2}$, and lose all knowledge of σ_x (analogous to regeneration of K^0 and $\bar{K}^0$ in an absorber). Finally, passage through a field H_x yields as eigenvalues $\sigma_x = \pm\frac{1}{2}$ (corresponding to reappearance of K_1). To make the analogy more exact, i.e., account for the different decay and nuclear absorption probabilities in the K^0-problem, one could arrange to absorb components of the beam deflected in two of the four directions.

The important feature of the Stern-Gerlach atomic-beam experiment is that it is impossible simultaneously to quantize the spin components of the

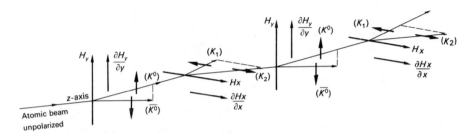

Figure 7.24 Analogy between deflection of atomic beam in two orthogonal planes, by means of inhomogeneous magnetic fields, and the K^0-decay and regeneration phenomena.

beam along two orthogonal axes. This may also be stated by saying that the spin operators σ_x and σ_y do not commute, or, of the three matrix operators σ_x, σ_y, and σ_z, only one may be diagonalized at a time (i.e., possess only diagonal elements, and have real, nonzero eigenvalues). Since an operator with real eigenvalues must commute with the Hamiltonian (energy) operator, the axis singled out in space is necessarily that defined by the magnetic field. In a similar fashion, the CP- and S-operators do not commute, so that one can have states which are eigenstates of CP or S, but not both.

7.14.3. CP-Violation in K^0-Decay

Following the discovery of parity violation in weak decay processes in 1957, it was for some time believed that the weak interactions were at least invariant under the CP-operation, and the ensuing description of the CP-eigenstates of neutral kaons, K_1 and K_2, has been given above. In 1964 however, an experiment by Christenson, Cronin, Fitch, and Turlay first demonstrated that the long-lived state we have called K_2 could also decay to $\pi^+\pi^-$ with a branching ratio of order 10^{-3}. The experimental arrangement of Christenson *et al.* is shown in Fig. 7.25. The nomenclature K_1 (for a state of $CP = +1$) and K_2 ($CP = -1$) has therefore been superseded by K_S (short-lived component) and K_L (long-lived component). The arguments quoted above on regeneration and mass difference, which follow from the superposition principle, remain essentially unchanged, although the formulae (7.78) need to be modified slightly if they are to describe K_S and K_L. The measure of the degree of CP-violation is usually quoted as the amplitude ratio

$$|\eta_{+-}| = \frac{\text{Ampl}(K_L \to \pi^+\pi^-)}{\text{Ampl}(K_S \to \pi^+\pi^-)} = (2.27 \pm 0.02) \times 10^{-3}. \qquad (7.91)$$

The state K_S consists principally of a $CP = +1$ amplitude, but with a little $CP = -1$, and K_L vice versa. Since both K_L and K_S decay to two pions, interference effects in the $\pi^+\pi^-$ signal are expected as a function of the time development of an initially pure K^0 beam. From (7.83) we expect the intensity to vary as

$$I_{2\pi}(t) = I_{2\pi}(0)[e^{-\Gamma_S t} + |\eta_{+-}|^2 e^{-\Gamma_L t} \\ + 2|\eta_{+-}|e^{-[(\Gamma_L + \Gamma_S)/2]t} \cos(\Delta mt + \phi_{+-})], \qquad (7.92)$$

where t measures proper time and ϕ_{+-} is an appropriate phase angle between the $K_S \to 2\pi$ and $K_L \to 2\pi$ amplitudes. Such interference effects are indeed observed (see Fig. 7.26). Similar CP-violating effects are found for the decay $K_L \to 2\pi^0$. The quantities η_{+-} and η_{00} depend on the amplitudes A_0 and A_2 for finding the final-state pions in a state of isospin $I = 0$ or 2 ($I = 1$ is

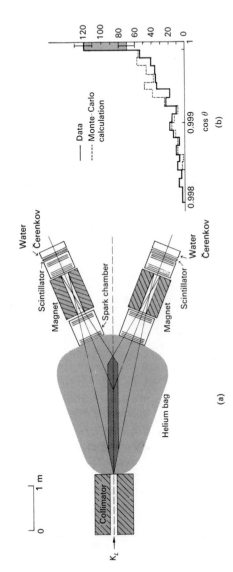

Figure 7.25 (a) Experimental arrangement of Christenson *et al.* (1964) demonstrating the decay $K_L \to \pi^+ \pi^-$. The K^0-beam is incident from the left and consists of K_L only, the K_S-component having died out. K_L-decays are observed in a helium bag, the charged products being analyzed by two spectrometers, consisting of bending magnets and spark chambers triggered by scintillators. The rare two-pion decays are distinguished from the common three-pion and leptonic decays on the basis of the invariant mass of the pair and the direction θ of their resultant momentum vector relative to the incident beam. (b) $\cos \theta$ distribution of events of $490 < M_{\pi\pi} < 510$ MeV. The distribution is that expected for three-body decays (dashed line), but with some 50 events (shaded) exactly collinear with the beam due to the $\pi^+ \pi^-$ decay mode.

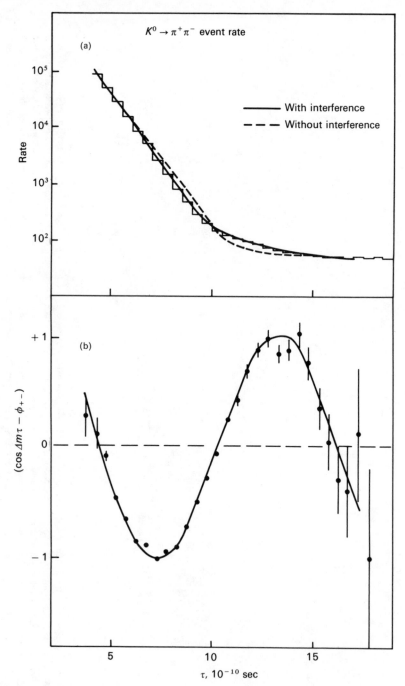

Figure 7.26 (a) Event rate for $\pi^+\pi^-$ decays from a neutral-kaon beam as a function of proper time, demonstrating that the best fit needs the existence of interference between K_L- and K_S-amplitudes. (b) The interference term extracted from the results in (a). From the fit one can obtain the $K_L - K_S$ mass difference Δm and the phase angle ϕ_{+-} between the two amplitudes. (After Geweniger *et al.* 1974.)

forbidden by Bose symmetry), and one obtains the expressions

$$\eta_{+-} = \varepsilon + \varepsilon',$$
$$\eta_{00} = \varepsilon - 2\varepsilon', \tag{7.93}$$

where

$$\varepsilon' = \frac{i}{\sqrt{2}} \frac{\operatorname{Im} A_2}{A_0} e^{i(\delta_2 - \delta_0)}.$$

In general, η_{+-} and η_{00} will be complex numbers and can be written as

$$\eta_{+-} = |\eta_{+-}|e^{i\phi_{+-}},$$
$$\eta_{00} = |\eta_{00}|e^{i\phi_{00}}, \tag{7.94}$$

where ϕ_{+-}, ϕ_{00} are related to ε' in (7.93).

If the $\Delta I = \frac{1}{2}$ rule holds, we expect $A_2 = 0$ and hence $|\eta_{+-}| = |\eta_{00}|$, $\phi_{+-} = \phi_{00}$. The experimental values are

$$\eta_{+-} = (2.274 \pm 0.022) \times 10^{-3}, \qquad \phi_{+-} = 44.6 \pm 1.2^0,$$
$$\eta_{00} = (2.29 \pm 0.04) \times 10^{-3}, \qquad \phi_{00} = 55 \pm 6^0, \tag{7.95}$$

and are thus compatible with the rule. The upper limit $\varepsilon'/\varepsilon \leq 0.02$ results from the above definitions and numbers.

CP-noninvariance is also demonstrated in the leptonic decay modes of K_L. These modes are

$$K_L \to e^+ v_e \pi^-, \tag{7.96}$$
$$K_L \to e^- \bar{v}_e \pi^+, \tag{7.97}$$

with similar ones for muons replacing electrons.

The decays (7.96) and (7.97) transform one into the other under the CP-operation, and if CP-invariance is violated, a small charge asymmetry is expected. The asymmetry is

$$\Delta = \frac{\operatorname{rate}(K_L \to e^+ v_e \pi^-) - \operatorname{rate}(K_L \to e^- \bar{v}_e \pi^+)}{\operatorname{rate}(K_L \to e^+ v_e \pi^-) + \operatorname{rate}(K_L \to e^- \bar{v}_e \pi^+)}$$
$$= (0.330 \pm 0.012) \times 10^{-2}. \tag{7.98}$$

The source of CP-violation has been a mystery ever since its discovery in 1964. In principle, the CP-violation can occur in the strong, electromagnetic, or weak interactions. If it occurred in the strong interactions, the decay of the neutral kaon would take place in two stages; first, a weak CP-conserving $\Delta S = 1$ transition from K_L to an intermediate state of $CP = -1$ (for example, 3π), followed by a strong CP-violating transition to 2π, at the level of 10^{-3} of the strong interaction. Since both CPT and P are conserved in strong interactions, both C- and T-violation is required at the 10^{-3} level. Direct experiments on C- and T-conservation just barely reach this level, so that a "millistrong" source cannot yet be ruled out—although it must be considered very improbable.

The search for neutral-current effects in atoms place stringent limits on C- and P-violation in electromagnetic interactions. Further, CP (and hence T) violation at the level required to explain the result from kaon decay would imply a neutron EDM of about $10^{-23} e$ cm, which is excluded by experiment. CP-violation to leading order in weak interactions at the 10^{-3} level ("milliweak") would imply T-violation at this level, again excluded experimentally from study of the decay of polarized neutrons and from Λ-decay.

One of the remaining possibilities is the famous "superweak" interaction, postulated by Wolfenstein (1964). This is a new, $\Delta S = 2$, CP-violating process which transforms $K_L \rightarrow K_S$. Because K_L and K_S are so closely equal in mass, the superweak coupling only has to be of order 10^{-10} of the normal weak coupling. In this case, the chance of observing CP-violation in any other system is essentially zero. All data available on CP-violation in K^0-decay are consistent with the superweak model. Among the predictions of this model are that $\varepsilon' = 0$ and thus $|\eta_{+-}| = |\eta_{00}|$, $\phi_{+-} = \phi_{00}$, and that if one neglects terms of order δ^2 compared with 1, one obtains the relation

$$\Delta = 2 \,\mathrm{Re}\, \eta = 2|\eta|\cos \phi = 0.322 \times 10^{-2} \qquad (7.99)$$

from (7.95), and in agreement with (7.98). There are, however, other models of CP-violation which are for practical purposes almost indistinguishable from the superweak theory. As discussed in Section 7.12, models involving six or more quark flavors and generalized Cabibbo mixing do include a possible finite CP-violating phase angle δ [Eq. (7.65)]. The value of the CP-violating parameter in K^0-decay in the six-flavor model is of order $\varepsilon \simeq s_1 s_3 \sin \delta$. A finite value for $\varepsilon'/\varepsilon \simeq 10^{-2}$ is also expected, which may be just about measurable in future experiments. The important point however is that, as distinct from the superweak theory, where CP-violation is detectable only in the narrow cul-de-sac of the neutral kaon system, such theories predict effects which are in principle detectable in other systems, for example in bottom meson decays, or in the form of a nonzero value for the electric dipole moment of the neutron. Unfortunately, the predicted EDM is extremely small ($10^{-30} e$ cm) and presumably unmeasurable.

On a much grander scale, CP-violation and baryon instability have been postulated in order to account for the mismatch between baryons and antibaryons and the ratio of baryons to photons in the universe (see Section 9.12).

A general remark may be made in conclusion. Until the advent of CP-violation, there was no unambiguous way of defining left-handed and right-handed coordinate systems, or of differentiating matter from antimatter, on a cosmic scale. Thus, aligned ^{60}Co nuclei emit negative electrons with forward-backward asymmetry and left-handed polarization (as we define left-handed). This is insufficient information to describe what we mean by a left-handed system, with the aid of light signals, to an intelligent being in a distant part of the universe, unless one can also uniquely define negative and positive charge or, equivalently, ^{60}Co and anti-^{60}Co. *CP*-violation, however, provides an unambiguous definition. Positive charge is now defined as that of the lepton associated with the more abundant leptonic decay mode of the long-lived K_L-particle, $(K_L \to \pi^- e^+ v)/(K_L \to \pi^+ e^- \bar{v}) > 1$. This lepton has the same (or opposite) charge as the local atomic nuclei of matter (or antimatter).

7.15. LEPTON FAMILIES, NEUTRINO MASSES, AND NEUTRINO OSCILLATIONS

Our discussion of leptons thus far has been largely in terms of the electron $(e^-, v_e; e^+, \bar{v}_e)$ and muon $(\mu^-, v_\mu; \mu^+, \bar{v}_\mu)$ leptons. In 1975, experiments by Perl and his coworkers at Stanford produced evidence for the existence of a new heavy lepton, named the tau and associated presumably with a new family $(\tau^-, v_\tau; \tau^+, \bar{v}_\tau)$. The discovery was based on the observation, in $e^+ e^-$ annihilation, of 24 $e\mu$ events consisting of a secondary high-energy electron and a high-energy muon and nothing else, i.e., attributable to

$$e^+ + e^- \to \tau^+ + \tau^-$$
$$\hookrightarrow \mu^- + \bar{v}_\mu + v_\tau \qquad (7.100)$$
$$\hookrightarrow e^+ + v_e + \bar{v}_\tau.$$

The mass of the τ-lepton is 1.78 GeV, so that the threshold for (7.100), 3.56 GeV, is rather close to that for charmed-meson production, 3.74 GeV, and the disentangling of this and other hadronic background effects constituted a superb piece of scientific detective work. There is not space here to describe the analysis of this and subsequent confirmatory experiments. Figure 7.27 shows just one example of an analysis which indicates spin-$\frac{1}{2}$ for the τ. The lepton momentum spectrum indicates a *V*-*A*-type coupling, as in the leptonic weak interactions: the relative leptonic branching ratios (i.e., to electrons and muons) are consistent with e/μ universality and the same (Fermi) coupling as describes μ-decay; and the semileptonic branching ratios are in agreement with the theory. For example, $\tau \to \pi v_\tau$ can be related to $\pi \to \mu v_\mu$, and $\tau \to \rho v_\tau$ to $e^+ e^- \to \rho$.

The mass value for the τ, and an upper limit to that for v_τ, are included in Table 7.5.

Our discussion of leptons thus far has been conducted almost exclusively under the assumption that neutrino masses are identically zero, so

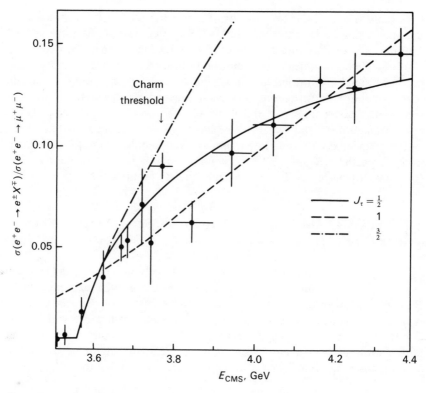

Figure 7.27   Analysis of τ-lepton production by DELCO collaboration at SLAC e^+e^- collider SPEAR (Bacino *et al.* 1978). Pair production of τ-leptons is recognized as events consisting of one electron track, and one other track due to a muon or hadron, of opposite sign of charge. The data clearly favor spin-$\frac{1}{2}$ for the τ.

that the distinction between neutrinos and antineutrinos—apart from their mode of production, for example $\pi^+ \to \mu^+ \nu_\mu$, $\pi^- \to \mu^- \bar{\nu}_\mu$—can be based on their helicity: neutrinos are left-handed ($H = -1$) and antineutrinos right-handed ($H = +1$), as in (7.18). This is a relativistic invariant description, just as for right and left circularly polarized light. If the neutrino mass were finite, this would not be true: one could always find a reference frame traveling faster than the neutrino, from which the sign of the helicity would be reversed.

TABLE 7.5   Lepton masses

	Electron family (e, ν_e)	Muon family (μ, ν_μ)	Tau family (τ, ν_τ)
Charged lepton (e, μ, τ)	0.511 MeV	105.66 MeV	1784 ± 4 MeV
Neutral lepton $(\nu_e, \nu_\mu, \nu_\tau)$	≤ 30 eV	< 0.5 MeV	< 70 MeV

The assumption that neutrinos are all massless and occur in several flavors is perhaps somewhat surprising, and is actively questioned at the present time. It has been necessary to invoke a quantum number (L_e, L_μ, L_τ) or internal degree of freedom to distinguish the three types, and such internal structure might be expected to manifest itself in a difference in masses, as it does for the charged leptons. There are indeed hints from astrophysics that massless neutrinos might lead to difficulties in our understanding of the expanding universe. Basically, endowing neutrinos with mass leads to dramatic differences in the gravitational potential energy, and would correct the apparent mismatch between the motional energy of the galaxies and their gravitational energy, the former being an order of magnitude larger than the latter. If one or more of the masses are in fact of the order of electron volts, as the astrophysicists suggest, then laboratory experiments may settle this matter in the near future, at least indirectly. There seems little hope of dramatically improving the present mass limits on v_μ and v_τ by direct measurement.

If lepton number is not absolutely conserved (and there are no compelling reasons for believing it should be) and neutrinos have finite masses, then mixing may occur between the different types of neutrino (v_e, v_μ, v_τ). This possibility was first envisaged many years ago (Maki *et al.* (1962), Pontecorvo (1968)). The weak-interaction eigenstates v_e, v_μ, v_τ are expressed as combinations of mass eigenstates, say v_1, v_2, v_3, which propagate with slightly different frequencies due to their mass differences. As a result, if one were to start off with a pure v_e-beam, for example, oscillations would occur and at subsequent times one would have admixtures of v_e with $v_\mu, v_{\tau'}$. In order to simplify the treatment we shall consider the case of two types of neutrino, say, v_e and v_μ. Each will be a linear combination of the two mass eigenstates, say v_1 and v_2, as given by the unitary transformation involving an arbitrary mixing angle θ:

$$\begin{pmatrix} v_\mu \\ v_e \end{pmatrix} = \begin{pmatrix} \cos\theta & \sin\theta \\ -\sin\theta & \cos\theta \end{pmatrix} \begin{pmatrix} v_1 \\ v_2 \end{pmatrix}, \tag{7.101}$$

so that the wavefunctions

$$v_\mu = v_1 \cos\theta + v_2 \sin\theta \quad \text{and} \quad v_e = -v_1 \sin\theta + v_2 \cos\theta$$

are orthonormal states. The states v_μ and v_e are those produced in a weak decay process, for example $\pi \to \mu + v_\mu$. However, propagation in space-time is determined by the characteristic frequencies of the mass eigenstates,

$$v_1(t) = v_1(0)e^{-iE_1 t},$$
$$v_2(t) = v_2(0)e^{-iE_2 t}, \tag{7.102}$$

where we set $\hbar = c = 1$. Since momentum is conserved we know that the states $v_1(t)$ and $v_2(t)$ must have the same momentum p. Then, if the mass $m_i \ll E_i$ $(i = 1, 2)$,

$$E_i = p + \frac{m_i^2}{2p}. \tag{7.103}$$

Suppose we start at $t = 0$ with muon-type neutrinos, so $v_\mu(0) = 1$ and $v_e(0) = 0$. Then from (7.101) we find

$$v_2(0) = v_\mu(0) \sin \theta,$$
$$v_1(0) = v_\mu(0) \cos \theta, \tag{7.104}$$

and

$$v_\mu(t) = \cos \theta \, v_1(t) + \sin \theta \, v_2(t).$$

Using (7.102) and (7.104), we then obtain

$$\frac{v_\mu(t)}{v_\mu(0)} = \cos^2 \theta \, e^{-iE_1 t} + \sin^2 \theta \, e^{-iE_2 t},$$

and the intensity is

$$\frac{I_\mu(t)}{I_\mu(0)} = \left| \frac{v_\mu(t)}{v_\mu(0)} \right|^2 = \cos^4 \theta + \sin^4 \theta + \sin^2 \theta \cos^2 \theta [e^{i(E_2 - E_1)t} + e^{-i(E_2 - E_1)t}]$$

$$= 1 - \sin^2 2\theta \sin^2 \left[\frac{(E_2 - E_1)t}{2} \right].$$

Writing $\Delta m^2 = m_2^2 - m_1^2$ and with the help of (7.103), we obtain the following form for the probability of finding v_μ or v_e after time t:

$$P(v_\mu \to v_\mu) = 1 - \sin^2 2\theta \sin^2 \left(\frac{1.27 \Delta m^2 L}{E} \right),$$
$$P(v_\mu \to v_e) = 1 - P(v_\mu \to v_\mu). \tag{7.105}$$

The numerical constant, 1.27, applies if Δm^2 is expressed in $(eV/c^2)^2$, while L, the distance from the source, is expressed in meters, and E, the beam energy, is in MeV. Equation (7.105) shows that the intensities of v_μ and v_e oscillate as a function of distance from the source. For example, at a reactor source (of antineutrinos), $E \simeq 1$ MeV, and for $\Delta m \simeq 1$ eV/c^2 the oscillation length will be a few meters.

The subject of neutrino oscillations acquired considerable impetus as a result of the so-called solar neutrino problem: The rate of detection of solar neutrinos in the reaction $v_e + {}^{37}\text{Cl} \to {}^{37}\text{Ar} + e^-$ is found to be about a factor 3 smaller than expected (Davis et al. (1968)). However, there are uncertainties in the solar model calculation and the discrepancy is hardly evidence for

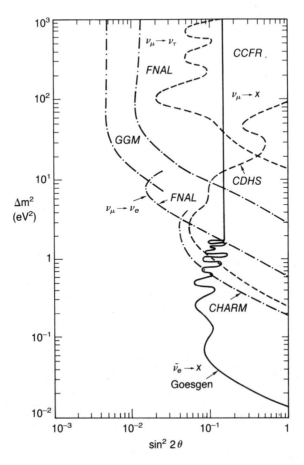

Figure 7.28 Limits on neutrino oscillations. The difference of mass squared of the mass eigenstates is plotted against $\sin^2 2\theta$, where θ is the mixing angle. The $\bar{\nu}_e$ result is from a reactor; the remainder are from accelerator experiments. The regions to the right of the curves are excluded at 90% confidence level. (After Boehm 1984.)

neutrino oscillations (although mixing ν_e with ν_μ and ν_τ would clearly give just such a factor). There is however at present (1984) no convincing evidence for oscillation phenomena from laboratory experiments using reactors or accelerators as the neutrino sources.

Figure 7.28 shows current limits on Δm^2 and $\sin^2 2\theta$ from reactor and accelerator experiments. At a reactor one measures the number of events of the type $\bar{\nu}_e + p \rightarrow n + e^+$ as a function of distance from the core and positron energy: It is sensitive to the *disappearance* of the $\bar{\nu}_e$ signal, i.e., to transformations $\bar{\nu}_e \rightarrow \bar{\nu}_X$, where X is another type of lepton. Because of the low energy ($\simeq 1$ MeV), these are not able to give the corresponding charged lepton in subsequent interactions. Accelerator neutrino experiments are at higher energy (1–50 GeV) and the beams are 99.5% pure v_μ; they search for the *appearance* of anomalously large numbers of interactions of the type $v_e + N \to e^- + \cdots$, $v_\tau + N \to \tau^- + \cdots$, arising from oscillations $v_\mu \to v_e, v_\tau$.

PROBLEMS

7.1. The ground states of $^{34}_{17}\text{Cl}$ and $^{34}_{16}\text{S}$ have $J^P = 0^+$ and belong to an $I = 1$ multiplet. The decay $^{34}_{17}\text{Cl} \to {}^{34}_{16}\text{S} + \beta^+ + v$ has a mean lifetime $\tau = 2.3$ s and a maximum positron energy of 4.5 MeV. The lifetime of the pion is 26 ns. Estimate the branching ratio for the decay $\pi^+ \to \pi^0 e^+ v$.

7.2. A "narrow-band" neutrino beam (see Fig. 7.12) is produced by bombarding a Be target with 400-GeV protons and forming a pencil secondary beam with a small spread in momentum centered at 200 GeV/c. This beam contains charged pions and kaons of one sign of charge and traverses an evacuated decay tunnel of length 300 m, where a fraction of them decay to muons and neutrinos. This is followed by an absorber of 200-m steel and 150-m rock. A cylindrical detector of radius 2 m is placed 400 m beyond the end of the decay tunnel, and aligned with the beam axis. (a) Find a relation between the laboratory energy of a neutrino and its angle relative to the beam axis, for neutrinos from pion and from kaon decays. (b) What are the maximum and minimum energies of neutrinos produced in the two cases? (c) Above what neutrino energy do all neutrinos from kaon decay pass through the detector? (d) If 10^{10} pions per burst enter the decay tunnel, how many neutrinos from pion decay traverse the detector? (The divergence of the pion beam may be neglected.) (e) If the detector has mass 100 tons, how many neutrinos from pion decay interact in it per burst, if the cross-section per nucleon at energy E is $0.6E \times 10^{-38}$ cm^2 with E in GeV? (f) Why is 350 m of absorber necessary? Refer to Problem 2.1 for estimates of straggling in range.

7.3. The neutron has a mean lifetime $\tau_n = 930$ s, and the muon $\tau_\mu = 2.2 \times 10^{-6}$ s. Show that the couplings involved in the two cases are of the same order of magnitude, when account is taken of the phase-space factors.

7.4. Obtain an estimate of the branching ratio $(\Sigma^- \to \Lambda e^- \bar{v})/(\Sigma^- \to n\pi^-)$, assuming that the matrix element for the electron decay is the same as that of the neutron, and that baryon recoil may be neglected. (See Table IV of Appendix for neutron and Σ-decay data.)

7.5. Strangeness-changing decays do not conserve isospin but appear to obey the $\Delta I = \frac{1}{2}$ rule. Use this rule to compute the ratios

(a) $K_S \to \pi^+\pi^-/K_S \to \pi^0\pi^0$,

(b) $\Xi^- \to \Lambda\pi^-/\Xi^0 \to \Lambda\pi^0$.

Compare your answers with the experimental values in Table IV. [*Hint*: The $\Delta I = \frac{1}{2}$ rule may be applied by postulating that a hypothetical particle of $I = \frac{1}{2}$—called a "spurion"—is added to the left-hand side in a weak decay process, and then treating the decay as an isospin-conserving reaction.]

7.6. What is the expected ratio $K_L \to 2\pi^0 / K_L \to \pi^+ \pi^-$ if the pions are in (a) an $I = 0$ state ($\Delta I = \frac{1}{2}$ rule), or (b) an $I = 2$ state ($\Delta I = \frac{3}{2}$ or $\frac{5}{2}$)?

7.7. Using the $\Delta I = \frac{1}{2}$ rule, find a relation between the amplitudes for

$$\Sigma^+ \to n\pi^+, \qquad \text{say } a_+,$$

$$\Sigma^- \to n\pi^-, \qquad \text{say } a_-,$$

$$\Sigma^+ \to p\pi^0, \qquad \text{say } a_0.$$

You will need to introduce $I = \frac{3}{2}$ and $I = \frac{1}{2}$ amplitudes. The triangle relation you should obtain is

$$a_+ + \sqrt{2} a_0 = a_-.$$

Experimentally, $|a_+| \simeq |a_0| \simeq |a_-|$, so that a right-angled triangle results.

7.8. Given that the width of the $\Delta(1234)$ πp resonance is 150 MeV, estimate the branching ratio for β-decay,

$$\frac{\Delta^{++} \to p e^+ \nu}{\Delta^{++} \to p\pi^+}.$$

7.9. How can one produce experimentally a beam of pure, monoenergetic K^0-particles? If a short pulse of such particles travels through a vacuum, compute the intensity of K^0 and $\bar{K}^0$ as a function of proper time, assuming the mass difference Δm equal to (a) $0.5/\tau_1$, (b) $2/\tau_1$. Display the phenomena on a graph.

7.10. An experiment in a gold mine in South Dakota has been carried out to detect solar neutrinos, using the reaction

$$\nu + {}^{37}\text{Cl} \to {}^{37}\text{Ar} + e^-.$$

The detector contained approximately 4×10^5 liters of tetrachloroethylene (C_2Cl_4). Estimate how many atoms of ${}^{37}\text{Ar}$ per day would be produced, making the following assumptions: (a) solar constant = 2 cal cm^{-2} min^{-1}; (b) 10% of thermonuclear energy of sun appears in neutrinos, of mean energy 1 MeV; (c) 1% of all neutrinos are energetic enough to induce the above reaction; (d) cross-section per ${}^{37}\text{Cl}$ nucleus for "active" neutrinos is 10^{-45} cm^2; (e) ${}^{37}\text{Cl}$ isotopic abundance is 25%; (f) density of C_2Cl_4 is 1.5 g ml^{-1}. Do you expect any difference between day rate and night rate? For a description of the experiment, see R. Davis, D. S. Harmer, and K. C. Hoffman, *Phys. Rev. Lett.* **20**, 1205 (1968).

7.11. Energetic neutrinos may produce single pions in the following reactions on protons and neutrons:

$$v + p \to \pi^+ + p + \mu^-, \tag{i}$$

$$v + n \to \begin{Bmatrix} \pi^0 + p \\ \pi^+ + n \end{Bmatrix} + \mu^-. \tag{ii}$$

Assume the process is dominated by the first pion-nucleon resonance $\Delta(1234)$, so that the π-nucleon system has $I = \frac{3}{2}$ only. As in weak decay processes of $\Delta S = 0$, the isospin of the hadronic state then changes by one unit ($\Delta I = 1$ rule). Show that this rule predicts a rate for (i) 3 times that of (ii). Also show that, on the contrary, for a $\Delta I = 2$ transition (for which there is no present evidence), the rate for (ii) would be 3 times that for (i).

7.12. Use the $\Delta I = \frac{1}{2}$ rule to demonstrate the following relations between the rate of three-pion decays of charged and neutral kaons:

$$\Gamma(K_L \to 3\pi^0) = \tfrac{3}{2}\Gamma(K_L \to \pi^+\pi^-\pi^0),$$

$$\Gamma(K^+ \to \pi^+\pi^+\pi^-) = 4\Gamma(K^+ \to \pi^+\pi^0\pi^0),$$

$$\Gamma(K_L \to \pi^+\pi^-\pi^0) = 2\Gamma(K^+ \to \pi^+\pi^0\pi^0).$$

Comment: These results depend on the (reasonable) assumption that the three pions are in a relative S-state. Then, any pair of pions must be in a symmetric isospin state, i.e., $I = 0$ and/or 2. The third pion ($I = 1$) must then be combined with the pair to yield a three-pion state of $I = 1$, $I_3 = 1$ for K^+, or $I = 1$, $I_3 = 0$ for K^0 (from the $\Delta I = \frac{1}{2}$ rule). It is necessary to write the three-pion wavefunction in a manner which is, as required for identical bosons, completely symmetric under pion label interchange. Thus, the $\pi^+\pi^+\pi^-$ state must be written

$$(+ + -) = \frac{1}{\sqrt{6}} (\pi_1^+\pi_2^+\pi_3^- + \pi_2^+\pi_1^+\pi_3^- + \pi_3^-\pi_2^+\pi_1^+ + \pi_3^-\pi_1^+\pi_2^+$$

$$+ \pi_2^+\pi_3^-\pi_1^+ + \pi_1^+\pi_3^-\pi_2^+).$$

Each term in this expression will be the product of an isospin state of $I = 0$ or 2 for the first two pions, and one of $I = 1$ for the third pion, with appropriate Clebsch-Gordan coefficients.

7.13. Show why observation of the process $\bar{v}_\mu + e^- \to e^- + \bar{v}_\mu$ constitutes unique evidence for neutral currents, whereas $\bar{v}_e + e^- \to e^- + \bar{v}_e$ does not. Prove that the maximum angle of emission of the recoil electron relative to the beam is $\sqrt{2m/E}$ where m, E are the electron mass and energy.

7.14. The muon lifetime is 2.197 μs. Calculate that of the τ-lepton, given that the branching ratio for $\tau^+ \to e^+ v_e \bar{v}_\tau$ is 18%, and that

$m_\tau = 1784$ MeV, $m_\mu = 105.7$ MeV. State any assumptions you need to make in the calculation and compare with the measured lifetime $\tau_\tau = (2.86 \pm 0.16) \times 10^{-13}$ s.

BIBLIOGRAPHY

Commins, E. D., *Weak Interactions*, McGraw-Hill, New York, 1973.

Ellis, J. *et al.*, "Physics of intermediate vector bosons", *Ann. Rev. Nucl. Part. Science* **32**, 443 (1982).

Feinberg, G., and L. Lederman, "Physics of muons and moun neutrinos", *Ann. Rev. Nucl. Science* **13**, 431 (1963).

Fritschi M., *et al. Phys. Lett.* **B** 173, 485 (1986).

Gasiorowicz, S., *Elementary Particle Physics*, John Wiley, New York, 1966.

Hung, P. Q., and C. Quigg, "Intermediate bosons; weak interaction carriers", *Science* **210**, 1205 (1980).

Kleinknecht, K., "CP violation and K^0 decays", *Ann. Rev. Nucl. Science* **26**, 1 (1976).

Konopinski, E. J., "Experimental clarification of the laws of β-radioactivity", *Ann. Rev. Nucl. Science* **9**, 99 (1959).

Lederman, L., "Neutrino physics", in Burhop (ed.), *Pure and Applied Physics*, Academic Press, New York, 1967, Vol. 25-II.

Lee, T. D., and C. S. Wu, "Weak interactions", *Ann. Rev. Nucl. Science* **15**, 381 (1965).

Okun, L. B., *Weak Interactions of Elementary Particles*, Pergamon, London, 1965.

Perl, M. L., and W. T. Kirk, "Heavy leptons", *Sci. Am.* **238**, 50 (Mar. 1978).

Perl, M. L., "The tau lepton", *Ann. Rev. Nucl. Part. Science* **30**, 299 (1980).

Reines, F., "Neutrino interactions", *Ann. Rev. Nucl. Science* **10**, 1 (1960).

Rubbia, C., "Experimental observation of the intermediate vector bosons W^+, W^-, and Z^0", *Rev. Mod. Phys.* **57**, 699 (1985).

CHAPTER 8

Quark-Quark Interactions: The Parton Model and QCD

A dynamical (rather than static) understanding of quark substructure had its origins in 1968, when new evidence for quarks started to come from experiments on deep-inelastic lepton-nucleon scattering. These showed firstly that the complicated process of leptoproduction of many hadrons in such a collision could be simply interpreted as (quasi) elastic scattering of the lepton by a pointlike, or nearly pointlike, constituent or *parton*, later to be identified with the quark; secondly that the analysis of precise and detailed scattering experiments could give valuable information on the quark-quark interactions. This evidence was strongly reinforced from the results of studies of e^+e^- annihilation to hadrons at high energy and of the production of lepton pairs in hadron-hadron collisions.

8.1. EVIDENCE FOR PARTONS: NEUTRINO-NUCLEON AND e^+e^- ANNIHILATION CROSS-SECTIONS AT HIGH ENERGIES

One of the most dramatic demonstrations of the constituent nature of hadrons is provided by the total cross-section, as a function of energy, for neutrino-nucleon scattering. Figure 8.1 shows an example of the interaction of a 200-GeV neutrino with a proton or neutron in the liquid neon-hydrogen mixture in a large bubble chamber,

$$\nu_\mu + N \rightarrow \mu^- + \text{hadrons}, \tag{8.1}$$

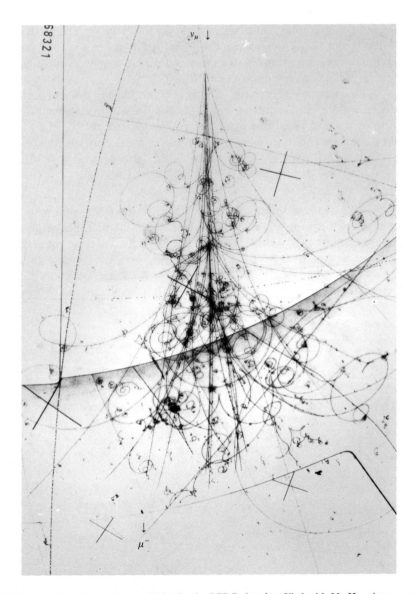

Figure 8.1 Neutrino-nucleon collision in the BEBC chamber filled with Ne-H$_2$ mixture, and exposed to the 200-GeV narrow-band neutrino beam from the CERN SPS (see Fig. 7.12). An incident muon-neutrino (ν_μ) transforms to a 100-GeV negative muon, appearing as a very straight track at an angle to the general "jet" of charged and neutral hadrons. The muon is identified by its penetration through iron and recorded in external proportional chambers (see Fig. 2.14). The hadrons produced carry off the remaining (100 GeV) energy. γ-rays from neutral-pion decay convert to pairs in the liquid, which point back to the main vertex. The development of electromagnetic cascades and secondary interactions of the hadrons are visible.

where about 10 secondary hadrons, in addition to the muon, are produced. The event looks very complicated, but if we just measure the total cross-section σ as a function of energy, the result is a simple one—an almost linear rise of σ with the neutrino energy (see Fig. 8.2).

    This is exactly the result we expect if we replace the complicated process of hadron production, as in Fig. 8.1, by the *elastic* scattering of the neutrino by a *single pointlike particle*, which can depend only on the Fermi coupling constant G and the phase space. Indeed, if the neutrino energy E is large compared with any of the masses involved, we expect, as in (7.14)

$$\sigma \simeq G^2 p^2 \simeq G^2 E, \tag{8.2}$$

where p is the momentum of the two particles in the CMS, and a simple calculation gives $p^2 = mE/2$, where m is the mass of the pointlike target. So the linear dependence of σ on E just tells us that the two particles undergo a contact interaction with a cross-section rising like phase space. In this discussion, we have ignored the effect on the cross-section associated with the $W^{\pm}$ propagator term, $(1 + q^2/M_W^2)^{-1}$, since it is negligible in the energy range shown.

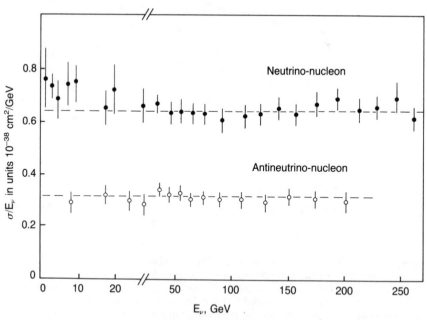

Figure 8.2 Total neutrino and antineutrino cross-sections on nucleons as a function of energy, from experiments at CERN, Fermilab, and Serpukhov. The ratio of cross-section per nucleon to incident energy is practically constant over two orders of magnitude and is a direct demonstration of pointlike constituents (partons) inside the nucleon. The gentle dependence of σ/E on energy is predicted in the field theory of interacting quarks (QCD).

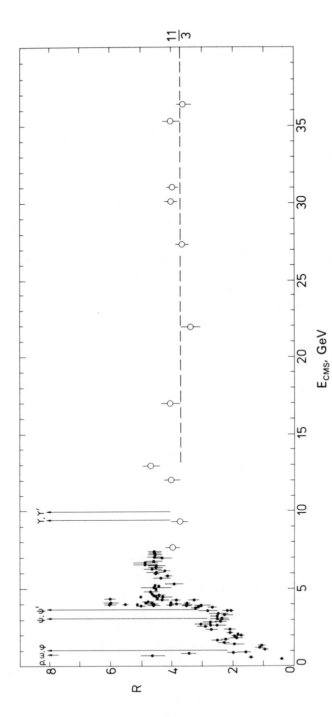

Figure 8.3 The ratio R of the cross-section for $e^+ e^- \rightarrow$ hadrons, divided by that for $e^+ e^- \rightarrow \mu^+ \mu^-$. The fact that R is constant above 10-GeV CMS energy is a proof of the pointlike nature of hadron constituents. The predicted value of R, assuming that the primary process is formation of a quark-antiquark pair, is $\frac{11}{3}$ if pairs of u, d, s, c, b quarks are excited and they have three color degrees of freedom. The data come from many storage-ring experiments. At high energy (> 10 GeV CMS) it is from the PETRA ring at DESY, Hamburg.

Figure 8.3 shows the results of measurements, carried out at high-energy e^+e^- colliders, of the cross-section ratio

$$R = \sigma(e^+e^- \rightarrow \text{hadrons})/\sigma(e^+e^- \rightarrow \mu^+\mu^-) \qquad (8.3)$$

as a function of the CMS energy, E. For $E > 10\,\text{GeV}$, clear of the ρ, ω, ϕ, ψ, and Υ resonances, R is practically constant, arguing that the process $e^+e^- \rightarrow \text{hadrons}$ is pointlike, just as is $e^+e^- \rightarrow \mu^+\mu^-$.

Before discussing these results in detail, and trying to establish the nature of the pointlike or parton constituents of the hadrons that are responsible for it, we shall first consider deep-inelastic electron-nucleon scattering, which provided the first evidence for partons.

8.2. DEEP-INELASTIC ELECTRON-NUCLEON SCATTERING

As indicated in (6.27) at high-momentum transfers, the elastic form factor is very small, and inelastic scattering of the incident electron is much more probable than elastic scattering. In a general inelastic scattering process, there is an extra variable because the space and time components $(\mathbf{q}, iv)$ of the momentum transfer q are no longer related by $q^2 = 2Mv$, as in (6.17b). If we denote the 3-momentum, energy, and invariant mass of the final hadron state by $\mathbf{p}^*$, E^*, and W, we obtain (see Fig. 8.4)

$$q^2 = (\mathbf{p}^* - 0)^2 - (E^* - M)^2,$$

$$v = E^* - M,$$

$$W^2 = E^{*2} - \mathbf{p}^{*2},$$

so

$$q^2 = 2Mv - W^2 + M^2. \qquad (8.4)$$

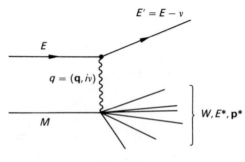

Figure 8.4

$W = M$ corresponds to elastic scattering, with $q^2 = 2Mv$ as before. In terms of q^2, v the cross-section can be written down in analogy with (6.24) as

$$\frac{d^2\sigma}{dq^2\,dv} = \frac{4\pi\alpha^2}{q^4}\frac{E'}{EM}\left[W_2(q^2, v)\cos^2\frac{\theta}{2} + 2W_1(q^2, v)\sin^2\frac{\theta}{2}\right], \quad (8.5)$$

where E and E' are the incident and emergent electron energies ($E, E' \gg mc^2$). W_1, W_2 are arbitrary structure functions corresponding to the two possible polarization states, transverse and longitudinal, of the mediating photon. The validity of this formula has been established from the linearity of the Rosenbluth plot, just as in the case of elastic scattering. From such a plot, the ratio W_1/W_2 of magnetic to electric scattering can be determined.

The variables q^2, $2Mv$ are plotted in Fig. 8.5(a), where the shaded area represents the physical region accessible by inelastic scattering, and we have defined the ratio

$$x = \frac{q^2}{2Mv} \quad (1 > x > 0). \quad (8.6)$$

Lines of constant W are indicated at $45°$ to the axes; those of constant x diverge from the origin with a slope equal to x. $q^2 < 2Mv$, $x < 1$ corresponds to the inelastic-scattering region, while the line $x = 1$ defines the kinematics for elastic scattering.

It is instructive to compare the dependence of $d^2\sigma/dq^2\,dv$ on the energy transfer $v = E - E'$ in different regions of q^2, for electron-nucleus and electron-nucleon scattering. Figure 8.5(b) shows typical results for scattering of energetic electrons by nuclei. At low q^2, one observes a strong elastic-scattering peak, at a unique value

$$E' = E - v = E - \frac{q^2}{2M_{nucleus}}. \quad (8.7)$$

At larger v values (smaller E'), a broad peak is observed with a maximum in the region

$$v \simeq \frac{q^2}{2M_{nucleon}}, \quad (8.8)$$

corresponding to *quasi-elastic scattering off individual nucleons*. If the nucleons were free, a sharp spike at $v = q^2/2M$ would be seen. The target nucleons are, however, bound in a potential well, of radius R, and thus have a Fermi momentum

$$p_F \sim \frac{\hbar}{R} = 200 \text{ MeV}/c.$$

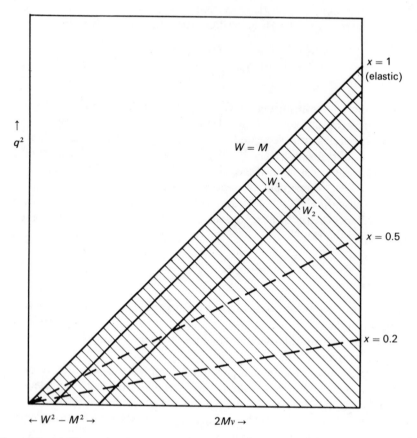

Figure 8.5 (a) Kinematic relations in inelastic lepton-nucleon scattering. The 4-momentum transfer squared, q^2, is plotted against $2M\nu$, where M is the nucleon mass and ν is the energy transfer measured in the rest frame of the target. The invariant mass of the final state of hadrons is given by $W^2 = M^2 + 2M\nu - q^2$, and the variable $x = q^2/2M\nu$. Elastic scattering corresponds to $x = 1$, $W = M$. Contours of fixed W are 45° lines; those of fixed x are radial lines from the origin with slope x.

The nuclear binding energy of 10 MeV or so is negligible in comparison with ν and hardly affects the elastic kinematics; but the Fermi momentum smears the elastic peak, with a symmetrical spread of order

$$\frac{\Delta\nu}{\nu} = \pm\frac{p_F}{M} \simeq 10\%. \tag{8.9}$$

(The proof is left as an exercise.) Put another way, the elastic kinematic relation $q^2 = 2M\nu$ defines the energy transfer ν in the rest frame of the struck nucleon. In the laboratory frame where the bound nucleon is moving, the

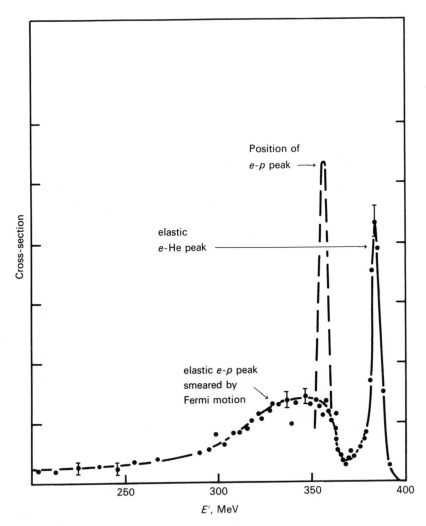

Figure 8.5 (b) Scattering of 400-MeV electrons in helium at 45°, showing a sharp elastic helium peak and a smeared peak due to elastic scattering from individual nucleons. (After Hofstadter 1956.)

measured energy transfer will be greater or less, depending on the direction of its velocity.

To summarize, in electron-nucleus scattering, as q^2 increases from low values, the coherent nuclear scattering dies away and the incoherent elastic scattering from individual nucleons becomes progressively more important. Increasing q^2 is equivalent to decreasing the wavelength or resolution of our "probe", and at sufficiently large q^2 we begin to "see" the constituents rather than the entire nucleus.

Turning now to electron-proton scattering, we observe the *same phenomena* as in the nuclear case (only at much higher values of q^2). For $q^2 < 1\,\mathrm{GeV}^2$, with v increasing, we observe first a strong elastic peak at $v = q^2/2M$, or $x = q^2/2Mv = 1$, followed by successive peaks due to the broad nucleon resonances $\Delta(1232)$, $N(1450)$, $\Delta(1688)$,... at $x = q^2/(q^2 + W^2 - M^2)$. In terms of the parameter x, as q^2 increases, these peaks move in toward $x = 1$ and are suppressed by the form factors, but the continuum excitation at smaller x-values remains large. In analogy with the nucleon case, we expect this to be due to *quasi-elastic scattering by nucleus constituents* which are more pointlike than the nucleon itself—and hence are not suppressed at large q^2 (see Figs. 8.6 and 8.7).

If one were to identify these constituents with pointlike quarks, of which the proton possesses three (u, u, d), one might expect each to have an effective mass $m = M/3$ and therefore $x = q^2/2Mv = (q^2/2mv)(m/M) = \frac{1}{3}$,

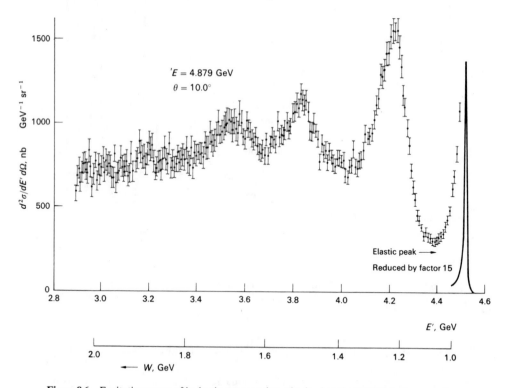

Figure 8.6 Excitation curve of inelastic *ep* scattering, obtained at the DESY electron accelerator (Bartel *et al.* 1968). E and E' are the energies of the incident and the scattered electron, and W is the mass of the recoiling hadronic state. The peaks due to the pion-nucleon resonances of masses 1.24, 1.51, and 1.69 GeV are clearly visible.

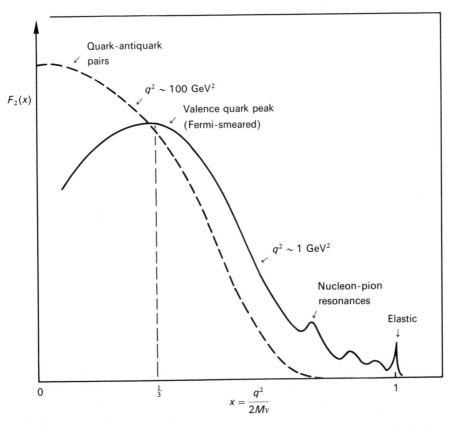

Figure 8.7 Schematic diagram of the dependence of the cross-section, as measured by the structure function $F_2(x, q^2)$, on q^2 as a function of $x = q^2/2Mv$. At low q^2, the elastic peak and those due to resonances, as in Fig. 8.6 appear at large x, followed by continuum excitation (lepton-quark elastic scattering) at lower x. At high q^2, the elastic and resonance peaks have been suppressed by form-factor effects and the whole distribution shifts to lower x.

that is, a sharp elastic peak. No such peak is observed. Again, the quarks are confined within a volume of dimensions R, the nucleon radius, and so have a Fermi momentum $p_F \simeq \hbar/R \simeq 250$ MeV/c. Thus, the spread of the elastic peak would be $\Delta x/x \simeq p_F/mc \simeq 1$, so it is essentially invisible.

8.3. SCALE INVARIANCE AND PARTONS

We now pursue more quantitatively the idea that inelastic lepton-nucleon scattering can be interpreted in terms of the elastic scattering of the lepton by pointlike parton constituents and how these can be identified with the quarks (if they can be). First we rewrite (8.5) in terms of related functions and variables:

$$F_2(q^2, v) = \frac{v W_2(q^2, v)}{M},$$

$$F_1(q^2, v) = W_1(q^2, v),$$

$$y = v/E,$$

$$E'/E = 1 - y,$$

$$q^2 = 2MExy.$$

Then (8.5) becomes

$$\frac{d^2\sigma}{dq^2\,dv} = \frac{4\pi\alpha^2}{q^4}\frac{E'}{Ev}\left[F_2(q^2, v)\cos^2\frac{\theta}{2} + \frac{2v}{M}F_1(q^2, v)\sin^2\frac{\theta}{2}\right], \qquad (8.10)$$

or, with $\cos^2(\theta/2) = 1 - q^2/(4EE') \simeq 1$ and $dv/v = dx/x$,

$$\frac{d^2\sigma}{dq^2\,dx} = \frac{4\pi\alpha^2}{q^4}\left[(1 - y)\frac{F_2(x, q^2)}{x} + \frac{y^2}{2}\frac{2xF_1(x, q^2)}{x}\right]. \qquad (8.11)$$

If the lepton-parton scattering is pointlike, then F_1 and F_2 cannot depend on q^2 and are purely functions of x. This assumption is known as the *Bjorken scaling hypothesis* (Bjorken 1967). In simple terms, this states that if, in the limit $q^2 \to \infty$, $v \to \infty$, the function $F(q^2, v)$ remains finite, it can depend only on the dimensionless and finite ratio of these two quantities, that is, on $x = q^2/2Mv$. Since x is dimensionless, there is no scale of mass or length; hence the term scale invariance. Figure 8.8 shows the values of $F_2(x, q^2)$ at $x = 0.25$ from SLAC electron-scattering experiments, where there is essentially no q^2-dependence. Even at larger or smaller values of x, the q^2-dependence is weak (see Fig. 8.28). For example, for $x = 0.5$, F_2 falls by 50% as q^2 increases from 1 to 25 GeV^2; this is to be contrasted with the square of the elastic form factor (Fig. 6.6), which falls by a factor 10^6 over the same interval. The scale invariance proposed by Bjorken in the limit $q^2 \to \infty$ seems then to hold approximately in the region of q^2 of a few times M^2.

The interpretation of scale invariance is given in physical terms by the *parton model* of Feynman (1969). Imagine a reference frame in which the target proton has very large 3-momentum—the so-called infinite-momentum frame (Fig. 8.9). The proton mass can be neglected, so it has 4-momentum $P = (p, 0, 0, ip)$ and is visualized as consisting of a parallel stream of partons, each with 4-momentum xP $(0 < x < 1)$. Again, if P is large, masses and transverse momentum components of the partons can be neglected. Suppose now that one parton of mass m is scattered elastically by absorbing the

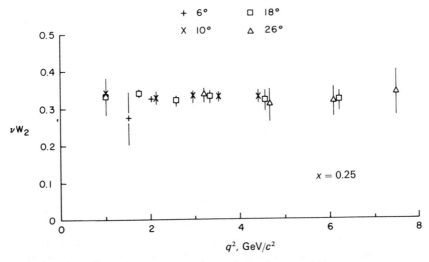

Figure 8.8 νW_2 (or F_2) as a function of q^2 at $x = 0.25$. For this choice of x, there is practically no q^2-dependence, that is, exact "scaling." (After Friedman and Kendall (1972).)

current 4-momentum q from the scattered lepton. Then

$$(xP + q)^2 = -m^2 \simeq 0,$$
$$x^2 P^2 + q^2 + 2xP \cdot q \simeq 0.$$

(8.12)

If $|x^2 P^2| = x^2 M^2 \ll q^2$, we obtain

$$x = \frac{-q^2}{2P \cdot q} = \frac{q^2}{2M\nu},$$

(8.13)

where the invariant scalar product $P \cdot q$ has been evaluated in the laboratory system in which the energy transfer is ν and the nucleon is at rest. Thus, x in

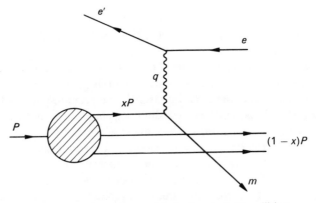

Figure 8.9 The parton model of a deep-inelastic collision.

(8.6) represents the fractional 3-momentum of the parton in the infinite-momentum frame. It is as if we had a hypothetical parton of mass m, *stationary* in the lab system, with the elastic relation $q^2 = 2mv$, so that, provided always that $q^2 \gg M^2$,

$$x = \frac{q^2}{2Mv} = \frac{m}{M}. \tag{8.14}$$

Then x is also the fractional mass of the nucleon carried by such a hypothetical free parton initially at rest in the lab system. The cross-section, or equivalently $F(x)$ in (8.11) therefore gives, in a sense, a measure of the effective mass distribution of parton constituents.

Of course, we do not observe partons in the final state, but hadrons. Somehow the scattered and unscattered partons have to recombine to form hadrons, and at present no one knows the mechanism involved in this final-state parton interaction. The basic assumption is that the collision occurs in two independent stages. First, one parton is scattered, the collision time being that required to define the energy transfer, i.e., $t_1 \simeq \hbar/v$. Over a much longer time, the partons recombine to form the final hadronic state, of mass W. Clearly the proper lifetime of this state must be $t_2 > \hbar/W$, or, transformed into the lab frame, $t_2 > \gamma\hbar/W = v\hbar/W^2$, so that since $W^2 \simeq 2Mv$, we have finally $t_2 \simeq \hbar/M \gg t_1$ for $v \gg M$. The recombination therefore takes place over a long time scale and can be treated separately from the initial collision. It is also surmised that the cross-section will depend first and foremost on the dynamics of the initial stage, and only weakly or not at all on the complexities of the final-state interaction. This turns out to be a good guess except in the low-energy region ($v \simeq M$), where there are significant resonance effects. Even then, when integrated over all W-values at fixed E, as in the total neutrino cross-sections in Fig. 8.2, the combined effect of several resonances averages out to give a linear cross-section, that is, the pure partonlike behavior, even in the low-energy domain.

8.4. NEUTRINO-NUCLEON INELASTIC SCATTERING

The expression for the neutrino-nucleon inelastic scattering cross-section is obtained by replacing the electromagnetic coupling factor $4\pi\alpha^2/q^4$ in (8.11) with $G^2/2\pi$, where G is the Fermi constant, and including a third structure function $F_3(x)$. In this treatment we shall neglect effects due to strangeness- and charm-changing transitions, which are suppressed by the Cabibbo factor $\tan^2 \theta_C$ relative to the $\Delta S = \Delta C = 0$ reactions. There are then three independent helicity states $(-1, +1, 0)$ for the mediating boson $W^\pm$, since in the weak interactions there is no conservation of parity which compels helicity -1 and $+1$ states to occur with equal probability as a coherent superposition, as in the electromagnetic case. Thus, *three* structure functions are needed to describe vp scattering, and three each for $\bar{v}p$, vn, and $\bar{v}n$ scattering. Since many experiments are carried out on nuclear targets containing approximately equal numbers of protons and neutrons, we consider here neutrino-nucleon (i.e., proton-neutron average) and antineutrino-nucleon structure functions. These are equal via the principle of charge symmetry:

$$\left.\begin{array}{l} F_i^{vn} = F_i^{\bar{v}p} \\ F_i^{vp} = F_i^{\bar{v}n} \end{array}\right\} \quad i = 1, 2, 3,$$

so that

$$F_i^{vN} = \tfrac{1}{2}(F_i^{vp} + F_i^{vn}) = \tfrac{1}{2}(F_i^{\bar{v}p} + F_i^{\bar{v}n}) = F_i^{\bar{v}N}, \tag{8.15}$$

except that the V-A interference term changes sign for antineutrinos: $F_3^{\bar{v}N} = -F_3^{vN}$. We then obtain, for exact scaling,

$$\frac{d^2\sigma^{vN,\bar{v}N}}{dx\,dq^2} = \frac{G^2}{2\pi}\left[(1-y)\frac{F_2^{vN}(x)}{x} + \frac{y^2}{2}\frac{2xF_1^{vN}(x)}{x} \pm y\left(1 - \frac{y}{2}\right)\frac{xF_3^{vN}(x)}{x}\right], \tag{8.16}$$

or, with $dq^2 = 2MEx\,dy$,

$$\frac{d^2\sigma^{vN,\bar{v}N}}{dx\,dy} = \frac{G^2ME}{\pi}\left[(1-y)F_2^{vN}(x) + \frac{y^2}{2}2xF_1^{vN}(x) \pm y\left(1 - \frac{y}{2}\right)xF_3^{vN}(x)\right]. \tag{8.17}$$

Upon integration over x and y from 0 to 1, Eq. (8.17) gives total cross-sections σ^{vN}, $\sigma^{\bar{v}N}$ proportional to E, as noted previously.

8.5. LEPTON-QUARK SCATTERING

The nature of the partons can be established by relating the data on electromagnetic and weak structure functions of nucleons, (F_1^{eN}, F_2^{eN} and F_1^{vN}, F_2^{vN}, F_3^{vN} respectively) to the quantum numbers of the partons.

8.5.1. Parton Spin

First we compare the expression (8.5) with the Dirac cross-section for scattering from spin-$\tfrac{1}{2}$ pointlike particles of charge ze and mass m. We have from (6.15 and 6.24)

$$\left(\frac{d\sigma}{dq^2}\right)_{\text{Dirac}} = \frac{4\pi\alpha^2 z^2}{q^4}\left(\frac{E'}{E}\right)^2\left(\cos^2\frac{\theta}{2} + \frac{q^2}{2m^2}\sin^2\frac{\theta}{2}\right)$$

and

$$\left(\frac{d^2\sigma}{dq^2\,dx}\right)_{\text{inelastic}} = \frac{4\pi\alpha^2}{q^4}\frac{E'}{E}\left(F_2(x)\cos^2\frac{\theta}{2} + \frac{q^2}{2M^2x^2}2xF_1(x)\sin^2\frac{\theta}{2}\right)\frac{1}{x}.$$

Comparing coefficients of $\cos^2(\theta/2)$ and $\sin^2(\theta/2)$ and with the identification $m^2 = M^2 x^2$ as in (8.14), we immediately obtain

$$\frac{2xF_1(x)}{F_2(x)} = 1. \qquad (8.18)$$

This is the relation expected between F_1 and F_2 (i.e., magnetic and electric scattering) in the scaling region (q^2 large), if the scattering is to be interpreted as due to *pointlike constituents of spin-$\frac{1}{2}$* and normal (Dirac) magnetic moments (i.e., $\mu = ze\hbar/2mc$). Equation (8.18) is called the *Callan-Gross relation* (1968). Figure 8.10 shows the observed ratios $2xF_1/F_2$ measured in

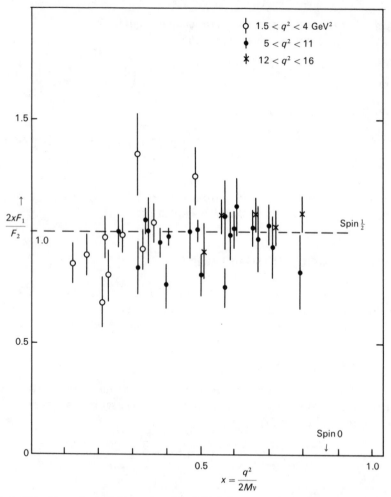

Figure 8.10 The ratio $2xF_1/F_2$ measured in SLAC electron-nucleon scattering experiments. For spin-$\frac{1}{2}$ partons, with $g = 2$, a ratio of unity is expected in the limit of large q^2—the Callan-Gross relation. (Data compiled from published SLAC data.)

electron-scattering experiments at SLAC for low values of q^2. They indicate spin-$\frac{1}{2}$ for the partons. Zero-spin partons $(2xF_1/F_2 = 0)$ are obviously excluded.

8.5.2. Parton Charges

Let us now compare the neutrino and antineutrino cross-sections in (8.17), for deep-inelastic scattering from nucleon targets with those expected for elastic scattering from pointlike targets of spin-$\frac{1}{2}$. In the V-A theory of weak interactions, neutrino-electron elastic scattering in reactions involving the so-called charged currents (i.e., via virtual $W^\pm$-boson exchange) is described by the following formulae (see Appendix F):

$$\frac{d\sigma^{\nu e}}{dy} = \frac{d\sigma^{\bar{\nu}\bar{e}}}{dy} = \frac{2G^2mE}{\pi}, \tag{8.19}$$

$$\frac{d\sigma^{\bar{\nu}e}}{dy} = \frac{d\sigma^{\nu\bar{e}}}{dy} = \frac{2G^2mE}{\pi}(1-y)^2, \tag{8.20}$$

where y is the fractional energy carried off by the recoiling electron target, and it is assumed that $E \gg m$, the electron mass. The cross-section (8.19) corresponds to isotropic scattering in the CMS, for a particle-particle or antiparticle-antiparticle scattering, with $J_z = 0$ for the net spin component along the collision axis z (see Fig. 8.11(a)). In (8.20), however, as shown in Fig. 8.11(b), $J_z = \pm 1$ and 180° scattering is impossible for a pointlike (S-wave) interaction, by angular-momentum conservation. Thus, either neutrino-positron or antineutrino-electron scattering contains a term $(1 - y)^2$, leading to a factor of $\frac{1}{3}$ when integrated over y. In simple terms, the reactions (8.20) proceed through a $J = 1$ state, but angular-momentum conservation allows only one of the 3 ($= 2J + 1$) substates, so the cross-section is reduced relative to that for (8.19).

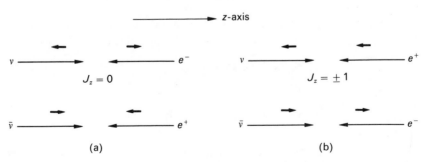

(a)   (b)

Figure 8.11

If we replace electrons by spin-$\frac{1}{2}$ partons and positrons by antipartons, the same formulae will apply, so that for neutrinos on a combination of partons Q and antipartons $\bar{Q}$ of mass $m = xM$ we get

$$\frac{d^2\sigma}{dy\,dx} = \frac{2G^2ME}{\pi}[xQ(x) + x\bar{Q}(x)(1 - y)^2], \qquad (8.21)$$

which is to be compared with (8.17) rewritten assuming $2xF_1 = F_2$:

$$\frac{d^2\sigma^{\nu N}}{dy\,dx} = \frac{G^2ME}{2\pi}\{[F_2(x) + xF_3(x)] + [F_2(x) - xF_3(x)](1 - y)^2\}. \quad (8.22)$$

Thus,

$$F_2^{\nu N}(x) = 2x[Q(x) + \bar{Q}(x)],$$
$$xF_3^{\nu N}(x) = 2x[Q(x) - \bar{Q}(x)]. \qquad (8.23)$$

The structure functions $F_2^{\nu N}$ and $F_3^{\nu N}$ describing neutrino scattering on nucleons are thus proportional to the sums and differences of the parton and antiparton densities at x, weighted by the mass fraction x, i.e., the fractional nucleon mass (or, strictly, 3-momentum) carried by partons (antipartons) at x.

If we identify the partons with the quarks u and d, the elementary charged-current reactions for muon-type neutrinos will be

$$\nu_\mu d \rightarrow \mu^- u, \qquad (8.24a)$$
$$\nu_\mu \bar{u} \rightarrow \mu^- \bar{d}, \qquad (8.24b)$$
$$\bar{\nu}_\mu u \rightarrow \mu^+ d, \qquad (8.24c)$$
$$\bar{\nu}_\mu \bar{d} \rightarrow \mu^+ \bar{u}. \qquad (8.24d)$$

If we let $u(x)$, $d(x)$, $\bar{u}(x)$, $\bar{d}(x)$ stand for the quark densities in the proton, then

$$F_2^{\nu p}(x) = 2x[d(x) + \bar{u}(x)]. \qquad (8.25)$$

From isospin invariance, we know that the quark populations in the neutron are given by $u(x)^n = d(x)$, $d(x)^n = u(x)$ and thus,

$$F_2^{\nu n}(x) = 2x[u(x) + \bar{d}(x)], \qquad (8.26)$$

or

$$F_2^{\nu N}(x) = x[u(x) + d(x) + \bar{u}(x) + \bar{d}(x)]. \qquad (8.27)$$

Similarly,

$$xF_3^{\nu N}(x) = x[u(x) + d(x) - \bar{u}(x) - \bar{d}(x)]. \qquad (8.28)$$

The value of F_2 in electron scattering can be easily obtained. From (8.11) we note that, in the limit $y \to 0$ ($\theta \to 0$),

$$\frac{d\sigma}{dq^2} = \frac{4\pi\alpha^2}{q^4} \int \frac{F_2^{eN}(x)\,dx}{x}, \qquad (8.29)$$

so that comparing with the formulae in Section (6.3), $\int F_2(x)\,dx/x$ is to be interpreted as the *sum of the squares of the charges of the partons.* Thus $F_2^{ep}(x)/x$ is given by the quark densities in the proton, weighted by the squares of the charges; i.e., if we consider u-, d-, and s-quarks,

$$F_2^{ep}(x) = x\{\tfrac{4}{9}[u(x) + \bar{u}(x)] + \tfrac{1}{9}[d(x) + \bar{d}(x) + s(x) + \bar{s}(x)]\}. \qquad (8.30)$$

F_2^{en} is obtained by replacing the symbols u with d and $\bar{u}$ with $\bar{d}$, so that for a nucleon

$$F_2^{eN}(x) = x\{\tfrac{5}{18}[u(x) + \bar{u}(x) + d(x) + \bar{d}(x)] + \tfrac{1}{9}[s(x) + \bar{s}(x)]\}. \qquad (8.31)$$

Thus, combining with (8.27), we find

$$F_2^{vN}(x) \leq \tfrac{18}{5} F_2^{eN}(x), \qquad (8.32)$$

where the number results directly from the fractional quark-charge assignment, and the equality holds if $s, c, \ldots$ quarks can be neglected (they in fact make only a few percent contribution). Figure 8.12 shows early neutrino data on F_2^{vN} compared with those from electron scattering. The comparison shows that neutrinos and electrons "see" the same type of substructure in the nucleon and that the parton constituents do indeed have the fractional charges previously ascribed to quarks.

The data on neutrino and antineutrino scattering can also be used to *count* the number of quarks in the nucleon. From (8.28) we expect that

$$I = \int_0^1 \frac{xF_3^{vN}(x)}{x}\,dx = \int_0^1 [u_v(x) + d_v(x)]\,dx = 3,$$

where $u_v = u - \bar{u}$ and $d_v = d - \bar{d}$ are the numbers of "valence" quarks in the nucleon (while $\bar{u}, \bar{d}$ correspond to the number of quark-antiquark pairs). The above prediction is called the Gross-Llewellyn-Smith sumrule. Experimentally, the main problem is that the above integrand is largest near $x = 0$, which implies $E_v \to \infty$ at fixed q^2. In practice the integral can be evaluated down to a lower limit $x_0 \simeq 0.03$, for which one obtains $I \simeq 2.8 \pm 0.5$, in agreement with expectations.

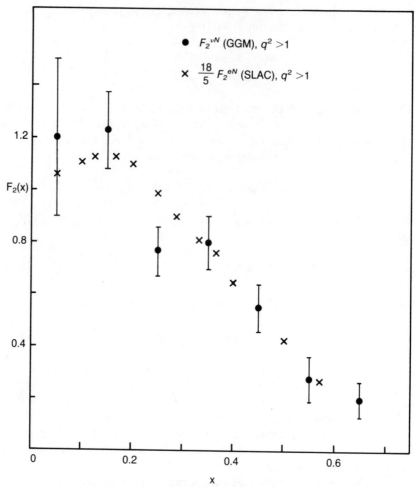

Figure 8.12 (a) First comparison of $F_2^{\nu N}$ measured in neutrino-nucleon scattering in the Gargamelle heavy-liquid bubble chamber in a PS neutrino beam at CERN, with SLAC data on F_2^{eN} from electron-nucleon scattering, in the same region of q^2. The two sets of data agree when the electron points are multiplied by the factor $\frac{18}{5}$, which is the reciprocal of the mean squared charge of u- and d-quarks in the nucleon. This is a confirmation of the fractional charge assignments for the quarks. Note that the total area under the curve, measuring the total momentum fraction in the nucleon carried by quarks, is about 0.5. The remaining mass is ascribed to gluon constituents, which are the postulated carriers of the interquark color field.

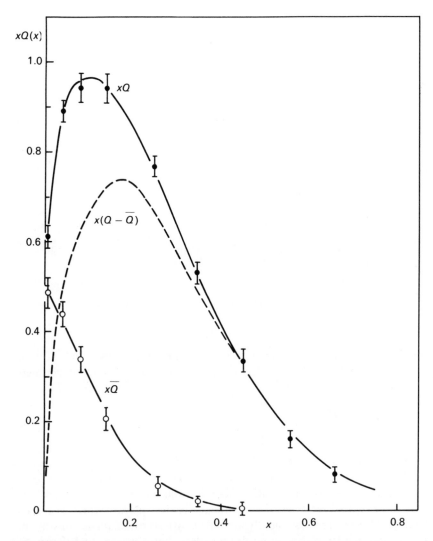

Figure 8.12 (b) Momentum distributions of quarks (Q) and antiquarks ($\bar{Q}$) in the nucleon, at a value of q^2 of order 10 GeV^2, obtained from results on neutrino and antineutrino scattering in experiments at CERN and Fermilab. The neutrino and antineutrino differential cross-sections measure the structure functions F_2 and F_3 in Eq. (8.17), and the difference and sum of these, through Eq. (8.23), give the quark and antiquark populations weighted by the momentum fraction x. The antiquarks ($\bar{Q}$) are concentrated at small x, the region of the so-called quark-antiquark "sea." The "valence" quarks of the static quark model ($Q - \bar{Q}$) are concentrated toward $x = 0.2$.

8.5.3. Antiquark Content of the Nucleon

According to the *V-A* theory of the weak interactions, the ratio of cross-sections for antineutrinos and neutrinos on spin-$\frac{1}{2}$ pointlike particles should be $R = \frac{1}{3}$ exactly. As pointed out above, this factor arises because in the antineutrino scattering on a fermion, the initial state has $J = 1$ (and $J_z = +1$), but in the final state, only one of the three possible $J = 1$ substates is allowed by angular-momentum conservation. However, if antipartons are involved as well, then $R > \frac{1}{3}$, and the ratio of antiparton to parton densities will be simply $\bar{Q}/Q = (3R - 1)/(3 - R)$. Figure 8.2 shows that $R \simeq 0.45$ in the energy range 10–100 GeV, so $\bar{Q}/Q \simeq 0.1$. The x-dependence is shown in Fig. 8.12(b).

8.5.4. Gluon Constituents

The identification of the partons with quarks leads us to expect that the structure-function integrals

$$\frac{18}{5} \int F_2^{eN}(x)\, dx = \int F_2^{vN}(x)\, dx = \int [u(x) + \bar{u}(x) + d(x) + \bar{d}(x)]x\, dx \simeq 1,$$

$$(8.33)$$

since the total fractional momentum summed over all constituents is necessarily unity. In 1972, the early electron and neutrino scattering experiments gave (see Fig. 8.12(a))

$$\frac{18}{5} \int F_2^{eN}(x)\, dx \simeq \int F_2^{vN}(x)\, dx = 0.50 \pm 0.05 \qquad (8.34)$$

for values of q^2 in the region 1–10 GeV^2. The partons responsible for the scattering of leptons, via their electric and weak charges, account for only *half* the nucleon mass. We must either abandon the model entirely or postulate that there is another type of constituent besides quarks, which is inert to leptons. These were subsequently identified with so-called *gluon* constituents. As we shall see later, the gluons are massless vector bosons and the specific quanta of the strong (color) force between quarks. They interact only with the strong charges of the quarks and have no interaction with electromagnetic or weak charges. There is, incidentally, no *a priori* reason why the fractional energy-momentum content of the gluons should be just equal to that of the quarks, as indicated in (8.34). The fact that the mass fraction just happens to be $\simeq 50\%$ in the region of q^2 a few GeV^2 is an indication of the mass scales associated with the strong quark-quark interaction and the number of degrees of freedom carried by the quarks and gluons, respectively. The important point is that we have postulated gluons as an important constituent of matter inside the nucleon.

At this stage it is worth trying to summarize the results so far from deep-inelastic lepton-nucleon scattering:

(a) The nucleon contains pointlike constituents, as evidenced from the approximate scale invariance of the structure functions, $F_2(x, q^2) \simeq F_2(x)$.
(b) The constituents have spin-$\frac{1}{2}\hbar$; $[2xF_1(x) \simeq F_2(x)]$.
(c) Electromagnetic and weak cross-sections are consistent with the identification of the "active" partons with fractionally charged quarks.
(d) The quarks account for only a fraction ($\simeq 50\%$) of the nucleon mass; the remainder is ascribed to gluon constituents which are responsible for the interquark binding.

8.6. ELECTRON-POSITRON ANNIHILATION TO HADRONS

Let us turn now to electron-positron annihilation at high energy, that is, above 10-GeV CMS energy, clear of the resonance region associated with ρ, ω, ϕ, and the ψ- and Υ-series of levels. The success of the parton model in deep-inelastic lepton-nucleon scattering naturally leads to the expectation that other processes involving hadrons and leptons can be described in similar fashion. The process of e^+e^- annihilation to hadrons can be considered as the two-stage reaction

$$e^+e^- \to Q\bar{Q}, \qquad Q, \bar{Q} \to \text{hadrons}, \tag{8.35}$$

in close analogy with lepton-quark scattering (Fig. 8.13). In (8.3) we already compared the process (8.35) with that for lepton-lepton scattering, $e^+e^- \to \mu^+\mu^-$:

$$R = \frac{\sigma(e^+e^- \to \text{hadrons})}{\sigma(e^+e^- \to \mu^+\mu^-)}$$

where, from (6.31) and with $s = E_{\text{CMS}}^2$,

$$\sigma(e^+e^- \to \mu^+\mu^-) = \frac{4\pi\alpha^2}{3s}. \tag{8.35a}$$

The magnitude of R, the total cross-section for $e^+e^- \to$ hadrons, in terms of that for $e^+e^- \to \mu^+\mu^-$, is easily calculated in the quark model. The (Rutherford) cross-section for electron-quark scattering (Fig. 8.13(a)) will be proportional to $\sum e_i^2$, the square of the quark charge summed over all contributing quark flavors i. If we replace incoming (outgoing) fermions by outgoing (incoming) antifermions on the legs of Fig. 8.13(a), we obtain the

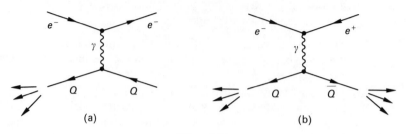

Figure 8.13

process $e^+e^- \to \gamma \to Q\bar{Q}$, Fig. 8.13(b). If follows that we expect, according to the Rutherford scattering formula,

$$R = \frac{\sum e_i^2}{1} \qquad (8.36)$$

summed over all contributing quark flavors.

At low s-values, below $c\bar{c}$ threshold, only u-, d-, s-quarks are involved, and thus we expect

$$R_{th}(\sqrt{s} < 3 \text{ GeV}) = (\tfrac{1}{3})^2 + (\tfrac{1}{3})^2 + (\tfrac{2}{3})^2 = \tfrac{2}{3}, \qquad (8.37)$$

while at high s-values, where u-, d-, s-, c-, b-quarks can contribute, we predict

$$R_{th}(\sqrt{s} > 10 \text{ GeV}) = (\tfrac{2}{3})^2 + (\tfrac{2}{3})^2 + (\tfrac{1}{3})^2 + (\tfrac{1}{3})^2 + (\tfrac{1}{3})^2 = \tfrac{11}{9}. \qquad (8.38)$$

It is obvious from Fig. 8.3 that these predictions are a long way below the data, and indeed this conflict was one of the primary reasons for introducing the *color* degree of freedom mentioned in Section 5.3. Quarks are endowed with three possible values of the strong color charge: red, green, and blue (R, G, B). Thus, any given $Q\bar{Q}$ flavor combination occurs in three substates $R\bar{R}$, $G\bar{G}$, and $B\bar{B}$, and the expected R-values must be multiplied by the factor 3, in good agreement with experiment.

At high energies, it is observed that the hadrons from the process $e^+e^- \to Q\bar{Q} \to$ hadrons are collimated into two oppositely directed "jets", as illustrated in the example in Fig. 2.23. Presumably the jet axis must be approximately that of emission of the primary $Q\bar{Q}$ pair. The angular distribution in the process $e^+e^- \to Q\bar{Q}$ will depend on the spin of the parton constituents. Again, in the prototype process $e^+e^- \to \mu^+\mu^-$ involving spin-$\tfrac{1}{2}$ particles, the angular distribution has the form (see (6.32) and Fig. 6.8):

$$\frac{dN}{d\Omega} \propto (1 + \cos^2 \theta), \qquad (8.39)$$

where θ is the angle between the $\mu^\pm$ and beam directions. Figure 8.14 shows the results for the angular distributions in the process $e^+e^- \to 2$ hadron jets; it is indeed consistent with the form (8.39) expected for spin-$\tfrac{1}{2}$ quark constituents.

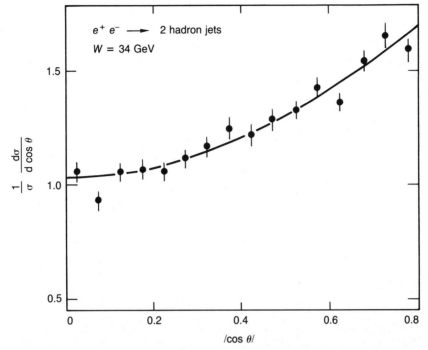

Figure 8.14 Center-of-mass angular distribution of the two hadron jets (as in Fig. 2.23) relative to the beam axis in e^+e^- annihilation at high energy. It is consistent with a $(1 + \cos^2 \theta)$ distribution, as expected if the fundamental process is $e^+e^- \rightarrow Q\bar{Q}$.

To summarize the results from the process $e^+e^- \rightarrow$ hadrons:

(a) The constancy of R (Fig. 8.3) is evidence for the pointlike (parton) constituents of hadrons.

(b) The angular distribution of the two hadron jets is proof of spin-$\frac{1}{2}$ for the partons.

(c) The value of R is equal to that expected if the charged partons are quarks, with fractional charges and the color quantum number.

8.7. LEPTON PAIR PRODUCTION IN HADRON COLLISIONS—THE DRELL-YAN PROCESS

The process of lepton pair production in hadron-hadron collisions, as first described by Drell and Yan in 1970, can also be simply interpreted in terms of the quark-parton model. Consider the reaction

$$p + p \rightarrow \mu^+\mu^- + X, \tag{8.40}$$

where X is any hadronic state. On the basis of the lepton-quark scattering diagram of Fig. 8.15(b), or that of e^+e^- annihilation to hadrons in Fig. 8.15(c), as discussed in the previous sections, it is natural to draw the related diagram of Fig. 8.15(a), in which a quark from one proton annihilates with an antiquark from the other, and the virtual photon produces a lepton pair. Note that in both (a) and (c), the 4-momentum of the photon is timelike (i.e., $q^2 < 0$), while in (b) it is spacelike ($q^2 > 0$).

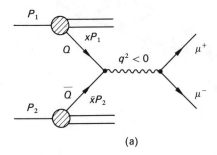

(a)

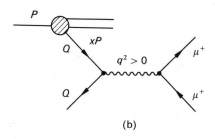

(b)

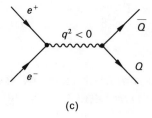

(c)

Figure 8.15 Diagrams for different types of quark-lepton interaction. (a) Production of a muon pair in a hadron-hadron collision (the Drell-Yan process) with $q^2 < 0$. (b) Muon-nucleon deep-inelastic scattering, with $q^2 > 0$. (c) Electron-positron annihilation to a quark-antiquark pair, with $q^2 < 0$.

In analogy with Fig. 8.15(c) and Eq. 8.35(a) we can write the cross-section in terms of pointlike scattering:

$$\sigma(Q\bar{Q} \rightarrow l^+ l^-) = \frac{4\pi\alpha^2}{3q^4} e_i^2, \tag{8.41}$$

where e_i, $-e_i$ are the electric charges of the $Q\bar{Q}$ pair. The differential cross-section is then

$$\frac{d\sigma}{dq^2}(Q\bar{Q} \rightarrow l^+ l^-) = \frac{4\pi\alpha^2}{3q^4} e_i^2. \tag{8.42}$$

The value of $-q^2$, the mass squared of the virtual photon, is related to the proton momenta and the momentum fractions x_i, $\bar{x}_i$ carried by the constituents. Let P_1, P_2 be the 4-momenta of the colliding protons: those of the constituents are $k_i = x_i P_1$ and $\bar{k}_i = \bar{x}_i P_2$, so the invariant masses squared of the photon and of the dilepton are

$$m^2 = -q^2 = -(k_i + \bar{k}_i)^2 = -(x_i P_1 + \bar{x}_i P_2)^2$$
$$= -(2P_1 P_2 x_i \bar{x}_i + x_i^2 P_1^2 + \bar{x}_i^2 P_2^2). \tag{8.43}$$

The total CMS energy squared of the proton-proton system is

$$s = -(P_1 + P_2)^2 = -(2P_1 P_2 + P_1^2 + P_2^2).$$

Now $P_1^2 = P_2^2 = -M^2$, and thus if $s \gg M^2$, $m^2 \gg x^2 M^2$, we obtain

$$m^2 = x_i \bar{x}_i s. \tag{8.44}$$

In this analysis, components of momentum of the constituents transverse to the proton beam direction have been neglected.

Introducing the dimensionless parameter

$$\tau = \frac{m^2}{s} = x_i \bar{x}_i, \tag{8.45}$$

we obtain for the lepton-pair cross-section in pp-collisions

$$\frac{d\sigma}{dq^2} = \frac{4\pi\alpha^2}{3q^2} F(\tau), \tag{8.46}$$

where, writing a δ-function to ensure the equality (8.44),

$$F(\tau) = \sum_i e_i^2 \frac{1}{N_c} \int_0^1 x_i dx_i \int_0^1 \bar{x}_i d\bar{x}_i f_i^1(x_i) f_i^2(\bar{x}_i) \delta(x_i \bar{x}_i - \tau) \tag{8.47}$$

and where the integration is performed over the densities $f_i^1(x_i)$, $f_i^2(\bar{x}_i)$ of quarks and antiquarks of flavor i in the protons, weighted by their fractional momenta x_i, $\bar{x}_i$ and by their squared charges e_i^2, in exact analogy with Eq. (8.30) for the structure function. The color factor $N_c = 3$ reduces the cross-section by the probability of pairing (for example) a red quark with an antired antiquark. According to (8.46), the cross-section

$$2q^4 \frac{d\sigma}{dq^2} = m^3 \frac{d\sigma}{dm} \propto F(\tau) \tag{8.48}$$

should scale, that is, depend only on the dimensionless ratio m^2/s. Direct tests of this prediction are difficult, since experiments tend to cover only a limited range of momentum of the dimuon pair. One can define the so-called rapidity of the dimuon pair,

$$y = \frac{1}{2} \ln \left(\frac{E + p_L}{E - p_L} \right), \tag{8.49}$$

where E and p_L are the total energy and net longitudinal momentum of the muon pair in the laboratory system (so $E^2 = p_L^2 + m^2$). Since y is dimensionless, we expect that the cross-section with respect to y will also scale; in particular, the dimensionless double differential cross-section

$$m^3 \frac{d^2\sigma}{dm\,dy}, \quad \text{or} \quad s \frac{d^2\sigma}{dy\,d\sqrt{\tau}},$$

should be a function of τ only. Figure 8.16 shows data on the invariant-mass distribution of muon pairs, obtained in an experiment at Fermilab by Yoh *et al.* (1978) using protons of 200–400 GeV ($\sqrt{s} = 20$–28 GeV) on a stationary target. Within the errors, the cross-section is a function of $\tau = m^2/s$ only. From such data it is possible to extract $F(\tau)$ defined in (8.46) and fit it to an analytical form, for example,

$$F(\tau) = \text{const} \cdot \exp(-18.6\sqrt{\tau}) \quad (0.2 > \sqrt{\tau} < 0.5). \tag{8.50}$$

This could then be used to predict the absolute cross-sections at the much higher energies of the CERN ISR, where oppositely circulating proton beams of up to 30 GeV collide head-on ($\sqrt{s}$ up to 62 GeV). Upon comparing the ISR data with the predictions of (8.50), it is found that the cross-sections do scale even though their absolute magnitudes in the two experiments differ by up to two orders of magnitude; this verification of the parton model holds for values of $q = m_{\mu\mu} > 3$ GeV.

The assumption that lepton pair production proceeds via quark-antiquark annihilation to a virtual photon is verified in these experiments by observing the polar angular distribution of either lepton in the dilepton rest frame, relative to the incident proton beam direction. As in the inverse process $e^+ e^- \to Q\bar{Q}$ via single photon exchange, we expect an angular distribution of the form

$$\frac{d\sigma}{d\Omega} = \text{const}(1 + \cos^2 \theta),$$

as is indeed observed (Fig. 8.16(b)).

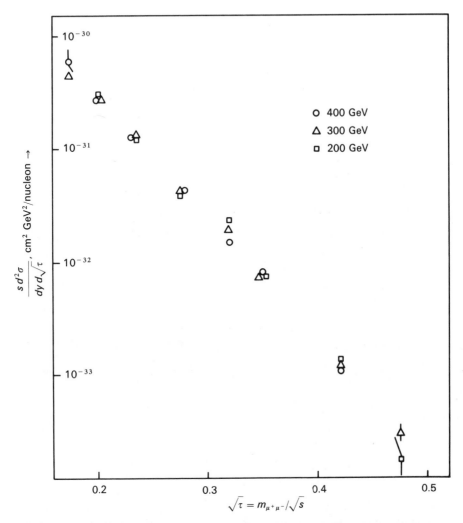

Figure 8.16 (a) Cross-section for muon pair production measured by the BFS collaboration (Yoh *et al.* 1978) in Fermilab experiments at three different proton energies. The cross-sections appear to "scale," being a function of the dimensionless variable $\tau = m_{\mu\mu}^2/s$, where s is the square of the CMS energy.

Equation (8.47) shows that $F(\tau)$ is related to integrals over the quark and antiquark distribution functions, which have been in principle determined independently in the process of lepton-nucleon scattering. We recall that in neutrino scattering, the antiquark distributions $\bar{u}(x)$, $\bar{d}(x)$, $\bar{s}(x)$,... are relatively difficult to measure, as their content in the nucleon is typically 5–10% only, and have to be determined using the neutrino-antineutrino cross-section differences (see for example Fig. 8.12). On the contrary, the

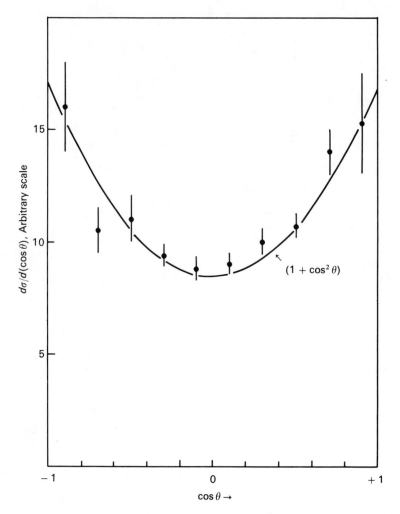

Figure 8.16 (b) Angular distribution of muons, measured in the dimuon rest frame, relative to the incident beam direction, in the process of dimuon production in hadron-hadron collisions (after Anderson *et al.* 1978). If the process follows the Drell-Yan mechanism of Fig. 8.15(a), involving spin-$\frac{1}{2}$ quarks ($Q\bar{Q} \to \mu^+\mu^-$), a distribution of the form $1 + \cos^2\theta$ is expected.

dilepton pair cross-section in hadronic interactions is directly proportional to the product of quark and antiquark distributions. The antiquark distributions obtained from such experiments are in reasonable agreement with those found from lepton scattering; in comparable regions of $|q^2|$, both for example give $x\bar{d}(x) \simeq 0.5(1-x)^7$.

A further important feature of massive dilepton production in hadronic reactions is that, using incident pion rather than proton beams, the cross-section is proportional to the antiquark distribution in the pion, where

$\bar{u}_\pi(x) = u_\pi(x)$, from symmetry. From such experiments, the pion structure function has been determined as

$$xu_\pi(x) \simeq 0.25(1 - x).$$

It is thus much harder than the valence or sea-quark momentum distributions in the nucleon, varying as $(1 - x)^3$ and $(1 - x)^7$, respectively.

8.8. QUANTUM CHROMODYNAMICS AND QUARK-QUARK INTERACTIONS

8.8.1. The Color Interaction between Quarks

The *color* quantum number has already been introduced as an extra degree of freedom in the quark model of hadrons and in lepton-quark interactions, particularly the cross-section for the process $e^+e^- \to$ hadrons (Section 8.6). Recall that the introduction of three colors for quarks increases the expected e^+e^- cross-section by a factor 3 and brings it into line with experiment. Quantum chromodynamics (QCD) is the formal gauge theory of the strong color interactions between quarks. The color charge of a quark has three possible values—say red, blue, or green. Antiquarks carry anticolor. The interquark interactions are assumed to be invariant under color interchange; in other words, they are described by the symmetry group SU(3). Since a quark can carry one of three possible colors, we can say that the quarks belong to the triplet (3) representation of SU(3). The bosons mediating the quark-quark interactions are called *gluons* and are postulated to belong to an octet (8) representation of SU(3). In analogy with the (flavor) octet of mesons in Section 5.5 we can write the color-anticolor states of the 8 gluons as follows:

$$r\bar{b}, r\bar{g}, b\bar{g}, b\bar{r}, g\bar{r}, g\bar{b}, \frac{r\bar{r} - b\bar{b}}{\sqrt{2}}, \frac{r\bar{r} + b\bar{b} - 2g\bar{g}}{\sqrt{6}}. \tag{8.51}$$

With 3 colors and 3 anticolors, we expect $3^2 = 9$ combinations, but one of these is a color singlet and has to be excluded. As an example, the color interaction between a red quark and a blue quark can proceed via exchange of a single $r\bar{b}$ gluon (Fig. 8.17).

The color charge of the strong quark interactions is analogous to the electric charge in electromagnetic interactions. Both forces are mediated by a massless, vector particle (a gluon or a photon). However, whereas in electromagnetism there are two types of charge and an uncharged mediating boson, in QCD there are six types of charge (color or anticolor) and a charged (i.e., colored) mediating boson. This difference turns out to be crucial in understanding the features of the quark interactions at short distances and the success of the parton model.

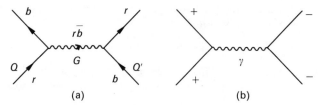

Figure 8.17 (a) QQ interaction via color (gluon) exchange. (b) Interaction between charges via (uncharged) photon exchange.

The color quantum number does not enter our description of hadrons, so that both baryons and mesons must be colorless, i.e., be singlets of the SU(3) of color. If we write down the various contributions due to exchange of gluons [Eq. (8.51)] between quarks, the quark configurations of lowest energy are found to consist of the color singlet QQQ state (baryon) and the color singlet $Q\bar{Q}$ state (meson). Other quark combinations (e.g., the $Q\bar{Q}$ color octet) have weaker binding, or even a repulsive interaction (see Appendix J). Thus, QCD correctly predicts that only two of all the possible multiquark combinations should exist in nature.

The actual form of the strong-interaction potential between quarks was discussed in Section 5.14. At small distances, the interaction is assumed to be of the Coulomb type, in analogy with electromagnetism, while at larger distances, the potential must increase indefinitely, so as to confine the quarks inside a hadron. The form

$$V = -\frac{4}{3}\frac{\alpha_s}{r} + kr \tag{8.52}$$

was quoted previously. We now discuss the experimental evidence for this form for the potential.

8.9. THE QCD POTENTIAL AT SHORT DISTANCES

Direct evidence supporting the potential (8.52) at short distances was obtained in experiments at the CERN $p\bar{p}$ collider (Arnison *et al.* 1984, Bagnaia *et al.* 1984). Rare events (of order 10^{-6} of all collisions) are found to consist of two high-energy jets of hadrons at large angle to the beam, as in Fig. 8.18. The selection of such events is based firstly on the requirement that the transverse energy $E_T = \sum E_i \sin \theta_i > 15$ GeV, where the sum is made over local energy depositions in the calorimeter cells, and θ_i is the angle to the incident beam direction. Most of these events ($\simeq 80\%$) consist of two jets[*]

---

[*] The definition of a jet is obviously arbitrary and found by a suitable algorithm. In the experiment, jets were found by combining signals from adjacent calorimeter cells, starting from the highest E_T cell, and separated from it by $(\Delta y^2 + \Delta\phi^2)^{1/2} < 1$, (where y is the rapidity and ϕ the azimuthal angle). Any cell of $E_T > 2.5$ GeV not assigned to an existing jet is considered as an initiator of a new jet.

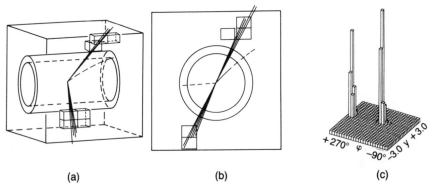

(a)   (b)   (c)

Figure 8.18 Displays of a two-jet event in the UA1 detector (Fig. 7.18) at the CERN $p\bar{p}$ collider (310-GeV protons on 310-GeV antiprotons). In (a) the event is shown in perspective. Activated calorimeter modules are shown as blocks. Low-energy hadrons are omitted. The cylinder encloses the inner tracking detector. (b) View down beam pipe, demonstrating the back-to-back nature of the jets in azimuth. (c) Plot of transverse energy as function of azimuth ϕ and rapidity y. The total energies of the two jets are 93 GeV and 84 GeV, their transverse energies 81 and 78 GeV.

with approximate momentum balance (as evidenced by the back-to-back configuration in azimuthal angle in Fig. 8.18). The remainder are principally three-jet events, although four-jet and one-jet events also occur. The two-jet events are interpreted in terms of elastic scattering of a parton (quark or gluon) from the proton with one from the antiproton, each scattered parton giving rise to a jet of hadrons in the manner familiar from the process $e^+e^- \to$ hadrons. The kinematics of the collision are shown in Fig. 8.19, where p_3, p_4 are the 4-momenta of the observed jets and p_1, p_2 are the (unknown) momenta of the partons before the collision. Then energy-momentum conservation gives

$$p_1 + p_2 = p_3 + p_4, \tag{8.53}$$

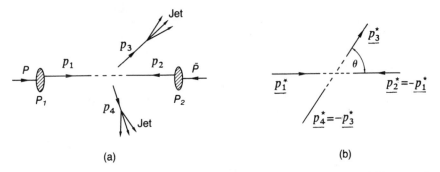

(a)   (b)

Figure 8.19 Kinematics of a parton-parton collision in (a) laboratory system and (b) center-of-momentum frame of the partons.

while 3-momentum conservation along the beam (z-axis) gives

$$\mathbf{p}_1 + \mathbf{p}_2 = p_{3z} + p_{4z} \tag{8.54}$$

The 4-momentum transfer is

$$q = p_3 - p_1 = p_2 - p_4. \tag{8.55}$$

In terms of the Bjorken x-variable (8.13),

$$p_1 = x_1 P_1; \qquad p_2 = x_2 P_2, \tag{8.56}$$

where P_1 and P_2 are the 4-momenta of the proton and antiproton. Let E be the energy in each beam. Then from (8.54), neglecting all particle masses, we obtain

$$x_1 - x_2 = (p_{3z} + p_{4z})/E = x_F, \tag{8.57}$$

whereas from (8.53)

$$(p_3 + p_4)^2 = (x_2 P_2 + x_1 P_1)^2 = -4x_1 x_2 E^2,$$

or

$$x_1 x_2 = -(p_3 + p_4)^2/4E^2 = \tau, \tag{8.58}$$

where the Feynman x_F and the τ variables have been used in earlier discussions (see Sections 4.12 and 8.7). From the above equations, we find

$$x_1, x_2 = \tfrac{1}{2}(x_F \pm \sqrt{x_F^2 + 4\tau}) \tag{8.59}$$

and therefore p_1, p_2 in terms of the known quantities E, p_3, and p_4. Thus the 4-momentum transfer q^2 in (8.55) can also be calculated. The direction of the equal and opposite 3-momentum vectors of the scattered partons in their common CMS frame (see Fig. 8.19(b)) is clearly given by $\mathbf{p}_3 - \mathbf{p}_4$, so that the CMS scattering angle of the partons relative to the beam direction is

$$\cos\theta = \frac{(\mathbf{p}_3 - \mathbf{p}_4)\cdot(\mathbf{p}_1 - \mathbf{p}_2)}{|\mathbf{p}_3 - \mathbf{p}_4||\mathbf{p}_1 - \mathbf{p}_2|}.$$

Note that since we do not know which incident parton (p_1 or p_2) belongs to which jet (p_3 or p_4), there is a $\theta \leftrightarrow \pi - \theta$ ambiguity. In the analysis, the smaller of the two values of θ was assumed, as it was also for the predicted distribution when comparing with theory. The expected cross-section for the proton-antiproton scattering to two jets will be

$$\frac{d^3\sigma(pp \to 2\,\text{jets})}{dx_1 dx_2 d(\cos\theta)} = \sum_{i,j} \left(\frac{F_i(x_1)}{x_1}\right)\left(\frac{F_j(x_2)}{x_2}\right)\frac{d\sigma_{ij}}{d(\cos\theta)}, \tag{8.60}$$

where $F_i(x_1)/x_1$, $F_j(x_2)/x_2$ are the densities of partons of type i, j in the proton and antiproton, with fractional momenta x_1 and x_2, and $d\sigma_{ij}/d(\cos\theta)$ is the elementary parton-parton cross-section. The contributions to the cross-

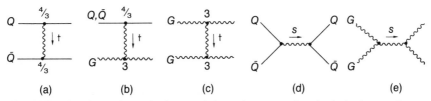

Figure 8.20 Quark-quark, quark-gluon, and gluon-gluon scattering via single gluon exchange, in the t- and s-channels.

section through the process of single gluon exchange are shown in Fig. 8.20. Scattering may proceed through momentum transfer in the so-called t-channel, where, as defined in Fig. 8.19,

$$t = -q^2 = -(p_1 - p_3)^2 \qquad (8.61)$$

or through the s-channel, where

$$s = -(p_1 + p_2)^2 = 4x_1 x_2 E^2 \qquad (8.62)$$

is the square of the CMS energy of the partons. Figure 8.20(a), (b), (c) show the t-channel exchanges, and (d), (e) show the s-channel processes. Broadly, the t-channel cross-sections vary as $d\sigma/dt \simeq 1/t^2$ (or $1/q^4$), whereas the crossed, s-channel cross-sections go as $d\sigma/ds \simeq 1/s^2$. Provided $t \ll s$, the t-channel exchanges dominate, which is the case in the experimental conditions. The QCD couplings of $Q\bar{Q}$, QG, and GG via one-gluon exchange are denoted by numbers in Fig. 8.20, giving the coefficients multiplying the strong coupling, α_s (see Appendix J). Clearly the GG coupling is the largest and in the approximation that all three processes have a similar angular distribution, we see that the cross-section effectively measures the combination of structure functions

$$F(x) = G(x) + \tfrac{4}{9}(Q(x) + \bar{Q}(x)), \qquad (8.63)$$

where now $Q(x)$, $\bar{Q}(x)$ and $G(x)$ represent momentum densities, i.e. the quark antiquark and gluon densities weighted by the momentum fraction, x. The angular distribution has the form (Combridge *et al* (1977)):

$$\frac{d\sigma}{d(\cos\theta)} = \frac{\pi\alpha_s^2}{s}\left(\frac{9}{16}\right)\frac{(3 + \cos^2\theta)^3}{(1 - \cos^2\theta)^2} \qquad (8.64)$$

or, for $\cos\theta \simeq 1$,

$$\frac{d\sigma}{d\Omega} \simeq \left(\frac{9}{8}\right)\frac{\alpha_s^2}{4p_0^2 \sin^4(\theta/2)}, \qquad (8.65)$$

where $p_0 = \sqrt{s}/2$ is the 3-momentum of each parton in the CMS. Note that this is just the Rutherford formula (6.14) for a $1/r$ potential, with the replacement of α_s for α and a different coefficient for the color coupling.

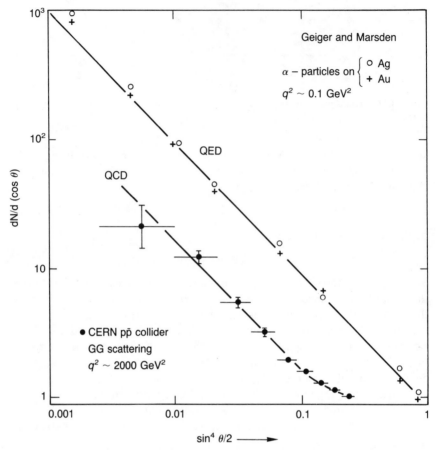

Figure 8.21 Examples of differential cross-sections for pointlike scattering via a $1/r$ potential. The upper plot is of the results of Geiger and Marsden (1911) for scattering of α-particles from radioactive sources by gold and silver foils, demonstrating the existence of a nucleus to the atom, scattering the α-particles through the Coulomb field. Their results involved momentum transfers $\simeq 0.1$ GeV^2 and are consistent with the Rutherford formula $dN/d\Omega \propto \sin^{-4}(\theta/2)$.

The lower plot is the two-jet angular distribution found at the CERN $p\bar{p}$ collider, at $q^2 \simeq 2000$ GeV^2, showing that the scattering of pointlike (quark or gluon) constituents of the nucleon also obeys the Rutherford law at small angles, and hence that the QCD potential varies as $1/r$ at small distances. At large angles, deviations from the straight line occur because of relativistic (spin) effects and because scatters of $\theta > \pi/2$ have to be folded into the distribution for $\theta < \pi/2$. The full-line curve is the QCD prediction for single vector gluon exchange. Scalar (spin-0) gluons are excluded, as they predict a very much weaker angular dependence.

The observed angular distribution is shown in Fig. 8.21. If we parameterize it in the form $(\sin \theta/2)^{-n}$ then the data give $n = 4.16 \pm 0.20$. We note that deviations from the straight line are expected near $\theta = \pi/2$, both because the full relativistic formula (8.64) for single vector gluon exchange must be used and because of the ambiguity problem mentioned before. It is instructive to compare these results for (predominantly) gluon-gluon scattering, typically at $q^2 \simeq 2000 \text{ GeV}^2$, with the results of Geiger and Marsden (1911) on scattering of α-particles by silver and gold nuclei, at $q^2 \simeq 0.1 \text{ GeV}^2$. The linearity of the plots in both cases is evidence that a $1/r$ potential is involved. In the case of the scattering of spinless, nonrelativistic α-particles, the Rutherford formula applies at all angles (if the nuclei act as point charges). In parton-parton scattering, the Rutherford formula is a small-angle approximation for the full relativistic formula including spin effects.

The measurement of the UA1 (Arnison *et al.* (1984)) and UA2 (Bagnaia *et al.* (1984)) experiments at the CERN $p\bar{p}$ collider constitute at present the best direct evidence that the QCD potential at small r has the $1/r$ variation assumed in (8.52). The small distances involved are of course guaranteed by the very high values of q^2.

The combination (8.63) of parton densities can be deduced from the differential $p\bar{p} \to 2$ jet cross-sections (8.60) up to a constant factor, since α_s is unknown. These results are discussed in Section 8.12.3. Alternatively, it is possible to deduce $F(x)$ from the lepton-nucleon deep-inelastic scattering data, and hence find a value for $\alpha_s \simeq 0.2$. Because of various corrections, this is considered to be an upper limit.

Clearly, the foregoing analysis can be extended to events with three, four, or more jets: The ratio of three-jet to two-jet events will be proportional to α_s, in the same way as described in Section 8.11 for $e^+ e^- \to$ two or three jets.

8.10. THE QCD POTENTIAL AT LARGE DISTANCES: THE STRING MODEL

One of the most remarkable empirical results of the study of baryon and meson resonances is that for states with a given isospin, C-parity, etc., there appears to be a linear dependence of the angular momentum of the states of highest angular momentum, on the square of the mass. Examples of the plots for the Δ- and Λ-baryon resonances are shown in Fig. 8.22. Such a plot is called a Chew-Frautschi plot and was originally of interest in Regge pole theory, which we do not discuss. Here we shall try to show its relevance to the form of the QCD potential at large distances.

In QCD, the characteristic feature of the gluon mediators of the color force is their strong self-interaction, because the gluons themselves are postulated to carry color charges. In analogy with the electric lines of force

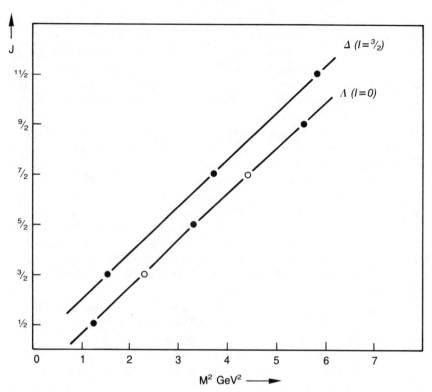

Figure 8.22   Plots of spin J against mass, squared, for baryon resonances of the Δ family ($S = 0$, $I = \frac{3}{2}$) and the Λ family ($S = -1, I = 0$). Positive and negative parity states are shown as full and open circles.

between two electric charges, as in Fig. 8.23(a), we can imagine that quarks are held together by color lines of force as in Fig. 8.23(b), but the gluon-gluon interaction pulls these together into the form of a tube or string.

Suppose k is the energy density per unit length of such a string, and that it connects together two massless quarks as in Fig. 8.23(c). The angular momentum of the quark pair will then be equal to the total angular momentum of the gluon tube, and we can calculate this if we assume that the ends of the tube rotate with velocity $v = c$. Then the local velocity at radius r will be

$$\frac{v}{c} = \frac{r}{r_0},$$

where r_0 is half the length of the string. The total mass is then (relativistically)

$$E = Mc^2 = 2 \int_0^{r_0} \frac{k\, dr}{\sqrt{1 - v^2/c^2}} = kr_0\pi,$$

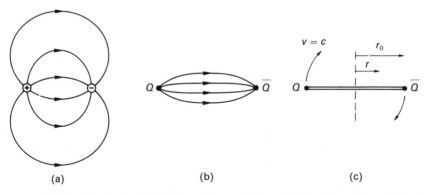

(a)   (b)   (c)

Figure 8.23 (a) Electric lines of force between two charges. (b) Color lines of force between quarks are pulled together into a tube or string, because of the strong self-interaction between the gluons, which are the carriers of the color field. (c) String model used in calculating the relation between angular momentum and mass of a hadron.

and its orbital angular momentum will be

$$J = \frac{2}{\hbar c^2} \int_0^{r_0} \frac{krv\,dr}{\sqrt{1 - v^2/c^2}} = \frac{kr_0^2 \pi}{2\hbar c}.$$

Eliminating r_0 between these equations gives the observed relation between the angular momentum and energy of a hadron state:

$$J = \alpha' E^2 + \text{const}, \tag{8.66}$$

This result holds for the case of a constant energy density k of the string, that is, to a potential of the form

$$V = kr. \tag{8.67}$$

Generally, for a potential of the form

$$V = kr^n$$

it is easy to show that the relation (8.66) acquires the form

$$J \propto E^{(1 + 1/n)},$$

so the observed linear dependence of J on M^2 is evidence for the linear potential (8.67). The value of k is obtained from the slope, α' in (8.66), of the plot, which in our model is

$$\alpha' = 1/(2\pi k\hbar c).$$

Inserting the observed value, $\alpha' = 0.93$ GeV^{-2}, we find

$$k = 0.85 \text{ GeV fm}^{-1} \tag{8.68}$$

This number also comes from consideration of the sizes of hadrons. A typical hadron mass is about 1 GeV, and its radius, as measured in electron scattering is about 1 fm, so the linear energy density will be $k \simeq 1$ GeV fm^{-1}.

8.11. MULTIJET EVENTS IN e^+e^- ANNIHILATION

Dramatic demonstrations of quark substructure are obtained in e^+e^- annihilation to hadrons at very high energy. As noted in Section 8.6, the elementary process is the annihilation to a $Q\bar{Q}$ pair, followed by "fragmentation" of the quarks to hadrons. At CMS energies of 30 GeV or more, typically about 10 hadrons (mostly pions) are produced. The average hadron momentum along the original quark direction is therefore large compared with its transverse momentum p_T, limited to $p_T \simeq 0.5$ GeV/c, that is, a magnitude $\sim 1/R_0$, where R_0 is a typical hadron size (~ 1 fm). Hence, the hadrons appear in the form of two "jets" collimated around the $Q\bar{Q}$ axis (see Fig. 8.24(a) and Fig. 2.23). Occasionally, one might expect a quark to radiate a "hard" gluon, carrying perhaps half of the quark energy, at a large angle (Fig. 8.24(b)), the gluon and quark giving rise to separate hadronic jets. Such processes seem to be observed (Fig. 8.25(a)). The measured rate of three-jet compared with two-jet events is clearly determined by α_s. We do not discuss here the complicated analysis procedures employed, but the final result, at CMS energy of 30–40 GeV, is that $\alpha_s = 0.16 \pm 0.03$ (an average value from several experiments).

The angular distribution in the three-jet events allows a determination of gluon spin. First the jets are ordered in energy, $E_1 > E_2 > E_3$. Then a transformation is made to the CMS frame of jets 2 and 3, and the angle $\tilde{\theta}$ calculated of jet 1 (of highest energy) with respect to the common line of flight of jets 2 and 3. Jet 3 (that of lowest energy) is most likely to be produced by the gluon, and the distribution in $\tilde{\theta}$ is sensitive to the gluon spin. Figure 8.25(b) shows data from the TASSO detector together with the predictions for scalar and vector gluons. The data strongly favor the vector gluon hypothesis.

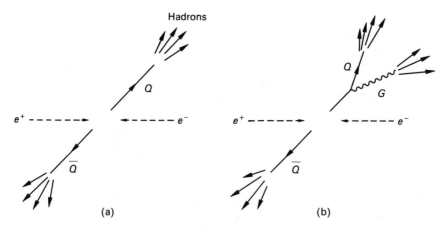

Figure 8.24

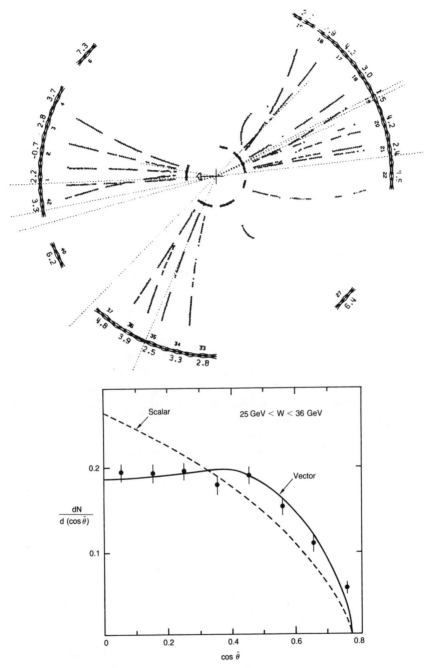

Figure 8.25 (a) Example of a three-jet event observed in the JADE detector at the PETRA e^+e^- collider (DESY, Hamburg). The total CMS energy in the event is 31 GeV. Such events (in comparison with the more common "two-jet" events of Fig. 8.24(a)) are suggestive of the process in Fig. 8.24(b). (b) Distribution in angle $\tilde{\theta}$ of the highest energy jet in three-jet events, with respect to the line of flight of the other two jets in their CMS frame. Curves show predictions for spin 1 (vector) and spin 0 (scalar) gluons. Data from TASSO detector at PETRA.

8.12. EFFECTS OF QUARK INTERACTIONS IN
DEEP-INELASTIC LEPTON-NUCLEON SCATTERING

Historically, the first experimental evidence in support of QCD (in the period 1973–78) came from detailed studies in deep-inelastic lepton-nucleon scattering. What was observed was "scaling violations," that is, small departures from the pure partonlike behavior of noninteracting, free quark constitutents. If such deviations can be measured accurately, they give information on the nature of the interquark forces.

Suppose that in a deep-inelastic scattering process, we vary q^2 (the 4-momentum transfer squared) between the incoming and the outgoing lepton. q^2 is defined as a positive quantity and is equal to $-m^2$, where m is the imaginary value of the mass of the virtual photon exchanged between lepton and nucleon. The spatial resolution of this virtual photon "probe" is $\Delta r = h/q$. At low q^2, such that $\Delta r > R$, the target-proton radius ($R \simeq 0.8$ fm), the entire proton recoils elastically from the collision; in other words it behaves like a pointlike particle (Fig. 8.26(a)). As q^2 is increased the virtual photon begins to "see" the parton structure of the nucleon as three pointlike "valence" quarks. In the parton model, where the quarks are assumed to be both pointlike and free, further increase in q^2 does not lead to revelation of any finer structure (see Fig. 8.26(b)).

On the other hand, in the interacting-quark model, significant changes are expected as q^2 is increased. The exchange of gluons between quarks implies that the quarks have a "structure". As our virtual photon probes smaller and smaller distances, what appeared as a single quark at low resolution is revealed as, for example, a quark which has emitted a gluon which in turn has formed a quark-antiquark pair. This process is analogous to that of vacuum polarization (virtual e^+e^- pair production) in QED (see Fig. 6.10). The pictures for the interacting quark model are shown in Fig. 8.26(b) and (c).

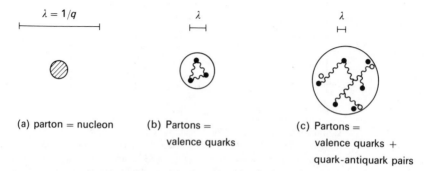

(a) parton = nucleon   (b) Partons =   (c) Partons =
                                 valence quarks   valence quarks +
                                                             quark-antiquark pairs

Figure 8.26   Symbolic picture of the nucleon, as probed in lepton scattering, as the wavelength $\lambda = 1/q$ of the probe is decreased and more structure is revealed.

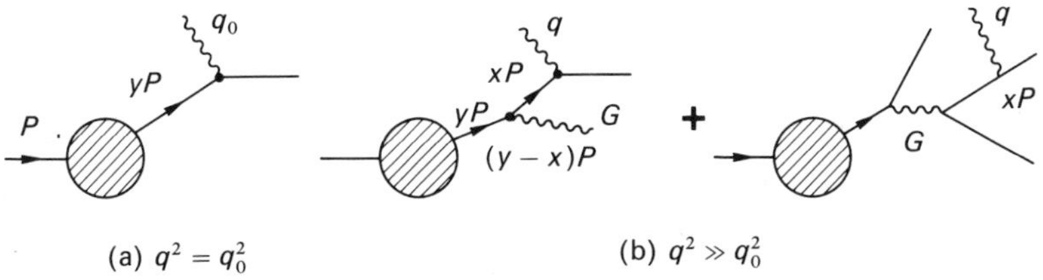

(a) $q^2 = q_0^2$ (b) $q^2 \gg q_0^2$

Figure 8.27

The effects of quark interactions on the behavior of nucleon-structure functions can easily be understood qualitatively from Fig. 8.27. In (a), a quark carrying a momentum fraction y of the nucleon momentum P absorbs a photon of momentum q_0. For $q^2 \gg q_0^2$ however, the quark has dissociated into a quark of momentum fraction $x < y$ and a gluon with fraction $y - x$. Thus, if x is large (>0.3 say), the effect of increasing q^2 is to lead to a shrinkage of the quark distribution (i.e., structure function) toward small x. At very small x, on the other hand, the increasing number of gluons can form quark-antiquark pairs, in the domain of small x, and one of these can absorb the photon and thus lead to an increase of the structure function in this region (Fig. 8.27(b)). This sort of behavior would be expected in any field theory of interacting quarks, and is not peculiar to QCD. Figure 8.28 shows practical examples of the evolution of structure functions with increasing q^2, bearing out the shrinkage in x as q^2 increases. Such deviations from the scale-invariant parton model were first observed in muon scattering experiments at Fermilab, Chicago in 1973 (Fox *et al.* 1974).

8.12.1. Running Coupling Constant: Quantitative Predictions of QCD

In Section 6.8 we described the radiative corrections to the electron and muon magnetic moments in QED, the expression for $(g - 2)/2$ appearing as a power series in the fine-structure constant α (measured, e.g., with a Josephson junction). To a certain degree of approximation, their effect is equivalent to a first-order correction (i.e., of order α), with α depending however on the masses or momentum transfers in the virtual processes involved. Thus, we can always write for the anomaly

$$\frac{g - 2}{2} = \frac{0.5}{\pi} \alpha_{\text{eff}}, \tag{8.69}$$

where α_{eff} is larger for the muon than for the electron. In fact we are treating α as a "running coupling constant" dependent on the masses or momentum transfers involved in any particular case. This is not so arbitrary as it sounds. We noted previously that attempts to define "bare" couplings, charges, or masses—α_0, e_0, m_0—in QED lead to infinities associated with self-energy

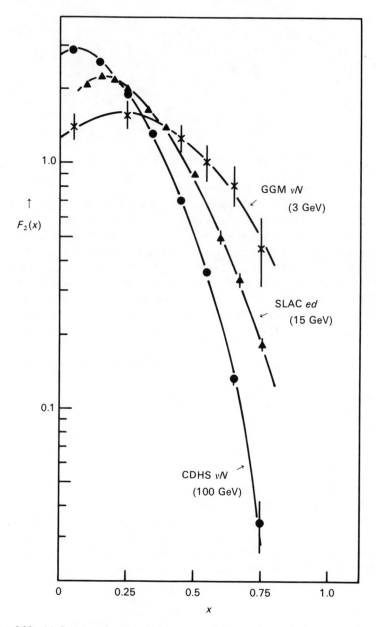

Figure 8.28 (a) Structure function $F_2(x)$ measured in neutrino and electron scattering on nucleons at different beam energies. The shrinkage towards smaller x with increasing average energy and increasing average q^2 is illustrated.

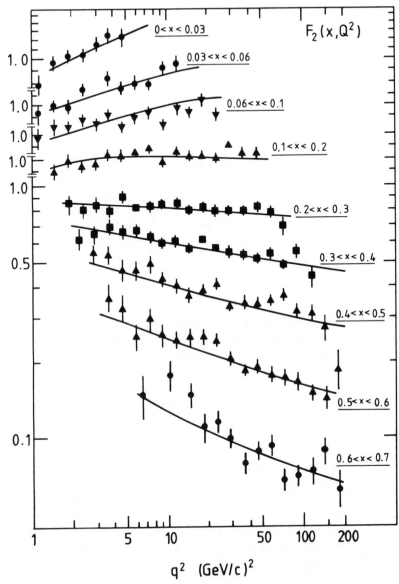

Figure 8.28   (b) $F_2^{\nu N}(x, q^2)$ plotted as a function of q^2 in different regions of x, from a counter experiment at CERN (see Fig. 2.24). Compare with Fig. 8.8, from electron scattering, at $x = 0.25$. The increase with q^2 at small x is shown, as well as the stronger decrease at large x. The curves are from empirical fits to quark momentum distributions, incorporating a q^2-dependence à la QCD, with $\Lambda = 0.3$ GeV. (After de Groot *et al.* 1979.)

Figure 8.29

terms, and we have to replace such quantities by the renormalized, physically measured values. The running coupling constant expresses the value of α at one value of q^2 in terms of that at another q^2 (incidentally avoiding bare couplings at $q^2 = \infty$). The relation can be approximated by

$$\alpha(q^2) = \frac{\alpha(q_0^2)}{1 - \frac{\alpha(q_0^2)}{3\pi} \ln\left(\frac{q^2}{q_0^2}\right)}, \tag{8.70}$$

so the change in the value of α, as we go from a process involving a typical momentum transfer q_0^2 to one involving $q^2 > q_0^2$, depends logarithmically on the ratio q^2/q_0^2. Equation (8.70) is obtained by summing the higher-order corrections, involving terms of the general form $\alpha^n[\ln(q^2/q_0^2)]^m$, but retaining only the leading logarithms, i.e., $m = n$. It must be emphasized that this is only an approximate procedure, which avoids the labor of summing enormous numbers of Feynman graphs (72, for the α^3-term in computing $(g - 2)$ of electron or muon) and which would be a quite impossible task in the case of QCD. We note that as q decreases, or the typical distance $r \simeq 1/q$ increases, the effective coupling α gets smaller. This is a well-known effect in a polarizable (dielectric) medium. A test charge immersed in the dielectric exerts a potential, at distances larger than or comparable with molecular dimensions, which is smaller than the Coulomb potential in free space; in other words, the dielectric introduces a *shielding* effect (Fig. 8.29). Actually, the medium is unnecessary. Even in a vacuum, the test charge continually emits and reabsorbs virtual photons which can temporarily produce e^+e^- pairs, again producing a shielding effect—the so-called vacuum polarization. At extremely small distances, the shielding effect becomes small and one obtains the potential due to the bare charge.[†]

In QCD, the quark interactions can also be represented by a running coupling constant, $\alpha_s(q^2)$. Again, quark-antiquark pairs produce a shielding

---

[†] Virtual electron pairs, according to the Uncertainty Principle, will be mainly confined to distances less than $\lambda = h/mc$, so it is only for $r < \lambda$ that shielding effects are important. For $r \gg \lambda$, shielding effects decrease dramatically and the potential rapidly approaches the Coulomb value (see Gell-Mann and Low 1954).

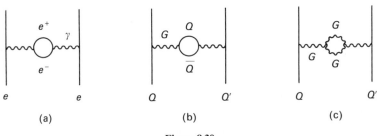

Figure 8.30

effect on the value of a test quark (color charge), Fig. 8.30(b). Because the field is non-Abelian, however, a gluon can also give rise to a gluon pair, and for certain states of polarization of this pair, an *antishielding* effect is produced (Fig. 8.30(c). $\alpha_s(q^2)$ has the form, in leading-logarithm approximation

$$\alpha_s(q^2) = \frac{\alpha_s(q_0^2)}{1 + B\alpha_s(q_0^2)\ln(q^2/q_0^2)} = \frac{1}{B\ln(q^2/\Lambda^2)}, \tag{8.71}$$

where $B = (33 - 2f)/12\pi$, and in the second expression we have made the substitution $\Lambda^2 = q_0^2 \exp[-1/B\alpha_s(q_0^2)]$. Thus, provided the number of quark flavors $f \leq 16$, it follows that $\alpha_s(q^2)$ decreases as q^2 increases. The effect of the self-coupling of the gluons is to "spread out" the color charge, and this effect becomes larger as q^2 increases. This decrease of α_s at large q^2 (or small distances) was first discussed in detail by Gross and Wilczek (1973) and by Politzer (1974). Clearly, (8.71) shows that at asymptotically large q^2 we have $\alpha_s(q^2) \to 0$, that is, the quarks behave as if free—a phenomenon called *Asymptotic freedom*, and precisely what is expected in the parton model of quasi-free quarks. At $q^2 \ll q_0^2$, such that $q \sim \Lambda$, $\alpha_s(q^2)$ becomes very large. Although at this point the theory becomes meaningless, large quark coupling may well be connected with the confinement of quarks at large distances, as for the potential (8.52).

The contrasting q^2 behavior of α and α_s is shown in Fig. 8.31.

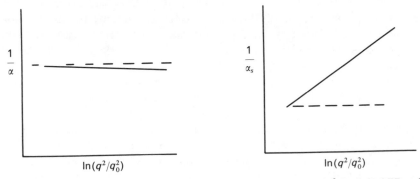

Figure 8.31   Reciprocal of the running coupling constant as a function of q^2, for (left) QED and (right) QCD.

8.12.2. q^2 Evolution of Structure Functions

The theory makes quantitative predictions about the q^2-evolution of the quark distributions (structure functions), in the limit where $\alpha_s(q^2) \ll 1$, so that the approximate expression in (8.71) may hold, requiring $q^2 \gg \Lambda^2$. Suppose $u(x, q_0^2)$ represents the quark distribution at $q^2 = q_0^2$. Then, as explained before, at $q^2 > q_0^2$ the quark distribution can be modified by the radiation of a gluon. Since there is no scale or mass in the problem, the fractional change in $u(x)$ will be proportional to the fractional change in q^2 and we can write (Altarelli and Parisi (1977))

$$\frac{q^2 \, du(x, q^2)}{dq^2} = \frac{du(x, q^2)}{d \ln q^2}$$

$$= \frac{\alpha_s(q^2)}{2\pi} \int_{y=x}^{y=1} u(y, q^2) P_{QQ}\left(\frac{x}{y}\right) \frac{dy}{y}. \tag{8.72}$$

The meaning of this equation is evident from Fig. 8.27. A quark with momentum fraction x which absorbs the current q is shown originating from a quark with fraction $y > x$ which has radiated a gluon with momentum fraction $y - x$. The probability for radiation to occur is measured by α_s, and the probability that a quark which radiates retains a fraction $z = x/y$ of its original momentum is described by the "splitting function" $P_{QQ}(z)$. The form of P_{QQ}, describing the radiation of a vector particle by a fermion, is

$$P_{QQ}(z) = \frac{4}{3} \frac{(1 + z^2)}{(1 - z)}. \tag{8.73}$$

This form of the z-dependence was actually derived by Weizsacker and Williams in 1934 for the radiation of a virtual photon by an electron: the coefficient of $\frac{4}{3}$ is a color factor taken over for QCD (see Appendix J).

Equation (8.72) holds for "valence" quarks, $u(x, q^2)$ or $d(x, q^2)$ in the nucleon, which can be measured from the differences of electron-proton and electron-neutron cross-sections or of neutrino and antineutrino-nucleon cross-sections, where the effects of the $Q\bar{Q}$ pairs of the "sea" cancel out. The evolution of the sea quarks themselves can be described by an equation analogous to (8.72), but containing two terms, since the gluons can transform to $Q\bar{Q}$ pairs. This leads to an increase of the sea-quarks at small x as q^2 increases.

The main point is that, given the measured quark distribution $u(x, q^2)$ over the range x to 1, the logarithmic derivative with respect to q^2, and hence the evolution of the distribution with q^2, is determined by α_s. A typical example of quark distributions (structure functions) from a recent experiment is given in Fig. 8.28(b). From such data, the value of α_s can be estimated, and hence the parameter Λ from Eq. (8.71). Typically, values of $\Lambda \simeq 0.1$–0.5 GeV are obtained. The smallness of Λ, together with the range of

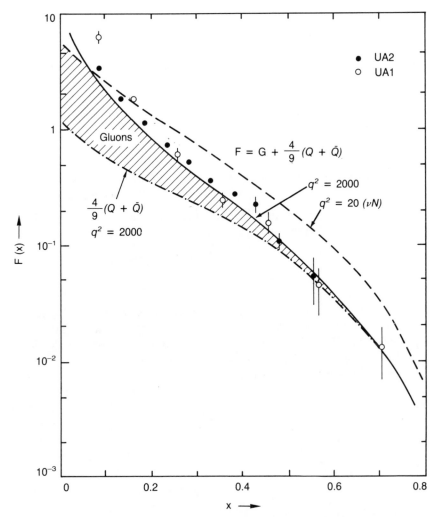

Figure 8.32   Points show structure function combination $G + \frac{4}{9}(Q + \bar{Q})$ of gluons, quarks, and antiquarks deduced from two-jet events at CERN $p\bar{p}$ collider ($q^2 = 2000$ GeV^2). The dashed curve is the same combination from neutrino-nucleon scattering data at $q^2 = 20$ GeV^2. Full curve is for this distribution evolved according to QCD to $q^2 = 2000$ GeV^2 (for $\Lambda = 0.2$ GeV). The dot-dash curve is that expected for quarks and antiquarks only.

q^2 under experimental investigation, shows that α_s is not large ($\simeq 0.2$) and that the corresponding q^2-dependence of the structure functions is small and difficult to measure.

From the experimental data on $Q(x)$ and $\bar{Q}(x)$ as a function of q^2, and the QCD evolution equations, it is possible to deduce the gluon distribution $G(x, q^2)$ in the nucleon. Very crudely, if we describe the momentum distributions by the form $(1 - x)^n$, then $n \simeq 3$ for the valence quarks, $n \simeq 5$ for gluons, and $n \simeq 7$ for sea quarks.

8.12.3. Comparison of Quark and Gluon Distributions in Lepton-Nucleon Scattering and the Process $p\bar{p} \to 2$ jets

In Section 8.9 we described how the two-jet events from $p\bar{p}$ collider experiments measure the combination of structure functions

$$F(x) = G(x) + \tfrac{4}{9}[Q(x) + \bar{Q}(x)] \tag{8.74}$$

where $G(x)$, $Q(x)$, and $\bar{Q}(x)$ denote the fractional momentum distributions of gluons, quarks and antiquarks. The points in Fig. 8.32 show the (normalized) $p\bar{p}$ collider data from the UA1 and UA2 experiments at CERN. The dashed curve is the combination (8.74) calculated from neutrino scattering data (for example, Fig. 8.28(b)) at $q^2 = 5$–150 GeV^2, with mean value $q^2 = 20$ GeV^2. The full-line curve is the result of evolving this distribution to $q^2 = 2000$ GeV^2, appropriate to the collider data, assuming $\Lambda = 0.2$ GeV. Considering the completely different nature of the two types of experiment, the measure of agreement in the shape of $F(x)$ is impressive. The discrepancy at small x is not unexpected: here the information on the gluon distribution from the neutrino data is very uncertain. The importance of the gluon contribution is emphasized by the dot-dash curve, for the Q, $\bar{Q}$ contribution only.

8.13. SUMMARY ON QCD

We have seen that the success of the parton model, of hadrons built from quasi-free quark and gluon constituents, can be understood in terms of a relatively weak color coupling between the quarks at high momentum transfer, i.e., $\alpha_s < 1$. The $p\bar{p}$ collider data is in agreement with a Coulombic ($1/r$) potential for close collisions (high q^2). The nature of three-jet events in $e^+ e^-$ annihilation, and the evolution of nucleon structure functions with q^2, are consistent with QCD predictions, and these data as well as those from $p\bar{p}$ collisions imply the existence of gluon-gluon scattering via gluon exchange and hence of the triple-gluon vertex, i.e., the non-Abelian character of QCD. All experiments confirm the vector nature of the gluon.

As a field theory, QCD has not been quantitatively verified in the same way as has QED, described in Chapter 6. The test particles of QED—the leptons and photons—exist as free particles, while quarks and gluons are confined in hadrons. As a result, interpretation of results is complicated by "cross-talk" between that quark in the nucleon participating in the hard collision (with a lepton, or quark from another hadron) and the other "spectator" quark constituents. These communicate via a gluon propagator and their effects therefore vary as $1/q^2$, $1/q^4$, etc., and should be less important at very high q^2, in comparison with the $\alpha_s \simeq 1/\ln q^2$ dependence of the primary process. Avoiding such "high-twist" contributions is one of the motivations for a high-energy ep collider (Table 2.1) able to reach $q^2 \simeq 20{,}000 \text{ GeV}^2$.

The non-Abelian nature of QCD implies that α_s must decrease logarithmically with q^2: the results to data are consistent with this behavior but do not yet prove it. Finally, we have to emphasize again that at low q^2, α_s becomes large and the perturbative approach (expansion in powers of α_s) discussed above become meaningless. In particular, the mechanism of color confinement at large distance is not understood.

Despite all the above deficiencies, no one doubts that QCD is enormously successful and will at least form an important part of the eventual complete field theory of the strong interactions.

PROBLEMS

8.1. Show that if in the parton model relation (8.12) we do not neglect the mass M of the nucleon or the mass m of the scattered parton, the fractional momentum of the nucleon carried by the initial parton is given by

$$\xi = x\left[1 - \frac{M^2 x^2 - m^2}{q^2} + \cdots\right],$$

where $q^2 \gg M^2 x^2$ or m^2, and as usual, $x = q^2/(2Mv)$.

8.2. Assuming that the quark and antiquark momentum distributions in the nucleon and pion are of the form $A(1 - x)^3$ and $B(1 - x)$ respectively, find an expression for the invariant cross-section for production of muon pairs of mass m in a pion-nucleon collision, in the limit where $\tau = m^2/s$ is small, where s is the square of the CMS energy.

8.3. The ratio of total cross-sections for antineutrinos and neutrinos on nucleon targets is found to be $R = 0.5$ in a particular experiment. Deduce the ratio of antiquark to quark momentum content in the nucleon and compute the average value of the quantity y (defined in (8.17)) in neutrino and antineutrino events.

8.4. Show that, at high q^2, the elastic form factor of the nucleon in Eq. (6.28) has the form naively expected in QCD, if the "struck" quark and the two "spectator" quarks are to recoil coherently, and their interactions are mediated by single gluon exchanges.

8.5. Assume that the momentum distributions of u quarks in the proton and $\bar{d}$-quarks in the antiproton have the forms

$$F_u(x) = xu(x) = a_1(1 - x)^3,$$

$$F_{\bar{d}}(x) = x\bar{d}(x) = a_2(1 - x)^3,$$

where x is the Bjorken variable (fractional momentum of nucleon carried by a quark). If the quarks account in total for half the nucleon momentum, find a_1 and a_2.

Using the value of the peak cross-section σ_0 for the process $u\bar{d} \to W^+$ as deduced from (7.69), integrate the cross-section over the above quark distributions to calculate the cross-section σ for the process $p\bar{p} \to W^+ + \dots$ as a function of the $p\bar{p}$ center-of-mass energy, $\sqrt{s}$. Express σ in terms of σ_0, $p = M_W^2/s$ and the total width Γ_W and mass M_W of the intermediate vector boson. Evaluate σ for $\sqrt{s} = 0.3$, 1.0, and 10 TeV. ($M_W = 83$ GeV, $\Gamma_W = 2.7$ GeV).

BIBLIOGRAPHY

Close, F. E., "The quark parton model", *Rep. Prog. Phys.* **42**, 1285 (1979).

Close, F. E., *An Introduction to Quarks and Partons*, Academic Press, New York, 1979.

Drees, J., and H. Montgomery, "Muon scattering", *Ann. Rev. Nucl. Part. Science* **33**, 383 (1983).

Feynman, R. P., *Photon-Hadron Interactions*, Benjamin, New York, 1972.

Fisk, H., and F. Sciulli, "Charged current neutrino interactions", *Ann. Rev. Nucl. Part. Science* **32**, 499 (1982).

Francis, W. F., and T. B. W. Kirk, "Muon scattering at Fermilab", *Phys. Rep.* **54**, 307 (1979).

Friedman, J. I., and H. W. Kendall, "Deep inelastic electron scattering", *Ann. Rev. Nucl. Science* **22**, 203 (1972).

Harari, H., "Quarks and leptons," *Phys. Rep.* **42**, 235 (1978).

Hofstadter, R., "Nuclear and nucleon scattering of high-energy electrons", *Ann. Rev. Nucl. Science* **7**, 231 (1957).

Jacob, M., and P. Landshoff, "Inner structure of the proton", *Sci. Am.* **243**, 46 (Mar. 1980).

Lederman, L. M., "Lepton production in hadron collisions", *Phys. Rep.* **26**, 149 (1976).

Perkins, D. H., "Inelastic lepton nucleon scattering", *Rep. Prog. Phys.* **40**, 409 (1977).

Schwitters, R. F., and K. Strauch, "The physics of e^+e^- collisions", *Ann. Rev. Nucl. Science* **26**, 89 (1976).

Tung-Mow Yan, "The parton model", *Ann. Rev. Nucl. Science* **26**, 199 (1976).

West, G. B., "Electron scattering from atoms, nuclei, and nucleons", *Phys. Rep.* **18c**, 264 (1975).

CHAPTER 9

The Unification of Electroweak and Other Interactions

As indicated in Chapter 1, we are faced in nature with several types of fundamental interaction or field between particles. Each field has its distinct characteristics, such as space-time transformation properties (vector, tensor, etc.), the set of conservation rules which are obeyed by the interaction, the characteristic coupling constant determining reaction cross-sections, and so forth (see Tables 1.4, 1.5 in Chapter 1).

The fact that the strength of the gravitational interaction between two protons, for example, is only 10^{-38} of their electrical interaction has always been a puzzle and a challenge, and many attempts have been made to try to understand the interrelation between the different fundamental fields. In recent years it has become fashionable to believe that the strong, weak, electromagnetic, and gravitational interactions are different aspects of a single universal interaction, which would be manifest at some colossally high energy. At much lower energies it is necessary to assume that this symmetry is for some reason badly broken, at mass or energy scales which are puny relative to the unification energy, but large enough to result in dramatic differences in the characteristics of the interactions which can be measured in laboratory experiments.

The first sucessful attempt to unify two interations was achieved by Clerk Maxwell in 1865. He showed that electricity and magnetism could be unified into a single theory involving a vector field (the electromagnetic field) interacting between charges and currents. The Maxwell equations involve the introduction of one arbitrary constant—called the velocity of light, c—which is not predicted by the theory and has to be determined by experiment. In the late 1960s Weinberg, Salam, and Glashow described how it might be possible to treat electromagnetic and weak interactions as different aspects of a single electroweak interaction, with the same coupling, e. This symmetry between electromagnetic and weak interactions would be manifest at very large momentum transfers ($q^2 > 10^4 \text{ GeV}^2$). At low energies it is a *broken symmetry*; of the four vector bosons involved, one (the photon) is massless and the others—W^+, W^-, Z^0—are massive. Thus, compared with electromagnetism, weak interactions are short-range and apparently feeble.

More recently, ambitious attempts have been made to carry unification further, to include strong and gravitational interactions as well. Although we discuss these so-called grand unified theories briefly in the closing pages of this chapter, the ideas are very speculative and have very little experimental support at the present time.

9.1. RENORMALIZABILITY IN QUANTUM ELECTRODYNAMICS

In Chapters 1 and 3 we noted the desirability of formulating quantum field theories which have the property of *renormalizability*, meaning that the amplitudes for different processes associated with an interaction should be well behaved, i.e., nondivergent at high energy and to high orders in the coupling constant. The prototype field theory, that of quantum electrodynamics (QED), does in fact contain divergent terms associated with integrals over intermediate states, but it is found that these divergences can always be absorbed into a redefinition of the "bare" lepton charges and masses, which are in any case arbitrary, as being equal to the physically measured values. Thus, we can say that a theory is renormalizable if, at the cost of introducing a *finite* number of arbitrary parameters (to be determined from experiment) the predicted amplitudes for physical processes remain finite at all energies and to all orders in the coupling constant. QED is an example of such a theory (and, for many years, was the only one). In it there is a small number of arbitrary constants; h, e, and m (the electron mass). The experimental tests of QED have been described in Chapter 6.

In contrast, early theories of weak interactions, while well behaved at low energy and to first order, involved divergences in higher orders, which could be canceled only at the price of introducing an indefinitely large number of arbitrary constants, thus losing essentially any predictive power. Good high-energy behavior and cancellation of divergent terms in higher order are thus sensible demands for any physical theory. The subject of renormalizability is highly technical and outside the scope of this text, but we just remark that it appears to be connected with the property of invariance of the interaction under gauge transformations, as discussed below (see also Section 3.6.1). Not surprisingly therefore, great emphasis has been placed in recent years on the development of generalized gauge theories, which also lead in a natural way to unification of the various fundamental interactions.

9.2. DIVERGENCES IN THE WEAK INTERACTIONS

In contrast with QED, the V-A (Fermi) theory of weak interactions is badly divergent. We recall that in the Fermi theory of β-decay, the four fermions involved are assumed to have a contact interaction specified by the Fermi constant G. An example is the process

$$v_e + e^- \rightarrow e^- + v_e. \tag{9.1}$$

The cross-section has the pointlike form given in (8.19) using (8.14) for the substitution $q^2 = 2Mv = 2MEy$;

$$\frac{d\sigma}{dq^2} = \frac{G^2}{\pi}, \tag{9.2}$$

where q^2 is the momentum transfer squared. The total cross-section is then

$$\sigma_{tot}(ve) = \frac{G^2}{\pi} q_{max}^2 = \frac{2G^2 mE}{\pi} = \frac{G^2 s}{\pi}, \tag{9.3}$$

where m is the electron mass, E the incident neutrino energy, and s is the CMS energy squared ($s = 2mE$). The cross-section depends only on G and the phase space, which rises with CMS momentum p^* as $p^{*2} = s/4$ (we assume $E \gg m$).

   The cross-section for such an elastic scattering process can also be found from wave theory (Eqs. (4.45) and (4.55)) which yields, for $\eta = 0$, spin $s = \frac{1}{2}$ for the target electron and pointlike scattering ($l = 0$),

$$\sigma_{max} = \pi \lambdabar^2 (2l + 1)/(2s + 1) = \pi \lambdabar^2/2, \tag{9.4}$$

where $\lambdabar = \hbar/p^*$ is the CMS wavelength. So Eq. (9.3) predicts a value for $\sigma_{tot} > \sigma_{max}$ when

$$\frac{4G^2 p^{*2}}{\pi} > \frac{\pi}{2p^{*2}}$$

or when

$$p^* > (\pi/G\sqrt{8})^{1/2} \simeq 300 \text{ GeV}/c, \tag{9.5}$$

   At sufficiently large energy, the Fermi theory therefore predicts a cross-section exceeding the wave-theory limit, which is determined by the condition that the scattered intensity cannot exceed the incident intensity in any partial wave: frequently this is called the unitarity limit. The basic reason for the bad high-energy behavior of the *V-A* theory (which becomes steadily worse if processes of higher order in G—i.e., G^2, G^4, etc.—are considered) is that G has the dimensions of an inverse power of the energy. Somehow, we have to redefine the weak interaction in terms of a dimensionless coupling constant.

The proposed intermediate vector boson $W^\pm$ of the weak interactions was discussed in Section 7.13. Its effect is to introduce a propagator term $(1 + q^2/M_W^2)^{-1}$ into the scattering amplitude, which "spreads" the interaction over a finite range, of order M_W^{-1}, so that σ_{tot} in (9.3) will tend to a constant value $G^2 M_W^2/\pi$ at high energy. The unitarity limit in a given partial wave is still broken, although only logarithmically. Quadratic divergences however still appear in other, more esoteric, processes; for example, $\nu\bar\nu \to W^+ W^-$ is a conceivable reaction for which $\sigma_{tot} \propto s$. What is needed therefore is a mechanism to cancel the weak-interaction divergences systematically and to all orders in G.

9.3. INTRODUCTION OF NEUTRAL CURRENTS

Figure 9.1 shows in (a) one of the recalcitrant diagrams giving a quadratic divergence (for longitudinally polarized Ws). One can in principle cancel this divergence by introducing *ad hoc* a neutral boson Z^0 *or* a new heavy lepton E^+ (with the same lepton number as e^- and ν_e)—or both—with suitably chosen couplings. Note that diagram (b) involves a neutral current, for which there is experimental evidence, as discussed in Section 7.10. There is no present evidence for a third, electron-type lepton E^+, and we do not consider this possibility (c) further.

Similarly the electromagnetic process $e^+ e^- \to W^+ W^-$ shown in Fig. 9.2(a) is also divergent, and cancellation in this order may again be effected by postulating a neutral vector boson Z_1^0 as in (b). Although in principle the cancellations in Figs. 9.1 and 9.2 could be effected with two different particles, Z^0 and Z_1^0, it is more economical to have just one. Nature seems to have chosen this course. In that case, it is clear that the Z^0 weak coupling, g_W, will have to be similar in magnitude to the electromagnetic coupling, e. In other

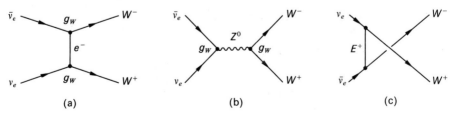

Figure 9.1

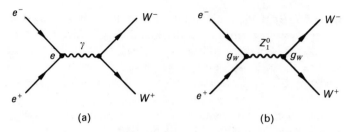

(a)   (b)

Figure 9.2

words, the *two fields are unified*, with (apart from numerical factors of order unity) the same intrinsic coupling strength $g_W \simeq e$ to leptons of the mediating bosons W, Z^0, and γ. Recalling from (7.56) and Fig. 9.3 that in the low-q^2 limit

$$\lim_{q^2 \to 0} \frac{g_W^2}{q^2 + M_W^2} \equiv \frac{G}{\sqrt{2}}, \tag{9.6}$$

we find, setting $g_W = e$,

$$M_W^2 = \frac{g_W^2 \sqrt{2}}{G} = \frac{e^2 \sqrt{2}}{G}. \tag{9.7}$$

Inserting $e^2 = 4\pi/137$, $G = 10^{-5}$ GeV^{-2} (in units $\hbar = c = 1$), we get

$$M_{W, Z^0} \simeq 100 \text{ GeV}. \tag{9.8}$$

This is the approximate mass required for the W- and Z^0-particles if they are to have the same coupling as in electromagnetism and to give an effective four-fermion coupling of magnitude G in weak interactions at low energy.

Actually we implied that by introducing a Z^0 with the right coupling, all the divergences in $e^+e^- \to W^+W^-$ would disappear. This is only true if the electron mass can be neglected. For a finite electron mass, a residual divergence exists and has to be canceled by introduction of further scalar particles, with special couplings proportional to the lepton mass. Such particles are called *Higgs scalars*.

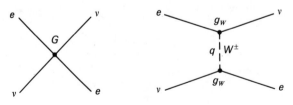

Figure 9.3

9.4. GAUGE INVARIANCE IN QED

Let us first recall the discussion of gauge invariance in Section 3.6.1. There we saw that the quantum-mechanical description of experiments on charged particles could be made invariant under local phase transformations on the particle wavefunction if we introduced a long-range field coupled to charge (the electromagnetic field) and we made a suitable simultaneous local gauge transformation on the electromagnetic potential. We can reproduce this argument in mathematical form as follows.

Consider a phase rotation on the charged-particle wavefunction $\psi(x)$, where x stands for a space-time coordinate, of the form

$$\psi(x) \rightarrow e^{ie\theta(x)}\psi(x), \tag{9.9}$$

where e is the electric charge of the particle. Then the gradient

$$\frac{\partial\psi(x)}{\partial x_\mu} \equiv \partial_\mu\psi(x) \rightarrow e^{ie\theta(x)}[\partial_\mu\psi(x) + ie\psi(x)\partial_\mu\theta(x)]$$

$$\neq e^{ie\theta(x)}\partial_\mu\psi(x),$$

where the subscript μ runs from 1 to 4 for the space and time components. The equations of motion involve terms of the form $\psi^*(x)\partial_\mu\psi(x)$, and this is clearly not invariant under this local phase transformation. However let us write a local gauge transformation on the potential A_μ of the electromagnetic field which is coupled to the charge:

$$A_\mu(x) \rightarrow A_\mu(x) + \partial_\mu\theta(x), \tag{9.10}$$

where the field equations of Maxwell are invariant if we add the derivative of any scalar $\theta(x)$ to the 4-vector potential. Let us also replace ∂_μ by the *covariant derivative*, defined by

$$D_\mu = \partial_\mu - ieA_\mu. \tag{9.11}$$

which is the same as (3.25), written in differential operator notation ($p = -i\hbar\partial/\partial x_\mu$). Under the simultaneous phase transformation on $\partial_\mu\psi(x)$ and gauge transformation on $A_\mu(x)$ we get

$$D_\mu\psi(x) \rightarrow e^{ie\theta(x)}[\partial_\mu\psi(x) + ie\psi(x)\partial_\mu\theta(x) - ie\psi(x)A_\mu(x) - ie\psi(x)\partial_\mu\theta(x)]$$

$$= e^{ie\theta(x)}D_\mu\psi(x), \tag{9.12}$$

and $\psi^*(x)D_\mu\psi(x)$ is now invariant. The fact that we must use D_μ instead of ∂_μ is already specifying the form of the interaction (eA_μ) of the charged particle with the field. As indicated in Chapter 3, the gauge invariance of the electromagnetic interaction leads to a conserved current (charge conservation) and the masslessness of the bosons* (photons) associated with A_μ. Most

---

* Gauge invariance fails if the bosons are massive, and can only be restored by including extra interactions (Higgs scalars).

importantly, the theory contains a high degree of symmetry, and as a result, it is renormalizable, with systematic cancellations of divergent terms, order by order in the coupling constant e.

9.5. GENERALIZED GAUGE INVARIANCE

The infinite set of phase transformations (9.9) form the unitary group called U(1). Since $\theta(x)$ is a scalar quantity, the U(1) group is said to be Abelian. More complex phase transformations are also possible, specified by noncommuting operators, and these belong to the so-called non-Abelian groups. Such gauge transformations were proposed by Yang and Mills in 1954 and involved fields containing both charged and neutral massless bosons. Specifically, they chose the group SU(2) of isospin, which involves the *noncommuting* Pauli matrices $\tau = \tau_1, \tau_2, \tau_3$ (see Appendix B). Recall that the conservation of isospin in strong interactions implies invariance under an isospin rotation (see (3.14)):

$$\psi \to e^{ig\tau \cdot \Lambda}\psi, \tag{9.13}$$

where Λ is an arbitrary vector about which the rotation in "isospin space" takes place and g is a constant. Again, we can require that $\Lambda(x)$ be chosen arbitrarily at different space-time points x. For example, we could choose the two states ψ of the nucleon as proton or neutron independently at different places. In the same way as for electromagnetism, a gauge-invariant description can be obtained by introducing a massless isovector field $\mathbf{W}_\mu$ with charged and neutral components, and a strong coupling constant g, analogous to e. Invariance of $\psi^* D_\mu \psi$ under the transformation (9.13) is obtained by introducing a covariant derivative of the form

$$D_\mu = \partial_\mu - ig\tau \cdot \mathbf{W}_\mu, \tag{9.14}$$

where an infinitesimal gauge transformation of the field $\mathbf{W}_\mu$ is given by

$$\mathbf{W}_\mu \to \mathbf{W}_\mu + \partial_\mu \Lambda - g\Lambda \times \mathbf{W}_\mu. \tag{9.15}$$

The extra term, as compared with (9.10), is associated with the fact that the isospin matrices $\tau = \tau^{(1)}, \tau^{(2)}, \tau^{(3)}$ do not commute.[†] The transformations (9.13) and (9.15) can be verified by writing down the expression for $\psi^* D_\mu \psi$ in the case where Λ is infinitesimal. Then, neglecting second-order terms in Λ,

$$\begin{aligned}
\psi^* D_\mu \psi &= \psi^*(\partial_\mu - ig\tau \cdot \mathbf{W})\psi \\
&\to \psi^*(1 - ig\tau \cdot \Lambda)[\partial_\mu - ig\tau \cdot (\mathbf{W} + \partial_\mu \Lambda - g\Lambda \times \mathbf{W})(1 + ig\tau \cdot \Lambda)]\psi \\
&= \psi^*(\partial_\mu - ig\tau \cdot \mathbf{W})\psi \\
&\quad + g^2\psi^*[-(\tau \cdot \Lambda)(\tau \cdot \mathbf{W}) + (\tau \cdot \mathbf{W})(\tau \cdot \Lambda) + i\tau \cdot \Lambda \times \mathbf{W}]\psi. \tag{9.16}
\end{aligned}$$

[†] Boldface type indicates isovector quantities, and superscripts the Cartesian components in "isospin space". The subscript μ stands for the space-time components of a 4-vector ($\mu = 1, \ldots, 4$).

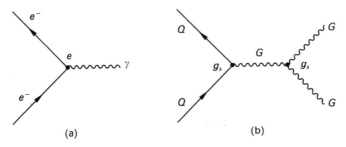

Figure 9.4

The second term, written out in isovector components i, j, k, is

$$\sum \tau^i W^i \sum \tau^j \Lambda^j - \sum \tau^j \Lambda^j \sum \tau^i W^i + i \sum \tau^k \sum \Lambda^j W^i$$

$$= \sum W^i \Lambda^j \sum (\tau^i \tau^j - \tau^j \tau^i) - i \sum W^i \Lambda^j \sum \tau^k = 0. \quad (9.17)$$

Hence, inclusion of the vector-product term in (9.15) leads to local gauge invariance. It implies an interaction of $\mathbf{W}_\mu$ with all particles carrying isospin and hence with itself. Thus the bosons $\mathbf{W}_\mu$ are at one and the same time the carriers and a part of the source of the isospin field. A parallel example is that of gravity, where the field quanta are the massless gravitons, mediating the interaction between masses; but since the gravitons carry energy and momentum, they are themselves part of the source of gravitational field. Similarly, we saw that in the gauge theory of strong interactions, involving the non-Abelian gauge group of color, SU(3), the quanta (gluons) of the color field themselves carry color and therefore are self-coupled (Fig. 9.4).

We note, however, that the hypothetical charged bosons of the isospin field are massless, just like the photon, and we infer that isospin symmetry in hadronic interactions cannot be (and is not) exact, since no massless charged particles exist.

9.6. THE WEINBERG-SALAM SU(2) × U(1) MODEL

In 1967–1968 Weinberg and Salam proposed a gauge theory unifying weak and electromagnetic interactions—the so-called electroweak interactions, based on an SU(2) group of "weak isospin" I and a U(1) group of "weak hypercharge" Y (phase transformations). They invoked a process called "spontaneous symmetry breaking" which endows the gauge bosons with mass, without spoiling the renormalizability of the theory. This may be achieved with the help of an isospin doublet of scalar mesons called Higgs scalars, which generate mass as a result of self-interaction, and which we mentioned above in connection with removal of divergences in $e^+e^- \to$ W^+W^-. This topic is outside the scope of this text, and the reader is referred to the chapter bibliography for details. At the price of extra scalar particles, the quanta of the "weak isospin" field are given a mass and the theory remains renormalizable. The crucial question of renormalizability was not settled until 1971 by 't Hooft, and until that time the theory was not taken too seriously. The fundamental vector bosons are a massless isovector triplet $\mathbf{W}_\mu = W_\mu^{(1)}, W_\mu^{(2)}, W_\mu^{(3)}$ (for SU(2)) and a massless isosinglet B_μ (for U(1)). As a result of spontaneous symmetry breaking, three bosons (denoted W_μ^+, W_μ^-, and Z_μ^0) acquire mass, and one (A_μ, the photon) remains massless. These four bosons are combinations of $\mathbf{W}_\mu$ and B_μ, as discussed below. There are no new leptons in this model, and the cancellation of divergences in Fig. 9.1 is accomplished by the neutral boson Z^0.

The interaction energy (usually represented by the so-called Lagrangian energy density $\mathscr{L}$) of fermions with the fields $\mathbf{W}_\mu$, B_μ is the product of the fermion currents with the fields, that is, of the form

$$\mathscr{L} = g\mathbf{J}_\mu \cdot \mathbf{W}_\mu + g' J_\mu^Y B_\mu, \tag{9.18}$$

where $\mathbf{J}_\mu$ and J_μ^Y represent respectively the isospin and hypercharge currents of the fermions (leptons or quarks), and g and g' are their couplings to $\mathbf{W}_\mu$ and B_μ, analogous to the charge e in (9.11). To avoid writing factors of two everywhere, we define the weak hypercharge as $Y = Q - I_3$ (rather than $2(Q - I_3)$ for the strong hypercharge, as in (5.1)), where Q is the electric charge and I_3 the third component of weak isospin. Then

$$J_\mu^Y = J_\mu^{\text{e.m.}} - J_\mu^{(3)}, \tag{9.19}$$

where $J_\mu^{\text{e.m.}}$ is the electromagnetic current, coupling to the charge Q, and $J_\mu^{(3)}$ is the third component of the isospin current $\mathbf{J}_\mu$. The physical bosons consist of the charged particles $W_\mu^\pm$ and the neutrals Z_μ and A_μ (the photon). The latter are taken as linear combinations of $W_\mu^{(3)}$ and B_μ. Thus, we can set (see Eqs. C.6 and C.7(c), Appendix C)

$$W_\mu^\pm = \frac{1}{\sqrt{2}} (W_\mu^{(1)} \pm i W_\mu^{(2)}) \tag{9.20}$$

and

$$W_\mu^{(3)} = \frac{g Z_\mu + g' A_\mu}{\sqrt{g^2 + g'^2}}, \tag{9.21}$$

$$B_\mu = \frac{-g' Z_\mu + g A_\mu}{\sqrt{g^2 + g'^2}}, \tag{9.22}$$

where $W_\mu^{(3)}$ and B_μ are orthogonal as required ($\langle W_\mu^{(3)}|B_\mu\rangle = 0$). Hence

$$
\begin{aligned}
\mathcal{L} &= g(J_\mu^{(1)}W_\mu^{(1)} + J_\mu^{(2)}W_\mu^{(2)}) + g(J_\mu^{(3)}W_\mu^{(3)}) + g'(J_\mu^{\text{e.m.}} - J_\mu^{(3)})B_\mu \\
&= (g/\sqrt{2})(J_\mu^- W_\mu^+ + J_\mu^+ W_\mu^-) + J_\mu^{(3)}(gW_\mu^{(3)} - g'B_\mu) + J_\mu^{\text{e.m.}}g'B_\mu,
\end{aligned}
$$

where $J_\mu^\pm = J_\mu^{(1)} \pm iJ_\mu^{(2)}$.

Inserting the expressions for $W_\mu^{(3)}$ and B_μ from (9.21) and (9.22) and setting

$$
g'/g = \tan \theta_w, \tag{9.23}
$$

we get the result

$$
\mathcal{L} = \frac{g}{\sqrt{2}}(J_\mu^- W_\mu^+ + J_\mu^+ W_\mu^-) + \frac{g}{\cos \theta_w}(J_\mu^{(3)} - \sin^2 \theta_w J_\mu^{\text{e.m.}})Z_\mu + g \sin \theta_w J_\mu^{\text{e.m.}} A_\mu.
$$

$$\uparrow \qquad\qquad\qquad\qquad \uparrow \qquad\qquad\qquad \uparrow$$

weak CC $\qquad\qquad\qquad$ weak NC $\qquad\qquad$ e.m. NC

$$\tag{9.24}$$

This equation shows that the interaction contains the weak charge-changing current, a weak neutral current, and the electromagnetic neutral current, for which we know (see (9.11)) the coupling to be e. Hence

$$
e = g \sin \theta_w. \tag{9.25}
$$

The angle θ_w is called the weak mixing angle (or Weinberg angle).

Let us now consider the electroweak coupling of the leptons. Just as for the gauge bosons, they will be endowed with weak isospin and weak hypercharge. Further, we know that the weak charged-current interaction (mediated by $W^\pm$) is parity-violating, and connects, for example, the left-handed states of neutrino and electron. On the other hand the electromagnetic interaction (the last term in (9.24)) is parity-conserving and involves both LH and RH states of the electron. Hence, we assign the lepton states to a LH doublet and a RH singlet, as follows:

$$
\psi_L = \frac{1+\gamma_5}{2}\begin{pmatrix} \nu_e \\ e^- \end{pmatrix} \quad \left.\begin{array}{l} I = \tfrac{1}{2}, I_3 = +\tfrac{1}{2}, Q = 0 \\ I = \tfrac{1}{2}, I_3 = -\tfrac{1}{2}, Q = -1 \end{array}\right\} \quad Y = -\tfrac{1}{2},
$$

$$
\psi_R = \frac{1-\gamma_5}{2}(e^-) \quad I = 0, Q = -1, \qquad\qquad Y = -1. \tag{9.26}
$$

The lepton wavefunctions in (9.26) are normalized to give unity for the electron state (both LH and RH components). With this normalization, the charge-changing leptonic currents in (9.24) are defined as

$$J_\mu^+(CC) = \bar{v}\gamma_\mu \frac{(1 + \gamma_5)}{2} e = \bar{v}_L\gamma_\mu e_L = \bar{\psi}_L\gamma_\mu \tau^+ \psi_L, \qquad (9.27a)$$

$$J_\mu^-(CC) = \bar{e}\gamma_\mu \frac{(1 + \gamma_5)}{2} v = \bar{e}_L\gamma_\mu v_L = \bar{\psi}_L\gamma_\mu \tau^- \psi_L, \qquad (9.27b)$$

where $\tau^\pm = \tau_1 \pm i\tau_2$ are the Pauli operators (B.5) suitable for describing $I = \frac{1}{2}$ systems. Then the third member of the isospin triplet of currents will be

$$J_\mu^{(3)}(NC) = \bar{\psi}_L\gamma_\mu \tau_3 \psi_L = \bar{\psi}\gamma_\mu \frac{(1 + \gamma_5)}{2} I_3\psi = \frac{1}{2}(\bar{v}_L\gamma_\mu v_L - \bar{e}_L\gamma_\mu e_L). \quad (9.27c)$$

Let us now compare the expressions (9.27(a),(b)) with the previous definition of the charged-current matrix element (7.27). Then using (9.24), and in the limit of $q^2 \ll M_W^2$—so that the W propagator gives a factor of $1/M_W^2$—we get

$$|M(CC)| = \left(\frac{g}{\sqrt{2}}\right)^2 \frac{1}{M_W^2} \left[\bar{e}\gamma_\mu \frac{(1 + \gamma_5)v}{2}\right]\left[\bar{v}\gamma_\mu \frac{(1 + \gamma_5)e}{2}\right]$$

$$\equiv \frac{G}{\sqrt{2}} [\bar{e}\gamma_\mu(1 + \gamma_5)v][\bar{v}\gamma_\mu(1 + \gamma_5)e],$$

so that

$$\frac{G}{\sqrt{2}} = \frac{g^2}{8M_W^2} = \frac{g_W^2}{M_W^2},$$

and from (9.25) it follows that

$$M_{W^\pm} = \left(\frac{g^2\sqrt{2}}{8G}\right)^{1/2} = \left(\frac{e^2\sqrt{2}}{8G \sin^2 \theta_w}\right)^{1/2} = \frac{37.4}{\sin \theta_w} \text{ GeV}, \qquad (9.28)$$

to be compared with our first rough guess, (9.8).

In order to find the predicted Z^0-mass, recall that in dealing with bosons (described by the Klein-Gordon equation (1.6)), the mass matrix will involve the squares of the boson masses (as for vector meson mixing in (5.19)). Inverting the relations (9.21) and (9.22), we find ($\theta = \theta_w$):

$$Z_\mu = W_\mu^{(3)} \cos \theta - B_\mu \sin \theta,$$

$$A_\mu = W_\mu^{(3)} \sin \theta + B_\mu \cos \theta,$$

so that, using the empirical fact that the photon is massless and orthogonal to the Z, we get

$$M_Z^2 = M_W^2 \cos^2 \theta + M_B^2 \sin^2 \theta - 2M_{BW}^2 \cos \theta \sin \theta,$$

$$M_\gamma^2 = 0 = M_W^2 \sin^2 \theta + M_B^2 \cos^2 \theta + 2M_{BW}^2 \cos \theta \sin \theta,$$

$$M_{Z\gamma}^2 = 0 = (M_W^2 - M_B^2) \sin \theta \cos \theta + M_{BW}^2(\cos^2 \theta - \sin^2 \theta),$$

where M_{BW} and $M_{Z\gamma}$ are the off-diagonal mass terms. Eliminating these, we easily find that

$$M_{Z^0} = \frac{M_{W^\pm}}{\cos \theta_w} = \frac{75}{\sin 2\theta_w} \text{ GeV}. \tag{9.29}$$

These predictions of boson masses are for the simplest (Salam-Weinberg) model of spontaneous symmetry-breaking, which involves one weak isospin doublet of scalar (Higgs) particles. More complex isospin structure for the Higgs (e.g., an isospin triplet) is possible, with the result that more physical Higgs mesons appear and (9.29) is no longer valid. The second term of (9.24), specifying the relative magnitude of the neutral-current coupling then acquires a factor ρ, and (9.29) is replaced by

$$M_{Z^0}^2 = \frac{M_W^2}{\rho \cos^2 \theta_w}. \tag{9.30}$$

However, all experiments to date find $\rho = 1$, consistent with the simplest model.

9.6.1. Neutral-Current Couplings of Fermions

From (9.24) and (9.26), the left-handed and right-handed couplings of the fermions to the Z^0 (neutral current) have the coefficients

$$g_L = I_3 - Q \sin^2 \theta_w,$$
$$g_R = -Q \sin^2 \theta_w, \tag{9.31}$$

where Q is the electric charge in units of $|e|$ and $I_3 = \pm\frac{1}{2}$ is the third component of the weak isospin. Since vector (axial vector) interactions couple LH and RH states with the same (opposite) sign, the V- and A- coefficients are

$$c_V = g_L + g_R = I_3 - 2Q \sin^2 \theta_w,$$
$$c_A = g_L - g_R = I_3. \tag{9.32}$$

The assignment for charged leptons and neutrinos was already given in (9.26). For the u-quark, or the c- and t-quarks of $Q = +\frac{2}{3}$, the value of $I_3 = +\frac{1}{2}$, whereas $I_3 = -\frac{1}{2}$ for the d-, s-, and b-quarks, of $Q = -\frac{1}{3}$. Thus, we can draw up the following table:

Fermion	$2c_V$	$2c_A$	
ν_e, ν_μ, ν_τ	1	1	
e, μ, τ	$-1 + 4 \sin^2 \theta_w$	-1	(9.33)
u, c, t	$1 - \frac{8}{3} \sin^2 \theta_w$	1	
d, s, b	$-1 + \frac{4}{3} \sin^2 \theta_w$	-1	

Note that for charged leptons and quarks and finite values of $\sin^2 \theta_w$, c_V and c_A have different magnitudes, that is, the neutral couplings have unequal contributions from the V- and A-amplitudes, as distinct from the charged current (V-A) coupling. The couplings of antiparticles are obtained by interchange of g_L and g_R in (9.31).

9.7 EXPERIMENTAL TESTS OF NEUTRAL CURRENTS IN THE WEINBERG-SALAM MODEL

The theory of Weinberg and Salam makes several predictions, in terms of the one free parameter, $\sin^2 \theta_w$. The first important prediction is of the existence of neutral weak currents, verified in 1973, as described in Chapter 7, and now discussed in more detail. The second and even more dramatic prediction is of the existence of the massive bosons $W^\pm$ and Z^0, with masses and decay characteristics determined in terms of $\sin^2 \theta_w$ and the structure of the weak currents as described above. Their discovery was also described in Chapter 7 and a more detailed discussion is given in a following section.

9.7.1. Neutrino-Electron Scattering

The pioneer experiment demonstrating the existence of the neutral current scattering process $\bar{v}_\mu e \to \bar{v}_\mu e$ has been described in Section 7.10. The various processes involving neutral (and charged) currents in neutrino-electron scattering are illustrated in Fig. 9.5.

In the limit of small q^2, the matrix element for neutral-current ve scattering via Z^0-exchange will be, from (9.24)

$$|M^{ve}(NC)| = \frac{g^2}{\cos^2 \theta_w} \cdot \frac{1}{M_Z^2} \cdot J^v(NC) \cdot J^e(NC),$$

where, using (9.27(c)),

$$J(NC) = J^{(3)} - \sin^2 \theta_w J^{em} = \tfrac{1}{2}\bar{\psi}\gamma_\mu(1 + \gamma_5)I_3\psi - \sin^2 \theta_w \bar{\psi}\gamma_\mu \psi Q$$
$$= \tfrac{1}{2}\bar{\psi}\gamma_\mu(c_V + \gamma_5 c_A)\psi,$$

with c_V and c_A defined in (9.32). Using (9.28) and (9.29) we therefore find

$$|M^{ve}(NC)| = \frac{8G}{\sqrt{2}} \cdot J^v(NC) \cdot J^e(NC)$$

$$= \frac{8G}{\sqrt{2}} \left[\tfrac{1}{2}\bar{v}\gamma_\mu(c_V^v + \gamma_5 c_A^v)v\right]\left[\tfrac{1}{2}\bar{e}\gamma_\mu(c_V^e + \gamma_5 c_A^e)e\right]$$

$$= \frac{G}{\sqrt{2}} \left[\bar{v}\gamma_\mu(1 + \gamma_5)v\right]\left[\bar{e}\gamma_\mu(c_V + \gamma_5 c_A)e\right],$$

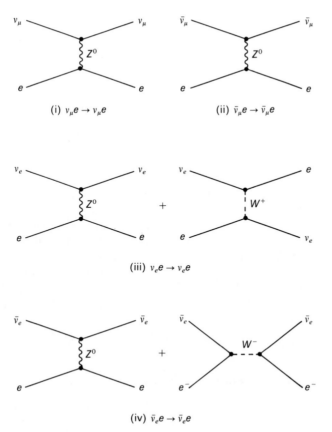

Figure 9.5 Diagrams for the scattering of electron-neutrinos and muon-neutrinos and the corresponding antineutrinos from electron targets.

where we have used the values $c_V^\nu = c_A^\nu = \frac{1}{2}$ from (9.33) and defined $c_V^e = c_V$, $c_A^e = c_A$. From (9.32) we obtain, in terms of the RH and LH electron couplings to the Z^0,

$$|M^{\nu e}(\text{NC})| = \frac{G}{\sqrt{2}} [\bar{\nu}\gamma_\mu(1 + \gamma_5)\nu][\bar{e}\gamma_\mu g_L(1 + \gamma_5)e + \bar{e}\gamma_\mu g_R(1 - \gamma_5)e].$$

From the form of this expression and the results (8.19) and (8.20)—see also Appendix F—we can immediately write the cross-section.

$$\frac{d\sigma^{\nu e}(\text{NC})}{dy} = \frac{2G^2 mE}{\pi} [g_L^2 + g_R^2(1 - y)^2], \qquad (9.34)$$

where E is the neutrino energy, m the mass of the electron target (with $E \gg m$), and yE is the electron recoil energy. The values of g_L and g_R are given in Table 9.1 for the various reactions of Fig. 9.5. Reactions (i) and (ii) proceed

TABLE 9.1 Neutrino-electron couplings in the Weinberg-Salam model

		g_L	g_R
(i)	$\nu_\mu e \to \nu_\mu e$	$-\frac{1}{2} + \sin^2 \theta_w$	$\sin^2 \theta_w$
(ii)	$\bar{\nu}_\mu e \to \bar{\nu}_\mu e$	$\sin^2 \theta_w$	$-\frac{1}{2} + \sin^2 \theta_w$
(iii)	$\nu_e e \to \nu_e e$	$\frac{1}{2} + \sin^2 \theta_w$	$\sin^2 \theta_w$
(iv)	$\bar{\nu}_e e \to \bar{\nu}_e e$	$\sin^2 \theta_w$	$\frac{1}{2} + \sin^2 \theta_w$

only via Z^0-exchange (neutral current). The antineutrino couplings are obtained from those for neutrinos from the interchange $g_L \leftrightarrow g_R$. Reactions (iii) and (iv) can proceed via both Z^0- and W^+-exchange, that is, both neutral and charged currents, the latter having $g_L = 1$, $g_R = 0$ for (iii), and $g_R = 1$, $g_L = 0$ for (iv). Adding these contributions, we obtain the results in the table.

The cross-sections for reactions (i) and (ii) have been measured in several accelerator neutrino experiments, using bubble-chamber and counter techniques. Based on only about 400 events altogether, which simply reflects the very low value of the cross-sections ($\sigma/E_\nu \simeq 10^{-42}$ cm^2 GeV^{-1}), the data provide an estimate

$$\sin^2 \theta_w = 0.22 \pm 0.03. \tag{9.35}$$

9.7.2. Electroweak Interference in $e^+ e^- \to \mu^+ \mu^-, \tau^+ \tau^-$

As explained in Section 6.6, the angular distribution for this process at high energy does not follow the pure QED prediction, since two diagrams are involved (Fig. 9.6), associated with photon and Z^0-exchange. Combining these and squaring, we find, with s the square of the CMS energy,

$$\frac{d\sigma}{d\Omega} = \frac{d\sigma}{d\Omega}(\text{QED}) + \frac{d\sigma}{d\Omega}(\text{interf.}) + \frac{d\sigma}{d\Omega}(\text{weak})$$
$$\qquad \downarrow \qquad\qquad \downarrow \qquad\qquad \downarrow$$
$$\qquad \alpha^2/s \qquad\quad \alpha G \qquad\quad G^2 s \tag{9.36}$$

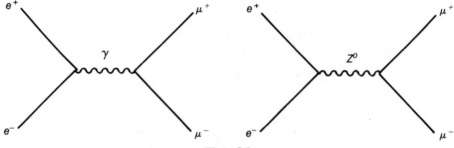

Figure 9.6

The first term is due to pure photon exchange, the last pure Z^0-exchange, and the middle one, the interference term. If $G^2s < \alpha^2/s$, which is true for $s < \alpha/G \simeq 10^3$ GeV^2, the relative magnitude of the interference term will be of order Gs/α. Since the neutral-current interaction is a mixture of V- and A-couplings, backward-forward asymmetry in the muon angular distribution is expected (a "forward" μ^+ means one traveling in the same direction as the incident e^+). The asymmetry is calculated to be

$$A = \frac{F - B}{F + B} = -6\chi c_A^2,$$

where

$$\chi = \frac{Gs\rho}{8\sqrt{2}\pi\alpha} \frac{M_Z^2}{(M_Z^2 + s)}, \qquad (9.37)$$

and the propagator factor for the mediating boson Z^0 has to be included if s is not negligible compared with M_Z^2. A depends only on c_A. Figure 6.8 shows the results obtained at the PETRA collider, from which it was found that $A = (-7.6 \pm 1.9)\%$ at $s \simeq 1000$ GeV^2. Assuming $\rho = 1$, c_A is well determined from A, although the vector coupling of leptons to the Z^0 is only poorly known (from the total cross-section). Figure 9.7 shows a plot of c_V against c_A from the data (Wu (1984)). Note that A determines c_A^2, so there are two solutions. Also shown on this graph are the results on c_A, c_V from neutrino-electron scattering and antineutrino-electron scattering previously described, also providing two solutions (which interchange c_V and c_A, as is clear from (9.34); see also Problem 9.2. The result (9.35) was found by *assuming* the couplings of the model, and fitting the cross-section to $\sin^2 \theta_w$.

The two sets of experiments taken together indicate one solution, with $c_V \simeq 0$ and $c_A = -\frac{1}{2}$, exactly consistent with the standard (Weinberg-Salam) model and with $\sin^2 \theta_w \simeq 0.25$, (see (9.33). The importance of the e^+e^- experiments is that they demonstrate the existence of electroweak interference, at the level predicted.

9.7.3. Deep-Inelastic Neutrino-Nucleon Scattering

The neutral-current cross-section for neutrino scattering by protons, via the quark constituents, is readily obtained from (9.34) if we replace m by xM, where x is the fractional nucleon momentum carried by the quark (see (8.14)), and use the results (9.33) for the neutral-current couplings of the quarks. For simplicity, we consider only u- and d-quarks (i.e., $\bar{u}, \bar{d}, \bar{s}, \ldots$

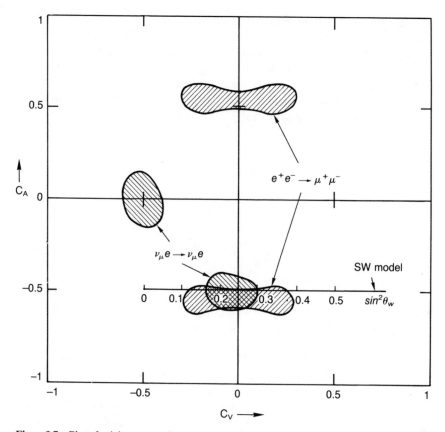

Figure 9.7 Plot of axial vector against vector coupling coefficients, c_A and c_V of charged leptons to the Z^0 (neutral weak current)—see Eq. (9.33). The regions allowed from the measured $\nu_\mu e$ and $\bar{\nu}_\mu e$ scattering cross-sections correspond to the two solutions $c_A \simeq -0.5$, $c_V \simeq 0$ and $c_V \simeq -0.5$, $c_V \simeq 0$. The allowed regions from the measurement of the asymmetry in $e^+ e^- \rightarrow \mu^+ \mu^-$ correspond to $c_A \simeq \pm 0.5$. The two experiments together provide a single solution in the doubly hatched region, with $\sin^2 \theta_w \simeq 0.25$. (After Wu 1984.)

constituents are neglected). Writing the recoil quark energy as yE, where E is the incident neutrino energy, we get, with $s = \sin^2 \theta_w$:

$$\frac{d^2\sigma^{\nu p}(\text{NC})}{dx\, dy} = \frac{G^2 M E x}{2\pi} [(1 - \tfrac{4}{3}s)^2 u(x) + (-1 + \tfrac{2}{3}s)^2 d(x)$$

$$+ (1 - y)^2 \{(-\tfrac{4}{3}s)^2 u(x) + (\tfrac{2}{3}s)^2 d(x)\}],$$

where $u(x)$, $d(x)$ are the quark densities at x. The expression for a neutron target is obtained by the replacement $u(x) \leftrightarrow d(x)$. Hence per nucleon the cross-section becomes

$$\frac{d^2\sigma^{vN}(NC)}{dx\,dy} = \frac{G^2MEx}{2\pi}[u(x) + d(x)][(1 + \tfrac{10}{9}s^2 - 2s) + \tfrac{10}{9}s^2(1 - y)^2]. \quad (9.38)$$

The result for antineutrinos is simply obtained by interchanging the $(1 - y)^2$ factor between the last two terms.

For charged-current reactions, we recall from (8.22), (8.27), and (8.28) that in the same approximation of neglecting antiquarks,

$$\frac{d^2\sigma^{vN}(CC)}{dx\,dy} = \frac{2G^2MEx}{2\pi}[u(x) + d(x)],$$

$$\frac{d^2\sigma^{\bar{v}N}(CC)}{dx\,dy} = \frac{2G^2MEx}{2\pi}[u(x) + d(x)](1 - y)^2. \quad (9.39)$$

Integrating these expressions over x and y, we obtain the ratios of neutral- to charged-current cross-sections,

$$R = \frac{\sigma^{vN}(NC)}{\sigma^{vN}(CC)} = \tfrac{1}{2} - \sin^2\theta_w + \tfrac{20}{27}\sin^4\theta_w, \quad (9.40)$$

$$\bar{R} = \frac{\sigma^{\bar{v}N}(NC)}{\sigma^{\bar{v}N}(CC)} = \tfrac{1}{2} - \sin^2\theta_w + \tfrac{20}{9}\sin^4\theta_w. \quad (9.41)$$

Recent results for the ratios R and $\bar{R}$ are given in Fig. 9.8. The variation predicted in the standard model is indicated in the curve, with minor corrections applied for sea-quark contributions, QCD effects, etc. The weighted average of these results is (Geweniger 1984):

$$\sin^2\theta_w = 0.223 \pm 0.010 \quad (9.42)$$

This number is determined mostly by the value of R, rather than $\bar{R}$, as is clear from Fig. 9.8. If the quantity ρ (the ratio of neutral- to charged-current coupling) is left as a free parameter (rather than the value of unity assumed in (9.42)), we find

$$\rho = 1.00 \pm 0.02, \quad (9.43)$$

determined mostly by $\bar{R}$. These, the most accurate determinations of ρ and $\sin^2\theta_w$ to date, are seen to be consistent with the previous results and with the standard model.

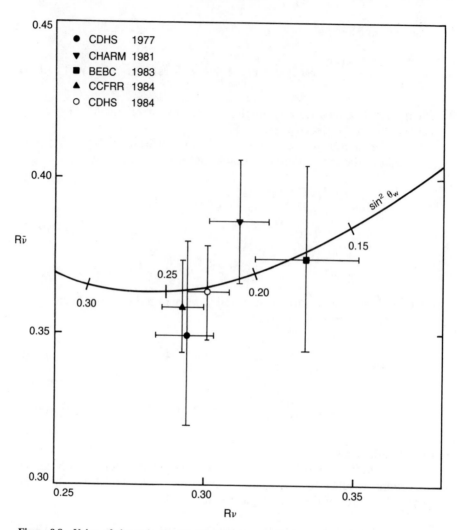

Figure 9.8 Value of the ratio $R_{\bar{\nu}}$ of antineutrino neutral-current cross-section to charged-current cross-section, plotted against the corresponding quantity R_{ν} for neutrinos, measured in deep-inelastic neutrino-nucleon scattering. The data points are from Fermilab (CCFRR) and CERN(CDHS, BEBC) experiments with beam energies of 50–150 GeV. The curve shows the prediction from the Weinberg-Salam model. The world data give a value $\sin^2 \theta_w = 0.22 \pm 0.01$. (After Geweniger 1984.)

9.7.4. Asymmetries in the Scattering of Polarized Electrons by Deuterons

Finally we discuss a very delicate experiment to detect tiny parity-violation effects (asymmetries) due to the interference between Z^0 and γ-exchange in inelastic scattering of polarized electrons by deuterons. The experiment was carried out with beams of electrons of 16–22-GeV/c momentum at SLAC, the reaction being

$$e^-_{L,R} + d_{unpolarized} \to e^- + X,$$

where X is any final hadron state. The weak and eletromagnetic exchanges are depicted in Fig. 9.9(a) and (b). The electromagnetic scattering amplitude is of order e^2/q^2, where q is the 4-momentum transfer, while the weak amplitude is of order G, the Fermi constant. The parity-nonconserving asymmetry, measured by the difference of cross-sections for LH and RH electrons, will then be

$$A = \frac{\sigma_R - \sigma_L}{\sigma_R + \sigma_L} \simeq \frac{Gq^2}{e^2} = \frac{137 \times 10^{-5}}{4\pi} \frac{q^2}{M_p^2}$$

$$\simeq 10^{-4}q^2 \qquad (q^2 \text{ in GeV}^2). \tag{9.44}$$

The method employed to measure polarization asymmetries as small as 10^{-5} is illustrated in Fig. 9.10(a). A source of electrons, either polarized or unpolarized, is accelerated in the SLAC linear accelerator to 16–22 GeV/c and impinges on a liquid deuterium target. The sign and degree of polarization $|P_e|$ of the beam (37% for the polarized source) was measured with a polarimeter, in which one observed the left-right asymmetry in the Møller (electron-electron) scattering from a magnetized iron foil. Inelastically scattered electrons were focused and momentum-analyzed in a spectrometer and recorded in a gas Čerenkov counter followed by a lead-glass shower counter.

The polarized-electron source consisted firstly of a dye laser producing linearly polarized light, which was transformed into circularly polarized light by means of a Pockels cell—a birefringent crystal in which the sign of

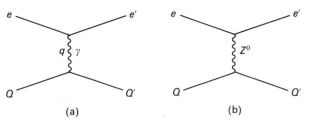

(a)   (b)

Figure 9.9   Weak and electromagnetic neutral-current couplings of electrons to quarks, relevant in the SLAC polarized electron-deuteron scattering experiment.

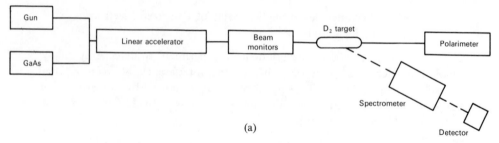

Figure 9.10 (a) Schematic layout of the SLAC experiment on scattering of polarized electrons on deuterons. (After Prescott et al. 1978.)

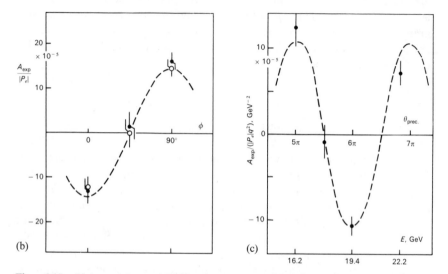

Figure 9.10 (b) Asymmetry, as defined in Eq. (9.44), as a function of the azimuth of the calcite prism in the experiment of (a). (c) Variation of asymmetry with electron-beam energy in the SLAC experiment of Prescott et al., showing the $g - 2$ rotation of the electron spin.

circular polarization could be switched by application of a high-voltage electric field of either sign. This circularly polarized light was then used to optically pump a gallium arsenide crystal, between valence and conduction bands, and thus provide a source of longitudinally polarized electrons. The magnitude and sign of the polarization could also be varied by rotating the plane of polarization of the laser light with a rotatable calcite prism.

The steps in the experiment were:

(a) Measurement of asymmetry using the unpolarized source (electron gun), yielding $A = (-2.5 \pm 2.2) \times 10^{-5}$, consistent with zero and proving the sensitivity of the method and reliability of beam monitoring.

(b) Setting the calcite prism at azimuthal angles of $0°$, $45°$, and $90°$. For $\phi = 45°$, the electrons from the GaAs source should be unpolarized, while for $\phi = 0°$ or $90°$, the polarization should be $R(L)$ or $L(R)$, respectively, according as the sign of the Pockels-cell voltage is $+$ $(-)$. Figure 9.10(b) shows the measured asymmetry as a function of ϕ.

(c) Varying the electron energy E_0 from 16 to 22 GeV/c. The beam, before hitting the target, has suffered a magnetic bending of $24.5°$, and in this process, the electron spin will "lead" over the momentum vector by an angle determined by the g-factor,

$$\theta_{\text{precession}} = \frac{E_0}{mc^2} \frac{g-2}{2} \theta_{\text{bend}}. \qquad (9.45)$$

Thus, as the beam energy E_0 is changed, the asymmetry should vary as the degree of longitudinal polarization is varied by the $g - 2$ effect. This is demonstrated by the results in Fig. 9.10(c).

The final results of this experiment was that a clear asymmetry was observed, of magnitude

$$\frac{A}{q^2} = -(9.5 \pm 1.6) \times 10^{-5}(\text{GeV}/c)^{-2}. \qquad (9.46)$$

In succeeding experiments, the variation of A was measured as a function of $y = (E_0 - E)/E_0$, the fractional energy loss of the electron in the collision. The y-dependence as well as the magnitude of A depends on the weak angle θ_w, according to

$$\frac{A}{q^2} = -\frac{9G}{20\sqrt{2}\pi\alpha} \left\{ a_1 + a_2 \frac{1 - (1 - y)^2}{1 + (1 - y)^2} \right\}, \qquad (9.47)$$

where

$$a_1 = 1 - \tfrac{20}{9} \sin^2 \theta_w, \qquad (9.48)$$
$$a_2 = 1 - 4 \sin^2 \theta_w.$$

These coefficients are derived from the quark-parton model, using the appropriate values of I_3 and Q in Eq. (9.31)—see Problem 9.4. The observed y-dependence was consistent with the prediction (9.47). The final result was

$$\sin^2 \theta_w = 0.22 \pm 0.02, \qquad (9.49)$$

agreeing with the previous estimates.

However, if the ratio ρ of neutral- to charged-current couplings is retained as a free parameter, the results are

$$\rho = 1.74 \pm 0.36,$$

$$\sin^2 \theta_w = 0.293 \begin{array}{c} +0.033 \\ -0.100 \end{array},$$

becaue the values of ρ and $\sin^2 \theta_w$ are strongly correlated. Thus, the neutrino results (9.42) provide by far the most stringent test of the Weinberg-Salam model.

Similar experiments, but at much higher energy, have been carried out at CERN using LH and RH positive and negative muon beams scattered from a carbon target (Argento *et al.* (1982)). The scattering asymmetry observed yields the result

$$\sin^2 \theta_w = 0.23 \pm 0.07. \tag{9.50}$$

9.7.5. Parity Violation in Atomic Transitions

A number of experiments have been carried out to detect parity-violating effects in atoms, again resulting from γ-Z^0 interference. For example, this can lead to a minute (10^{-7} rdn) optical rotation of polarized light traversing bismuth vapor. We do not have space to describe these experiments and refer the reader to a review article by Fortson and Lewis (1984). They quote for the combined result of all the atomic experiments

$$\sin^2 \theta_w = 0.21 \pm 0.05.$$

In summary, it is found that a wide variety of experiments, at low and at high energy, and involving both purely leptonic and lepton-quark interactions, are consistent with a single value of the parameter $\sin^2 \theta_w \simeq 0.22$.

9.8. COMPARISON OF OBSERVED AND PREDICTED W- AND Z-BOSON MASSES

From the various neutral-current experiments, on νe, νN, and ed scattering etc, the "world average" value of $\sin^2 \theta_w$ is given as

$$\sin^2 \theta_w = 0.227 \pm 0.015. \tag{9.51}$$

Any particular experiment is subject to small corrections dependent on experimental cuts in the data, as well as "radiative corrections," particularly for charged-current neutrino events, where the secondary muon may radiate a photon which gets counted in the hadronic energy. The different experiments are also carried out in different q^2 ranges; what would have been the results of the measurements if carried out at a single value of $|q^2|$, conventionally chosen as M_W^2, estimated using a particular renormalization scheme? The result (Marciano and Sirlin (1981), Llewellyn-Smith and Wheater (1981)) is

$$\sin^2 \theta_w = 0.215 \pm 0.015. \tag{9.52}$$

The masses of the W- and Z-particles in the standard (Weinberg-Salam) model are given by (9.27) and (9.29):

$$M_W = \left(\frac{\pi\alpha}{\sqrt{2}G}\right)^{1/2} \frac{1}{\sin \theta_w} = \frac{37.281}{\sin \theta_w} \text{ GeV}, \tag{9.53}$$

$$M_Z = M_W/\cos \theta_w,$$

using $\alpha^{-1} = 137.036$, $G = 1.1663 \times 10^{-5} \text{ GeV}^2$. This formula neglects radiative corrections to the W- and Z-masses themselves, arising from the fact that the W- "self-energy" depends on corrections from virtual fermion and boson loops as in Fig. 9.11. The most important correction is from QED (photon emission and absorption), equivalent to the renormalization of α in (9.53). Finally, one obtains the result that the value of 37.28 in (9.53) should be replaced by 38.50. The comparison between expected and observed mass values is then as follows:

	M_W, GeV	M_Z, GeV
Theory, no corrections	78.2	89.0
Theory, corrected	83.0 ± 2.7	93.8 ± 2.2
Experiment (1985)	$81.1 \pm 0.8 \pm 1.2$	$94.0 \pm 1.0 \pm 1.4$

$$\tag{9.54}$$

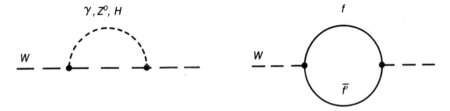

Figure 9.11 Diagrams showing leading-order radiative corrections to the W-boson mass for (left) boson exchanges and (right) exchange of a fermion pair.

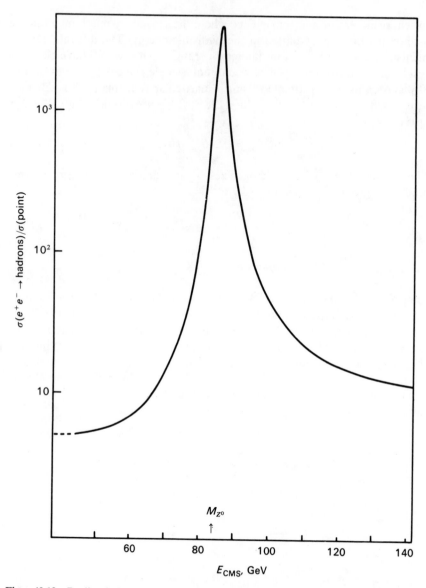

Figure 9.12 Predicted dependence of the cross-section for the process $e^+e^- \to$ hadrons, compared with the pointlike cross-section ($M_{Z^0} = \infty$), near the Z^0 resonance, for $\sin^2 \theta_W = 0.25$.

The first quoted experimental error is statistical, the second the systematic uncertainty of the energy scale. The error in the predicted value reflects that in (9.52).

In time, measured W- and Z-masses will become more precise, with the advent of large e^+e^- colliders (LEP, SLC—see Table 2.1). Figure 9.12 shows the resonance in the process $e^+e^- \rightarrow Z^0$. For a further discussion of the width and other properties of the W and Z^0, see Appendix H and Problem 9.6.

In the standard model, the width of the Z^0 is about 2.8 GeV, assuming it can decay to three generations of quark and lepton pairs, as in (9.33). Each neutrino flavor (i.e., $Z^0 \rightarrow \nu\bar{\nu}$) contributes 180 MeV to the width, so that a good measurement of the width counts the number of neutrino flavors, N_ν. There is already some astrophysical argument to indicate that $N_\nu \not> 4$ (see Section 9.12)

9.9. THE HIGGS BOSON

We have already mentioned that in the Weinberg-Salam model, the bosons consist initially of a weak isospin triplet **W** of vector particles, an isospin singlet vector **B**, and an isospin doublet of scalar Higgs fields, which are complex, denoted by ϕ^+, ϕ^0 with a total of four real components. All these bosons are massless. In the process of spontaneous symmetry-breaking, the Higgs generate mass by self-interaction. Three of the four components, ϕ^+, $\phi^- = \bar{\phi}^+$, and $(\phi^0 - \bar{\phi}^0)/\sqrt{2}$, are "eaten" by the W's, giving them mass and appearing as a third degree of freedom (longitudinal as well as transverse polarization). There is also W/B mixing, resulting in the massless photon and the massive Z^0, as well as the massive W^+ and W^-. This leaves one physical Higgs scalar, $H = (\phi^0 + \bar{\phi}^0)/\sqrt{2}$. (More complex Higgs structure, e.g., an $I = 1$ triplet, is possible, predicting charged as well as neutral physical Higgs bosons, but $\rho < 1$, contrary to experiment).

The properties of the Higgs are ordained by the job it was invented to do: a coupling to fermions proportional to fermion mass and to the bosons (W, Z) proportional to boson mass, squared, in order to avoid, respectively, divergences in $e^+e^- \rightarrow W^+W^-$ and in $WW \rightarrow WW$. The mass is not predicted. However, using the same argument as for the width of the W in (7.70) on dimensional grounds, we expect that for large M_H

$$\Gamma_H \sim GM_H^3.$$

If the Higgs self-coupling is not to be strong, we should require that $\Gamma_H < M_H$, that is,

$$M_H < 1/\sqrt{G}.$$

A more exact calculation gives for the applicability of perturbation theory, from the unitarity limit in the WW scattering amplitude

$$M_H < (4\pi\sqrt{2}/G)^{1/2} = 1.2 \text{ TeV}. \tag{9.55}$$

The argument for the upper limit to the Higgs mass is similar to that for the breakdown of the pointlike Fermi theory at the unitarity limit $\sqrt{s} = 0.6$ TeV in (9.5). Indeed the electroweak theory overcomes the problems of the Fermi theory—and predicts a whole new range of phenomena borne out by experiment—but effectively transfers the problems to the Higgs sector. Some new phenomenon in physics, like the existence of the Higgs boson or another type of interaction which fulfils the role of the Higgs, has to happen in the 1-TeV energy range. That is the main motivation for proposing pp super-colliders to operate in the multi-TeV energy range.

9.10. BEYOND THE STANDARD MODEL I—GRAND UNIFICATION

A major problem in our understanding of the fundamental interactions is in their number (four, if we include gravity together with strong, electromagnetic, and weak interactions) and disparate strengths and properties. The electroweak theory postulated a single interaction to describe electromagnetic and weak processes, and spontaneous symmetry breaking to account for their different apparent strengths in the energy domain well below the masses of the mediating bosons concerned. The so-called grand unified theories (called GUTs for brevity) appeal to further symmetry-breaking processes in order to reconcile the relatively great strength of strong interactions at low energies with a unique intrinsic coupling (approximately α, the fine-structure constant) for all three interactions at the unification energy.

The basic idea of the approach to a universal coupling is illustrated in Fig. 9.13. The graph indicates the dependence on momentum transfer, or mass scale, of the running coupling constants for the Abelian U(1) field described by g' in (9.18), for the non-Abelian SU(2) field denoted by g in (9.18), and for the non-Abelian SU(3) color field which we denote g_s (where $\alpha_s = g_s^2/4\pi$). As q increases, g' increases slightly while g decreases slightly and g_s decreases more quickly (compare (8.70) and (8.71)). It is a remarkable fact that when all three interactions are extrapolated, they appear to meet at a single point at the colossal value of $q \simeq 10^{15}$ GeV. (There are numerical constants of order unity multiplying g, g', and g_s, which we have omitted from Fig. 9.13). Implicit in this enormous extrapolation is the bold assumption that Nature has run out of ideas and has no new surprises in the range from $q \simeq 30$ GeV/c of present accelerator experiments, all the way to $q \simeq 10^{15}$ GeV/c. There may instead be new internal degrees of freedom, quark or

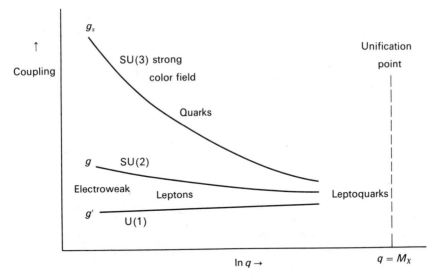

Figure 9.13 Running coupling constants g, g', and g_s of the electroweak and strong interactions appear to extrapolate to a single value at $q \sim 10^{15}$ GeV.

lepton substructure, only just around the corner and to be revealed at the next generation of accelerators. On the basis of past experience, one would think that this is rather likely—see also Section 9.11.

There are many ways in which the SU(2), U(1), and SU(3) symmetries could be incorporated into a more global gauge symmetry. The simplest grand unifying symmetry is that of the group SU(5) (Georgi and Glashow 1974). This incorporates the known fermions (leptons and quarks) in multiplets, inside which quarks can transform to leptons, and quarks to antiquarks, via the mediation of very massive (10^{15}-GeV) bosons Y and X, with electric charges $-\frac{1}{3}$ and $-\frac{4}{3}$ respectively. There are a total of 24 gauge bosons in the model. These consist of the 8 gluons of SU(3) and the $W^\pm$, Z^0, and photon of SU(2) $\times$ U(1), plus the 12 varieties of X and Y boson (each carrying 3 colors and existing in particle and antiparticle states). The fermions (quarks and leptons) are assigned to different "generations". The first generation consists of 15 states: the u- and d-quarks, each in 3 color and 2 helicity states; the e^- with 2 helicity states; and the ν_e, with 1 helicity only. By convention, we write them down as LH states, since, for example, the e^-_{RH} state and the e^+_{LH} state are equivalent (by CP). The 15 states comprise "$\bar{5}$" and "10" representations as follows:

$$\bar{5} = \begin{bmatrix} \nu_e \\ e^- \\ \bar{d}_R \\ \bar{d}_B \\ \bar{d}_G \end{bmatrix}_{\text{LH}} \qquad \begin{array}{l} \rightrightarrows W^- \\ \rightleftharpoons X \\ \rightrightarrows G_{BG} \end{array} \qquad . \tag{9.56}$$

The arrows indicate a gluon mediating the color force between quarks, the $W^\pm$ mediating the charged weak current, and an X "leptoquark" boson transforming a quark to a lepton. The quantum numbers of members of the "10" are obtained from the antisymmetric combinations $q_i q_j - q_j q_i$ of the members q_i of the "5" (conjugates of the $\bar{q}_i$ of the "$\bar{5}$"). Thus,

$$
10 = \begin{bmatrix}
0 & e^+ & d_R & d_B & d_G \\
-e^+ & 0 & u_R & u_B & u_G \\
-d_R & -u_R & 0 & \bar{u}_G & \bar{u}_B \\
-d_B & -u_B & -\bar{u}_G & 0 & \bar{u}_R \\
-d_G & -u_G & -\bar{u}_B & -\bar{u}_R & 0
\end{bmatrix}_{LH} . \tag{9.57}
$$

These multiplets have the property that the total electric charge $\sum Q_i = 0$. The heavier leptons (μ, v_μ, τ, v_τ) and quarks (s, c, b) have to be assigned to separate quark-lepton generations.

Among the attractive features of this (and other) grand unifying symmetries are:

(a) The fractional charges ($\frac{2}{3}$ and $\frac{1}{3}$) of the quarks occur because the quarks come in three colors and the electron is colorless (and $\sum Q_i = 0$).

(b) The equality of the baryon (proton) and electron charge—a historic puzzle—is accounted for.

(c) The strong similarity between the weak lepton and quark doublet patterns, for example, $(v_e, e)_L$ and $(u, d_c)_L$, and the fact that $Q(v) - Q(e) = Q(u) - Q(d)$, occur as natural consequences of leptoquark unification.

The model makes a number of predictions, some at least of which are accessible in present or future experiments. We discuss here two of the predictions.

9.10.1. Prediction of the Weak Mixing Angle

The bringing together of quarks and leptons into multiplets allows one to estimate the weak mixing angle, as follows. Consider the diagram of Fig. 9.14 depicting the neutral boson Z^0 mixing with a photon, γ, via an intermediate fermion loop (analogous to the $K^0 \bar{K}^0$ mixing of Fig. 7.22). We

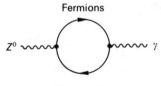

Figure 9.14

know that Z^0 and γ are orthogonal states, $\langle Z^0 | \gamma \rangle = 0$; hence there should be no net coupling when summed over all fermions (leptons and quarks) which can contribute.

The coupling of fermions to Z^0 is proportional to $I_3 - Q \sin^2 \theta_w$, and to the photon is simply Q—see (9.31). For the "$\bar{5}$" representation, using (9.31), we obtain from (9.56)

$$
\begin{bmatrix} \nu_e \\ e^- \\ \bar{d}_R \\ \bar{d}_B \\ \bar{d}_G \end{bmatrix}_{\text{LH}}
\quad
\begin{array}{cc}
I_3 & Q \\
\hline
+\tfrac{1}{2} & 0 \\
-\tfrac{1}{2} & -1 \\
0 & +\tfrac{1}{3} \\
0 & +\tfrac{1}{3} \\
0 & +\tfrac{1}{3}
\end{array}
\tag{9.58}
$$

where the LH states of antifermions (or RH states of fermions) have $I_3 = 0$, as explained previously. The net coupling, summed over all fermions, to both Z^0 and photon has to vanish, giving us the relation

$$\sum Q(I_3 - Q \sin^2 \theta_w) = 0$$

or

$$\sin^2 \theta_w = \frac{\sum Q I_3}{\sum Q^2} = \frac{3}{8}. \tag{9.59}$$

The above prediction for $\sin^2 \theta_w$ is clearly much larger than the observed value (9.52). However, it applies to the ratio $g'/g = \tan \theta_w \, (= \sqrt{3/5})$ at the *unification point*, $M_X \simeq 10^{15}$ GeV. From Fig. 9.13 we see that corrections to the running coupling constants g and g' are required to bring them to the values expected at accelerator energies, and the effect of these corrections is clearly to decrease g'/g and $\sin^2 \theta_w$. When these corrections are made, a value $\sin^2 \theta_w = 0.21 \pm 0.01$ is obtained, in remarkable agreement with the observed value of $0.215 \pm .015$.

9.10.2. Prediction of Proton Decay

There is no doubt that protons are very stable objects, and the mere existence of life on Earth sets a lower limit on the proton lifetime more than one million times longer than the age of the solar system (see Problem 9.3). Not surprisingly, baryon number was for many years considered to be an absolutely conserved quantity in all interactions, following the suggestion of Wigner and others. What we learn from gauge theories however is that absolute conservation laws are connected with gauge invariance and the existence of an appropriate long-range (massless) field. No fields are known which could be connected with baryon or lepton conservation (see Problem 9.5).

The grand unified models, on the contrary, predict that protons must decay, violating baryon-number conservation. Some possible diagrams for proton decay via single X- or Y-boson exchange are given in Fig. 9.15. This exchange transforms a u-quark into a positron, and a d-quark into $\bar{u}$, for example (see (9.56)), corresponding to the decay $p \to e^+ \pi^0$.

The decay probability can be crudely estimated from such diagrams. The dominant factor will be the X-boson propagator, $(q^2 + M_X^2)^{-1}$. Since the momentum transfer is $q^2 \simeq 1$ GeV^2 only, the decay rate will then be proportional to M_X^{-4}. Thus, we can write for the lifetime

$$\tau = \frac{A}{\alpha^2} \frac{M_X^4}{M_p^5}, \tag{9.60}$$

where A is dimensionless and, in units $\hbar = c = 1$, the right-hand side has dimensions of mass^{-1}. Our choice of the proton mass as the scale in the denominator is arbitrary but quite appropriate. The constant A involves a great deal of theory, which we do not discuss, but it turns out to be of order unity. If we take $A = 1$, then $M_X = 3 \times 10^{14}$ GeV gives $\tau_p = 10^{31}$ years.

Experimental lower limits on proton decay have been set over the last years in giant underground water Cerenkov and other detectors. In particular, the limit for the decay mode $p \to e^+ + \pi^0$ is more than 2×10^{32} years (Bionta et al. (1985)), in definite disagreement with the SU(5) prediction, which is at least an order of magnitude smaller (see Langacker (1981) and Perkins (1984) for recent reviews of the theoretical and experimental situation). Other GUT theories, involving more complex groups or supersymmetry, emphasize other decay modes or make less definite lifetime predictions.

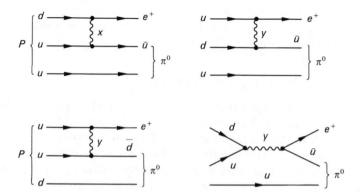

Figure 9.15 Some diagrams relevant to nucleon decay, in which quarks transform to leptons and to antiquarks via massive vector bosons X and Y of charge $\frac{4}{3}$ and $\frac{1}{3}$.

9.10.3. Magnetic Monopoles

In 1929 Dirac proposed that magnetic monopoles might exist with values of the magnetic charge[‡]

$$g = n \frac{\hbar c}{2e} \tag{9.61}$$

where n is an integer. The existence of a monopole of charge g would in a sense "explain" why charge was quantized in units of $|e|$. In grand unified theories, the electric charge operator is one of the generators of the symmetry, and all particles must have the same unit of charge (or fraction of it, in the case of quarks), so charge quantization occurs naturally. In that case, monopoles of charge g and mass of order M_X (in fact, 10^{16}–10^{17} GeV) are definitely predicted. Searches for such GUT monopoles have been made with superconducting coils, based on observing the change in flux $\Delta\phi = 4\pi g$ when a monopole traverses the coil; and by detecting the ionization or excitation of atoms resulting from the magnetic interaction of the pole with the moments of the electrons. Present upper limits on monopole flux are of the order of the so-called Parker bound, of 10^{-15} monopoles cm^{-2} sr^{-1} s^{-1}. This is the maximum flux that could be tolerated if monopoles are not to destroy the galactic magnetic field, of order a few microgauss.

9.11. BEYOND THE STANDARD MODEL II—SUPERSYMMETRY

The SU(5) model of grand unification described above possesses some basic weaknesses of principal. For example, neutrinos are considered to be massless and occur in only one helicity state ($H = -1$). Although this is consistent with all experiments to date, we can ask why the weak interactions of neutrinos should possess the mysterious property of parity violation when other interactions are parity-conserving. Perhaps parity violation is a low-energy phenomenon, a manifestation of spontaneous symmetry breaking: The missing right-handed couplings do exist but are associated with massive neutrinos and/or very massive mediating bosons. This is part of a very much larger problem in understanding the very broad range and the origin of fermion masses in general.

Another problem arises from differences in the energy scales of the various interactions: from the QCD scale of order 1 GeV, through the electroweak scale of $M_W \simeq 100$ GeV, to the GUT scale of $M_X \simeq 10^{15}$ GeV. This range of scales constitutes the so-called hierarchy problem. Higgs scalars in the electroweak theory were specifically introduced to generate mass by

[‡] For a derivation, see for example J. D. Jackson, *Classical Electrodynamics*, 2nd ed., John Wiley, New York, 1975, p. 254.

self-interaction and, as a result, spontaneous breaking of the gauge symmetry, differentiating the massless gauge boson (photon) from the massive bosons (W, Z). Equally, Higgs scalars—but of mass $\simeq M_X$ rather than $\simeq M_W$—will be needed to account for SU(5) symmetry breaking. The light Higgs will inevitably receive radiative corrections to the mass from the GUT Higgs, with the result that the masses of H, W, and Z of the electroweak sector become unstable, unless one can arrange clever cancellations working to a precision of order $M_W/M_X \simeq 10^{-13}$.

The hypothesis of supersymmetry (SUSY) does provide just such cancellations. It introduces the attractive idea of symmetry between fermions and bosons. New boson (fermion) partners are postulated for all known pointlike fermions (bosons), as indicated in Table 9.2. The radiative corrections to the Higgs mass from virtual boson and fermion loops have opposite signs. Since the particles and their SUSY partners have the same couplings, the mass divergences are kept under control provided $|M_F^2 - M_B^2| < 1$ TeV^2. So supersymmetry invokes a host of new particles (or sparticles) on roughly the Fermi mass scale, $G^{-1/2}$.

In most SUSY models, the new particles must be produced in association, that is in pairs with an R quantum number ($+1$ for particles, -1 for antiparticles). For example a quark-antiquark pair could make a pair of squarks (bosons): $Q + \bar{Q} \rightarrow \tilde{Q} + \bar{\tilde{Q}}$. Because of R-symmetry, at least one SUSY particle (the lightest) must be stable. If this is the photino $\tilde{\gamma}$ (the spin-$\frac{1}{2}$ partner of the photon), then production of a squark could be manifest in the decay $\tilde{Q} \rightarrow Q + \tilde{\gamma}$. Since the $\tilde{Q}$-mass is of order 1 TeV, the missing photino among the decay products would leave a dramatic momentum unbalance in the event.

To date, there is no evidence for SUSY particles and lower limits on the masses are more than 4 GeV for gluinos and more than 20 GeV for squarks and sleptons.

There are other proposed models to deal with the Higgs problem —for example, that it is a composite rather than an elementary particle. All these models point to new types of particles and interactions on roughly the 1-TeV energy scale. Whether there is any substance to these ideas is a matter for future experiments.

TABLE 9.2 Supersymmetric particles

Particle	Spin	Sparticle	Spin
Quark Q	$\frac{1}{2}$	Squark $\tilde{Q}$	0
Lepton l	$\frac{1}{2}$	Slepton $\tilde{l}$	0
Photon γ	1	Photino $\tilde{\gamma}$	$\frac{1}{2}$
Gluon G	1	Gluino $\tilde{G}$	$\frac{1}{2}$
W	1	Wino $\tilde{W}$	$\frac{1}{2}$
Higgs H	0	Shiggs $\tilde{H}$	$\frac{1}{2}$

9.12. PARTICLE PHYSICS AND COSMOLOGY

The grand unified theory described above may be relevant to our understanding of the temporal development of, and present distribution of matter and radiation in, the universe. This is a vast subject, and the interested reader is referred to the chapter bibliography for recent reviews and papers. We only mention here a few of the ideas and facts, simply to indicate the relevance of results from particle physics to cosmology.

The isotropic cosmic background radiation in the microwave region, discovered in 1965, has a spectrum equivalent to that from a black body at $3°K$ and is interpreted as the cooled radiative remnant of the "big bang" preceding the expansion of the universe. From the present distribution of matter, the ratio of baryons to photons averaged over the universe is estimated to be

$$\frac{n_b}{n_\gamma} = 10^{-9 \pm 1}$$

and should be time-independent in an adiabatically and isotropically expanding universe. In the initial hot stage of the big bang ($t < 10^{-4}$ s, $kT > 1$ GeV), baryon pairs should have been created with a number density comparable with that of photons. As the universe expanded and cooled, nucleon pair creation would no longer compensate for annihilation to photons, and the baryon number would therefore fall catastrophically until a baryon residue was "frozen out" as the expansion rate exceeded the annihilation rate. This final ratio of baryons and antibaryons to photons can be calculated without too much uncertainty and is found to be

$$\frac{n_b}{n_\gamma} \sim 10^{-18},$$

in violent contrast with the observed ratio. So there are two problems: The observed baryon density is nine orders of magnitude bigger than expected, and baryons preponderate over antibaryons. We know this is so in our own galaxy because heavy primary cosmic-ray nuclei—which are, so to speak, messengers from the most distant parts of the Milky-Way—are invariably nuclei rather than antinuclei. Nor is there any evidence for the intense γ-ray emission that would follow annihilation between, for example, the matter of a galaxy and intergalactic antimatter. So, with all the usual reservations about the average isotropy of the universe, we can be sure that the antibaryon-to-baryon ratio, now, is $< 10^{-4}$.

It appears therefore that, in the early universe, some processes occurred which resulted in a net baryon-antibaryon asymmetry of order 10^{-9}. All but this tiny fraction of matter annihilated to photons, leaving a residue of baryons. Clearly, if one starts off with a symmetrical situation, an asymmetry can only arise through a baryon-number-nonconserving process, but there are two other requirements as first emphasized by Sakharov in 1967: the baryon-creating processes must be CP-nonconserving, and the baryons must be out of thermal equilibrium. (In thermal equilibrium, the baryon density can depend only on the temperature and on the particle mass, which is the same for particle and antiparticle, by the CPT theorem.) One possibility is the initial generation of the X, $\bar{X}$ bosons mentioned above, which by the CPT theorem will have identical total decay rates, but for which some partial decay rates into quarks, antiquarks, and leptons may violate CP, just as in the leptonic decay of K_L (Eq. (7.98)). Such decays could generate a baryon asymmetry as a second-order process, and it thus appears possible to account for the preponderance of baryons in the universe as well as the observed baryon-to-photon ratio. The effects predicted in the SU(5) model of grand unification are too small, and more general and larger unification groups—involving more free parameters and hence less predictive power—seem to be required. Although, in principle, CP-violation is possible with only four quark flavors, the asymmetry becomes appreciable only with at least six flavors (see Section 7.12).

Astrophysical arguments can be used the other way round, to set interesting limits on the properties of elementary constituents. The density of light neutrinos (mass $\ll 1$ MeV) created in the big bang should be comparable with the photon density, now of order 10^9 times the nucleon density. Thus, a neutrino mass of only a few eV (i.e., 10^{-9} of the nucleon mass), would increase the gravitational potential energy by a substantial factor; indeed, this is one of the reasons for postulating nonzero neutrino masses, in order to correct the order-of-magnitude mismatch between motional and gravitational energy in the universe.

Limits on the number of flavors of the light neutrino have been set from considerations of the observed helium/hydrogen mass ratio (about 0.29), which must be a little greater (via stellar synthesis) than the primordial ratio attained in thermonuclear fusion following the big bang. The primordial helium content from nucleosynthesis (at $t \simeq 100$ s, $kT \simeq 0.1$ MeV) depends on the value of the n/p ratio, which is determined by the reversible reactions $e^- + p \rightleftharpoons v_e + n$. The back reaction is exothermic by 0.8 MeV, so $n/p < 1$ at nucleosynthesis and is smaller (larger) the slower (faster) the cooling rate. The latter is in turn proportional to the number of fundamental fermions produced in the earlier stages, and each extra neutrino flavor is expected to increase the helium mass fraction by about 1 %. These ideas suggest not more than four light-neutrino flavors; so perhaps the total eventual number of lepton (and presumably quark) flavors is finite and small, and not, as some people fear, embarrassingly large. The question of the number of possible types of neutrino will be independently and finally settled in a few years, from observation of the branching ratio for $Z^0 \rightarrow v\bar{v}$ at giant e^+e^- colliders.

PROBLEMS

9.1. Using Eq. (9.34) and the results from Table 9.1, plot the total cross-sections σ/E_v for v_e, $\bar{v}_e$, v_μ, and $\bar{v}_\mu$ scattering from stationary electron targets as a function of $\sin^2 \theta_w$. Where do σ_{v_μ}, $\sigma_{\bar{v}_\mu}$ have minimum values? For what value of $\sin^2 \theta_w$ is the $\bar{v}_\mu$ coupling to electrons purely axial-vector?

9.2. Using (9.34) and integrating over the electron recoil energy E, plot the relation between g_V and g_A assuming that $\sigma(v_\mu e) = 3 \times 10^{-42} E_v$ cm^2 GeV. Take $G = 1.02 \times 10^{-5}/M_p^2$. Show that if $\sigma(\bar{v}_\mu e)$ is also known (without error), there are four possible solutions for g_A and g_V.

9.3. The unit of radiation dosage is the rad, which corresponds to an energy liberation (in ionization) of 100 erg g^{-1}. The annual permissible body dose for a human is cited as 5 rad. Assuming that 100 times this dose would lead to extinction of advanced life forms, what limit does this set on the proton lifetime? (Assume that in proton decay, a substantial fraction of the total energy released is deposited in body tissue.)

9.4. Verify Eq. (9.47) for the assymmetry in scattering of polarized electrons by deuterons. Use the expressions in Eqs. (9.31) and (9.33) for the couplings of electrons and quarks to the Z^0 and photon. The cross-section results from the sum of the photon- and Z^0-exchange amplitudes. In summing these, only the pure photon-exchange term and the Z^0-photon interference term will contribute, assuming $q^2 \ll M_Z^2$. Assume also equal numbers of u- and d-quarks in the deuteron, and neglect antiquarks. (In case of difficulty, consult for example R. N. Cahn and F. J. Gilman, *Phys. Rev. D* **17**, 1313 (1977), where the problem is fully worked out.)

9.5. The Equivalence Principle states that the ratio R of inertial to gravitational mass is the same for all substances. It has been tested by comparing the centrifugal force due to the earth's rotation on a body with the gravitational force of the earth (or sun). R is found to be the same for Al and Pt within 1 part in 10^{12}. These experiments also set a limit on the coupling, K_B of any long-range $(1/r^2)$ field coupling to baryon number. By considering nuclear binding energies and neutron/proton ratios, show that the difference in baryon number per unit mass in Al and Pt is 4×10^{-4}. Hence show that $K_B/K < 10^{-9}$, where K is the gravitational constant. (For further details, see, for example, Perkins (1984)).

9.6. Using the expressions (9.33) for the Z^0-fermion couplings, estimate the total width for Z^0-decay, for three fermion generations, $\sin^2 \theta_w = 0.22$ and $M_Z = 93$ GeV. (The partial width $\Gamma(Z \to v\bar{v}) = GM_Z^3/12\pi\sqrt{2}$—see Appendix H).

9.7. Calculate the value of the ratio $R = \sigma(e^+e^- \to Z^0 \to \text{anything})/\sigma(\text{point})$ at the peak of the Z^0-resonance (see Fig. 9.12), where $\sigma(\text{point}) = \sigma(e^+e^- \to \mu^+\mu^-)$. Show that $R = (9/\alpha^2)(\Gamma_{ee}/\Gamma)$ where Γ is the total Z^0-width and Γ_{ee} is the partial width for $Z^0 \to e^+e^-$. Using the results of Problem 9.6, show that $R = 5180$.

BIBLIOGRAPHY

Aitchison, I. J., and A. J. Hey, "Gauge theories in particle physics" (Adam Hilger Ltd., Bristol 1981).

Bahcall, J. N., "Solar neutrino experiments", *Rev. Mod. Phys.* **50**, 881 (1978).

Barrow, J. D., "The baryon asymmetry of the universe", *Surveys in High En. Phys.* **1**, 182 (1980).

Beg, M. A., and A. Sirlin, "Gauge theories of weak interactions", *Ann. Rev. Nucl. Science* **24**, 379 (1974).

Glashow, S. L., "Towards a unified theory: Threads in tapestry", *Rev. Mod. Phys.* **52**, 539 (1980).

Goldhaber, M., P. Langacker, and R. Slansky, "Is the proton stable?", *Science* **210**, 851 (1980).

't Hooft, G., "Gauge theories of the forces between elementary particles", *Sci. Am.* **243**, 90 (June 1980).

Hung, P. Q., and C. Quigg, "Intermediate bosons: weak interaction couriers", *Science* **210**, 1205 (1980).

Langacker, P., "Grand unified theories and proton decay", *Phys. Rep.* **72**, 185 (1981).

Perkins, D. H., "Proton decay experiments", *Ann. Rev. Nucl. Part. Science* **34**, 1 (1984).

Primakoff, H., and S. P. Rosen, "Baryon number and lepton number conservation laws", *Ann. Rev. Nucl. Science* **31**, 145 (1981).

Pullia, A., "Structure of charged and neutral weak interactions at high energy", *Riv. del Nuov. Cim.* **7**, Ser. 3, Nos. 1–6 (1984).

Salam, A., "Gauge unification of fundamental forces", *Rev. Mod. Phys.* **52**, 525 (1980); *Science* **210**, 723 (1980).

Steigman, G., "Cosmology confronts particle physics", *Ann. Rev. Nucl. Science* **29**, 313 (1979).

Weinberg, S., "Conceptual foundations of the unified theory of weak and electromagnetic interactions", *Rev. Mod. Phys.* **52**, 515 (1980); *Science* **210**, 1212 (1980).

Weinberg, S., "Unified theories of elementary particle interactions", *Sci. Am.* **231**, 50 (July 1974).

APPENDIX A

Relativistic Transformations

A.1. LORENTZ TRANSFORMATIONS

One of the fundamental symmetries to be required of any theory of particle interactions is that of relativistic invariance; the equations describing the system should have the same form in all inertial frames. This invariance is achieved by formulating the equations in 4-*vector notation*. Recall that the relativistic relation between total energy E, 3-momentum $\mathbf{p}$, and rest mass m of a free particle, in units $c = 1$ *is*

$$E^2 = p^2 + m^2. \tag{A.1}$$

The Cartesian components of $\mathbf{p}$, E can be written as components of an energy-momentum 4-vector p_μ, where $\mu = 1\text{--}4$. In one convention we write these as

$$p_1 = p_x, \qquad p_2 = p_y, \qquad p_3 = p_z, \qquad p_4 = iE,$$

so that

$$p^2 = \sum_\mu p_\mu^2 = p_1^2 + p_2^2 + p_3^2 + p_4^2 = \mathbf{p}^2 - E^2 = -m^2 \tag{A.2}$$

is invariant, with the same value in all reference frames. If E, $\mathbf{p}$ refer to values measured in the laboratory frame Σ, those in a different frame Σ' moving, for instance, along the x-axis with velocity βc are found from the Lorentz transformations, which in matrix form are given by

$$p'_\mu = \sum_{v=1}^{4} \alpha_{\mu v} p_v, \tag{A.3}$$

where

$$\alpha_{\mu v} = \begin{vmatrix} \gamma & 0 & 0 & i\beta\gamma \\ 0 & 1 & 0 & 0 \\ 0 & 0 & 1 & 0 \\ -i\beta\gamma & 0 & 0 & \gamma \end{vmatrix}$$

and $\gamma = (1 - \beta^2)^{-1/2}$. So in terms of energy and momentum we find

$$p'_1 = \gamma p_1 + i\beta\gamma p_4 \quad \text{or} \quad p'_x = \gamma(p_x - \beta E)$$

$$p'_2 = p_2 \qquad\qquad\qquad p'_3 = p_3 \tag{A.4}$$

$$p'_4 = -i\beta\gamma p_1 + \gamma p_4 \quad \text{or} \quad E' = \gamma(E - \beta p_x).$$

If the particle is at rest in $\sum$, then in the frame $\sum'$ moving at velocity $-\beta$ along x with respect to $\sum$ we have, since $E = m$ and $p_x = p_y = p_z = 0$,

$$E' = \gamma m \qquad \mathbf{p}' = p'_x = \gamma\beta m = \sqrt{\gamma^2 - 1}\, m, \tag{A.5}$$

and, of course,

$$\mathbf{p}'^2 - E'^2 = -m^2.$$

These transformations are for reference frames in relative motion along the x-axis. They can be generalized to relative motion in arbitrary directions, but it is simpler always to redefine the x-axis as that of the relative motion.

The above transformation laws apply equally to space-time coordinates. They are found by replacing $p_1 \rightarrow x_1 (= x)$, $p_2 \rightarrow x_2 (= y)$, $p_3 \rightarrow x_3 (= z)$, $p_4 \rightarrow x_4 (= it)$.

A.2. LORENTZ SCALARS

The 4-momentum squared in (A.2) is an example of the invariant scalar product of two 4-vectors, i.e., $\sum p_\mu p_\mu$. It is referred to as a Lorentz scalar, since it is invariant under a Lorentz transformation. The phase of a plane wave—whether it is at a crest or a trough—must be the same for all observers and can be written as a Lorentz scalar. With $\mathbf{k}$ as the propagation vector and ω the angular frequency, we have

$$\text{Phase} = \mathbf{k} \cdot \mathbf{x} - \omega t = \frac{1}{\hbar}(\mathbf{p} \cdot \mathbf{x} - Et) = \sum p_\mu x_\mu \qquad \text{(in units of } \hbar = 1\text{)}.$$

The *metric* used in this book is defined above, namely, a 4-vector momentum is $p = (\mathbf{p}, iE)$ so that

$$(\text{4-momentum})^2 = (\text{3-momentum})^2 - (\text{energy})^2 \qquad (\text{A.6})$$

$$\underset{\text{space}}{\llcorner} \qquad \underset{\text{time}}{\llcorner}$$

In analogy with the space-time coordinates x_μ, the components $p_{1,2,3}$ are said to be *spacelike* and the energy component, E, *timelike*. Thus, if q denotes the 4-momentum transfer in a reaction, i.e., $q = p - p'$, where p, p' are initial and final 4-momenta, then

$$
\begin{array}{lll}
q^2 > 0 & \text{is spacelike} & \text{e.g., in scattering process} \\
q^2 < 0 & \text{is timelike} & \text{e.g., (mass)}^2 \text{ of free particle.}
\end{array}
\qquad (\text{A.7})
$$

A.3. OTHER NOTATIONS AND CONVENTIONS

The Minkowski notation (A.2) for 4-vectors is used in this text because it is the one most commonly used in books on special relativity at the undergraduate level. In texts on field theory, other notations are in more general use. These avoid the imaginary value of the fourth component ($p_4 = iE$) and introduce the negative sign in (A.6) via the metric tensor, $g_{\mu\nu}$. The scalar product of 4-vectors A and B is then defined as

$$AB = g_{\mu\nu}A_\mu B_\nu = A_0 B_0 - \mathbf{A} \cdot \mathbf{B}, \qquad (\text{A.8})$$

where all components are real and $\mu, \nu = 0$ stands for the time (or energy) component; $\mu, \nu = 1, 2, 3$ for the space (momenta) components, and where

$$g_{00} = +1, \quad g_{11} = g_{22} = g_{33} = -1, \quad g_{\mu\nu} = 0, \quad \text{for } \mu \neq \nu. \quad (\text{A.9})$$

Note that the metric employed here results in Lorentz scalars with sign opposite to that in (A.6). The correspondence between the notation in this text and in standard field theory is therefore as follows:

	This text	Field theory (e.g., Bjorken and Drell)
Momentum (space) coordinate	p_1, p_2, p_3	p_1, p_2, p_3
Energy (time) coordinate	$p_4 = iE$	$p_0 = E$
Metric tensor	—	$g_{11} = g_{22} = g_{33} = -1; g_{00} = +1$
(4-momentum)^2	$p^2 = \sum p_\mu^2 = \mathbf{p}^2 - E^2$	$p^2 = \sum g_{\mu\nu}p_\mu p_\nu = E^2 - \mathbf{p}^2$
Spacelike 4-momentum	$q^2 > 0$	$q^2 < 0$
Timelike 4-momentum	$q^2 < 0$	$q^2 > 0$

A.4. EXAMPLE OF USE OF 4-VECTORS

The 4-vector notation is invaluable in kinematic calculations. As an example, let us calculate the available energy for new-particle creation when an incident particle of mass m_A, total energy E_A, and momentum $\mathbf{p}_A$ hits a target particle of mass m_B, energy E_B, and momentum $\mathbf{p}_B$, all quantities being measured in the laboratory system (LS). The total (4-momentum)^2 of the system is

$$p^2 = (\mathbf{p}_A + \mathbf{p}_B)^2 - (E_A + E_B)^2 = -m_A^2 - m_B^2 + 2\mathbf{p}_A \cdot \mathbf{p}_B - 2E_A E_B.$$

The center-of-momentum system (CMS) is defined as a reference frame in which the total 3-momentum is zero. If the total energy in this system is E^*, then also

$$p^2 = -E^{*2}.$$

Let us consider two interesting cases:

(i) The target particle (m_B) is at rest in the LS, $E_B = m_B$, and $p_B = 0$. Then

$$E^{*2} = -p^2 = m_A^2 + m_B^2 + 2m_B E_A.$$

The threshold energy required to create a state of mass m^* from m_A and m_B will be

$$E_A \text{ (threshold)} = \frac{m^{*2} - m_A^2 - m_B^2}{2m_B} \simeq \frac{m^{*2}}{2m_B} \quad \text{for } m^* \gg m_A, m_B.$$

$$\text{(A.10)}$$

(ii) The incident and target particles m_A and m_B travel in opposite directions (as in an e^+e^- or $p\bar{p}$ collider). Then with $|\mathbf{p}_A| = p_A$, $|\mathbf{p}_B| = p_B$,

$$E^{*2} = -p^2 = 2(E_A E_B + p_A p_B) + (m_A^2 + m_B^2)$$
$$\simeq 4E_A E_B \quad \text{if } E_A, E_B \gg m_A, m_B. \quad \text{(A.11)}$$

We note that the total CMS energy for new particle creation in a collider with equal beam energies E rises linearly with beam energy, $E^* \simeq 2E$, while for a fixed-target machine, the available CMS energy rises as the square root of the incident energy, $E^* \simeq \sqrt{2m_B E_A}$.

A.5. LORENTZ INVARIANT CROSS-SECTIONS

Suppose that, in some reaction, any one particle is produced with momentum $\mathbf{p}$ and energy E, both measured in the laboratory system. The phase-space factor in this system will be proportional to

$$d^3\mathbf{p} = dp_x \, dp_y \, dp_z \quad \text{(A.12)}$$

For a transformation to an inertial frame moving with velocity $+\beta$ along the x-axis, we have (in units $c = 1$) from (A.4)

$$p'_x = \gamma(p_x - \beta E) \qquad p'_y = p_y, \qquad p'_z = p_z,$$

$$E' = \gamma(E - \beta p_x).$$

Then

$$dp'_x = \frac{\partial p'_x}{\partial p_x} dp_x + \frac{\partial p'_x}{\partial E} dE$$

$$= \gamma \, dp_x \quad - \gamma\beta \, dE$$

$$= \gamma(1 - \beta p_x/E) \, dp_x,$$

using the fact that $p_x^2 + p_y^2 + p_z^2 - E^2 = -m^2$ and that for p_y, p_z fixed, $p_x \, dp_x = E \, dE$. Hence,

$$\frac{dp'_x}{E'} = \frac{\gamma(1 - \beta p_x/E) \, dp_x}{\gamma E(1 - \beta p_x/E)} = \frac{dp_x}{E}$$

so

$$\frac{d^3\mathbf{p}'}{E'} = \frac{dp'_x \, dp'_y \, dp'_z}{E'} = \frac{dp_x \, dp_y \, dp_z}{E} = \frac{d^3\mathbf{p}}{E}. \qquad (A.13)$$

This result must hold for a Lorentz transformation in any direction, since one can always suitably redefine the coordinate axes. Thus, the differential cross-section for producing a particle of momentum $\mathbf{p}$ and energy E can be written in the form

$$E d^3\sigma/d^3\mathbf{p} \qquad (A.14)$$

and is Lorentz invariant.

The physical origin of the factor E in the expression $d^3\mathbf{p}/E$ for the phase space can be understood as follows:

First let one of the particles in the interaction be at rest relative to a box of volume V. Then the particle density is $1/V$ in this frame. Now suppose the particle plus box moves at velocity $\mathbf{v}$ relative to the chosen reference frame, $\sum$, in which W is to be calculated. If the particle mass is m, its energy in $\sum$ is E, where $E/m = (1 - v^2/c^2)^{-1/2}$. As a consequence of the Lorentz transformation, a length measured in the direction of $\mathbf{v}$ in the particle rest frame is reduced by a factor m/E when measured in the frame $\sum$, so that the volume of the box is contracted to a value $V' = Vm/E$. If the wavefunction is integrated over a volume V in $\sum$, then $\int \psi^*\psi \, dV = E/m$; i.e., the particle density is increased by a factor E/m. Therefore, to ensure that the particle density is correctly normalized in the same way in all frames, we should incorporate a factor $\sqrt{m/E}$ with the wavefunction ψ, for each particle, where

E is the energy in the chosen frame $\sum$. Several conventions are in use: $\sqrt{m/E}$, $1/\sqrt{2E}$, etc., depending on the choice of normalization factors for fermion and boson wavefunctions. Here we shall take a $1/\sqrt{2E}$ factor for each of the particles taking part in an interaction. The relativistically invariant form of the expression (4.4) for the transition rate in an interaction then becomes (with $V = 1$ and $\hbar = c = 1$)

$$W = \frac{2\pi |M|^2}{\prod_{\text{initial}} 2E_j} \rho_f, \tag{A.15}$$

where the final-state phase-space factor, for n final state particles and $(n-1)$ independent 3-momenta, is

$$\rho_f = \frac{d}{dE_{\text{total}}} \frac{\int d^3\mathbf{p}_1 \, d^3\mathbf{p}_2 \cdots d^3\mathbf{p}_{n-1}}{(2\pi)^{3(n-1)} \prod_{\text{final}} 2E_k}. \tag{A.16}$$

APPENDIX B

Pauli Matrix Representation of the Isospin of the Nucleon

Just as for the description of particles of spin $\frac{1}{2}$, we can formally express the isospin wavefunction of a nucleon as a two-component matrix

$$\chi = \begin{bmatrix} 1 \\ 0 \end{bmatrix} = p,$$

$$\chi = \begin{bmatrix} 0 \\ 1 \end{bmatrix} = n,$$

(B.1)

where, for brevity, p and n signify the proton (or isospin-up) state and neutron (or isospin-down) state, respectively. As in the formalism for spin $\frac{1}{2}$, we introduce an isospin operator τ which has Cartesian components given by the 2×2 matrix operators

$$\tau_1 = \frac{1}{2}\begin{bmatrix} 0 & 1 \\ 1 & 0 \end{bmatrix}, \quad \tau_2 = \frac{1}{2}\begin{bmatrix} 0 & -i \\ i & 0 \end{bmatrix}, \quad \tau_3 = \frac{1}{2}\begin{bmatrix} 1 & 0 \\ 0 & -1 \end{bmatrix}, \quad (B.2)$$

which, apart from the factor $\frac{1}{2}$, are identical with the Pauli spin matrices (see (D.7)) and obey the usual commutation relations of the angular momentum operators, as in (C.5). Thus,

$$\tau_1 \tau_2 - \tau_2 \tau_1 = i\tau_3, \text{ etc.}$$

If we apply τ_3 to the wave functions (B.1), we obtain

$$\tau_3 p = \frac{1}{2}\begin{bmatrix} 1 & 0 \\ 0 & -1 \end{bmatrix}\begin{bmatrix} 1 \\ 0 \end{bmatrix} = \frac{1}{2}\begin{bmatrix} 1 \\ 0 \end{bmatrix} = \frac{1}{2}p, \tag{B.3}$$

and

$$\tau_3 n = -\tfrac{1}{2}n,$$

expressing the fact that the eigenvalues of τ_3 are $+\frac{1}{2}$ and $-\frac{1}{2}$ for the two charge states of the nucleon. When we say that a nucleon has an isospin $\tau = \frac{1}{2}$, we really mean that there are $2\tau + 1$ eigenvalues of the *operator* τ_3. The eigenvalue of τ^2 should then be $\tau(\tau + 1) = \frac{3}{4}$. We verify this from (B.2):

$$\tau^2 = \tau_1^2 + \tau_2^2 + \tau_3^2, \qquad \text{where} \quad \tau_1^2 = \tau_2^2 = \tau_3^2 = \frac{1}{4}\begin{bmatrix} 1 & 0 \\ 0 & 1 \end{bmatrix}, \tag{B.4}$$

so

$$\tau^2 \chi = \frac{3}{4}\begin{bmatrix} 1 & 0 \\ 0 & 1 \end{bmatrix}\chi = \tfrac{3}{4}\chi.$$

Other combinations of the Pauli operators are the "raising" and "lowering" operators (see (C.6)):

$$\tau_+ = \tau_1 + i\tau_2 = \begin{bmatrix} 0 & 1 \\ 0 & 0 \end{bmatrix},$$

$$\tau_- = \tau_1 - i\tau_2 = \begin{bmatrix} 0 & 0 \\ 1 & 0 \end{bmatrix}. \tag{B.5}$$

Applying these to (B.1), we obtain

$$\tau_+ p = 0, \qquad \tau_+ n = p,$$

$$\tau_- p = n, \qquad \tau_- n = 0. \tag{B.6}$$

Thus, the operator τ_+ transforms a neutron into a proton state, and τ_- transforms a proton into a neutron—in other words, the operators $\tau_\pm$ flip the sign of the third component of isospin of the nucleon.

APPENDIX C

Clebsch-Gordan Coefficients and Rotation Matrices

C.1. CLEBSCH-GORDAN COEFFICIENTS: THE ADDITION OF ANGULAR MOMENTA OR ISOSPINS

Suppose we have two particles of angular momenta $\mathbf{j}_1$ and $\mathbf{j}_2$ with z-components m_1 and m_2. The total z-component is

$$m = m_1 + m_2.$$

The total angular momentum is

$$\mathbf{j} = \mathbf{j}_1 + \mathbf{j}_2,$$

and may therefore lie anywhere inside the limits

$$|j_1 - j_2| \leq j \leq |j_1 + j_2|.$$

We wish to find the weights of the various allowed j-values contributing to the two-particle state, that is,

$$\phi_1(j_1 m_1)\phi_2(j_2 m_2) = \sum_j C_j \psi(j, m), \quad \text{with} \quad m = m_1 + m_2. \quad \text{(C.1)}$$

The C_j are called Clebsch-Gordan coefficients (or Wigner, or vector addition, coefficients). Alternatively, we may want to express $\psi(j, m)$ as a sum of terms of different j_1 and j_2 combinations. We can do this by the use of angular-momentum (or isospin) *shift operators* (also known as "raising" and "lowering" operators).

First let us recall the definition of the x-, y-, and z-component angular-momentum operators, in terms of the differential Cartesian operators:

$$J_x = -\frac{ih}{2\pi}\left(y\frac{\partial}{\partial z} - z\frac{\partial}{\partial y}\right),$$

$$J_y = -\frac{ih}{2\pi}\left(z\frac{\partial}{\partial x} - x\frac{\partial}{\partial z}\right), \tag{C.2}$$

$$J_z = -\frac{ih}{2\pi}\left(x\frac{\partial}{\partial y} - y\frac{\partial}{\partial x}\right).$$

These Cartesian operations can also be interpreted in terms of rotations. A rotation in azimuthal angle in the xy-plane has Cartesian components

$$\delta y = r\cos\phi\delta\phi \quad = x\delta\phi$$

$$\delta x = -r\sin\phi\delta\phi = -y\delta\phi$$

Thus, the effect of a small rotation on a function $\psi(x, y, z)$ will be

$$R(\phi, \delta\phi)\cdot\psi(x, y, z) = \psi(x + \delta x, y + \delta y, z) = \psi(x, y, z) + \delta x\frac{\partial\psi}{\partial x} + \delta y\frac{\partial\psi}{\partial y}$$

$$= \psi\left[1 + \left(x\frac{\partial}{\partial y} - y\frac{\partial}{\partial x}\right)\delta\phi\right] = \psi\left(1 + \delta\phi\frac{\partial}{\partial\phi}\right)$$

and from (C.2),

$$J_z = -ih\left(x\frac{\partial}{\partial y} - y\frac{\partial}{\partial x}\right) = -ih\frac{\partial}{\partial\phi}. \tag{C.3}$$

So

$$R = 1 + \delta\phi\frac{\partial}{\partial\phi} = 1 + \frac{iJ_z}{\hbar}\delta\phi. \tag{C.4}$$

It is readily verified that the operators

$$J_x, J_y, J_z \quad \text{and} \quad J^2 = J_x^2 + J_y^2 + J_z^2$$

obey the commutation rules

$$J^2J_x - J_xJ^2 = 0, \quad \text{etc.}$$

and

$$J_x J_y - J_y J_x = iJ_z,$$
$$J_y J_z - J_z J_y = iJ_x, \qquad (C.5)$$
$$J_z J_x - J_x J_z = iJ_y,$$

where we have used units $\hbar = c = 1$ for brevity. The eigenvalues of the operators J^2 and J_z are given in Eq. (C.7) below.

The *shift operators* are defined as

$$\left. \begin{aligned} J_+ &= J_x + iJ_y \\ J_- &= J_x - iJ_y \end{aligned} \right\}, \qquad \text{whence} \qquad \begin{cases} J_z J_+ - J_+ J_z = J_+, \\ J_z J_- - J_- J_z = -J_-. \end{cases} \qquad (C.6)$$

Thus,

$$J_z(J_- \phi) = J_z J_- \phi = J_-(J_z - 1)\phi = (m - 1)J_- \phi.$$

Similarly,

$$J_z(J_+ \phi) = (m + 1)(J_+ \phi).$$

This last equation shows that the wavefunction $J_+ \phi$ is an eigenstate of J_z with eigenvalue $m + 1$. We can therefore write it as

$$J_+ \phi(j, m) = C_+ \phi(j, m + 1),$$

where C_+ is an unknown (and generally complex) constant. If we multiply both sides of this equation by $\phi^*(j, m + 1)$, and integrate over volume, we get

$$\int \phi^*(j, m + 1) J_+ \phi(j, m) \, dV = C_+ \int \phi^*(j, m + 1)\phi(j, m + 1) \, dV,$$

where * indicates complex conjugation.

We choose the normalization of ϕ so that the last integral is unity, and all allowed m-values have unit weight. So

$$C_+ = \int \phi^*(j, m + 1) J_+ \phi(j, m) \, dV.$$

Similarly,

$$C_- = \int \phi^*(j, m) J_- \phi(j, m + 1) \, dV$$

$$= \int \phi^*(j, m) J_+^* \phi(j, m + 1) \, dV$$

$$= C_+^*,$$

from Eq. (C.6).

If we neglect arbitrary and unobservable phases, we must have

$$C_+ = C_- = C \qquad \text{(a real number)}.$$

Also, from Eq. (C.6),

$$J_+ J_- = J_x^2 + J_y^2 - i(J_x J_y - J_y J_x) = J_x^2 + J_y^2 + J_z = J^2 - J_z^2 + J_z.$$

Then

$$J_+ J_- \phi(j, m+1) = [j(j+1) - m^2 - m]\phi(j, m+1) = C^2\phi(j, m+1).$$

So

$$C = \sqrt{j(j+1) - m(m+1)}$$

is the coefficient connecting states $(j, m) \leftrightarrow (j, m+1)$.

To summarize, the angular-momentum operators have the following properties:

$$J_z \phi(j, m) = m\phi(j, m), \tag{C.7a}$$

$$J^2 \phi(j, m) = j(j+1)\phi(j, m), \tag{C.7b}$$

$$J_+ \phi(j, m) = \sqrt{j(j+1) - m(m+1)}\phi(j, m+1), \tag{C.7c}$$

$$J_- \phi(j, m) = \sqrt{j(j+1) - m(m-1)}\phi(j, m-1). \tag{C.7d}$$

Example As an example, we consider two particles of j_1, m_1, and j_2, m_2, forming the combined state $\psi(j, m)$, and we take the case where $j_1 = 1, j_2 = \frac{1}{2}$, and $j = \frac{3}{2}$ or $\frac{1}{2}$.

Obviously the states of $m = \pm\frac{3}{2}$ can be formed in only one way:

$$\psi(\tfrac{3}{2}, \tfrac{3}{2}) = \phi(1, 1)\phi(\tfrac{1}{2}, \tfrac{1}{2}), \tag{C.8}$$

$$\psi(\tfrac{3}{2}, -\tfrac{3}{2}) = \phi(1, -1)\phi(\tfrac{1}{2}, -\tfrac{1}{2}). \tag{C.9}$$

Now we use the operators $J_\pm$ to form the relations:

$$J_- \phi(\tfrac{1}{2}, \tfrac{1}{2}) = \phi(\tfrac{1}{2}, -\tfrac{1}{2}), \qquad J_- \phi(\tfrac{1}{2}, -\tfrac{1}{2}) = 0$$

$$J_- \phi(1, 1) = \sqrt{2}\phi(1, 0), \qquad J_- \phi(1, 0) = \sqrt{2}\phi(1, -1), \qquad J_- \phi(1, -1) = 0,$$

using (C.7c, d).

Now operate on (C.8) with J_- on both sides:

$$J_- \psi(\tfrac{3}{2}, \tfrac{3}{2}) = \sqrt{3}\psi(\tfrac{3}{2}, \tfrac{1}{2}) = J_- \phi(1, 1)\phi(\tfrac{1}{2}, \tfrac{1}{2})$$
$$= \sqrt{2}\phi(1, 0)\phi(\tfrac{1}{2}, \tfrac{1}{2}) + \phi(1, 1)\phi(\tfrac{1}{2}, -\tfrac{1}{2}).$$

So

$$\psi(\tfrac{3}{2}, \tfrac{1}{2}) = \sqrt{\tfrac{2}{3}}\phi(1, 0)\phi(\tfrac{1}{2}, \tfrac{1}{2}) + \sqrt{\tfrac{1}{3}}\phi(1, 1)\phi(\tfrac{1}{2}, -\tfrac{1}{2}). \tag{C.10}$$

Similarly, for (C.9),

$$\psi(\tfrac{3}{2}, -\tfrac{1}{2}) = \sqrt{\tfrac{2}{3}}\phi(1, 0)\phi(\tfrac{1}{2}, -\tfrac{1}{2}) + \sqrt{\tfrac{1}{3}}\phi(1, -1)\phi(\tfrac{1}{2}, \tfrac{1}{2}). \qquad (C.11)$$

The $j = \tfrac{1}{2}$ state can be expressed as a linear sum:

$$\psi(\tfrac{1}{2}, \tfrac{1}{2}) = a\phi(1, 1)\phi(\tfrac{1}{2}, -\tfrac{1}{2}) + b\phi(1, 0)\phi(\tfrac{1}{2}, \tfrac{1}{2})$$

with $a^2 + b^2 = 1$. Then

$$J_+\psi(\tfrac{1}{2}, \tfrac{1}{2}) = 0 = a\phi(1, 1)\phi(\tfrac{1}{2}, \tfrac{1}{2}) + b\sqrt{2}\phi(1, 1)\phi(\tfrac{1}{2}, \tfrac{1}{2}).$$

Thus, $a = \sqrt{\tfrac{2}{3}}$, $b = -\tfrac{1}{3}$, and so

$$\psi(\tfrac{1}{2}, \tfrac{1}{2}) = \sqrt{\tfrac{2}{3}}\phi(1, 1)\phi(\tfrac{1}{2}, -\tfrac{1}{2}) - \sqrt{\tfrac{1}{3}}\phi(1, 0)\phi(\tfrac{1}{2}, \tfrac{1}{2}). \qquad (C.12)$$

Similarly,

$$\psi(\tfrac{1}{2}, -\tfrac{1}{2}) = \sqrt{\tfrac{1}{3}}\phi(1, 0)\phi(\tfrac{1}{2}, -\tfrac{1}{2}) - \sqrt{\tfrac{2}{3}}\phi(1, -1)\phi(\tfrac{1}{2}, \tfrac{1}{2}). \qquad (C.13)$$

Expressions (C.8) to (C.12) give the coefficients appearing in Table III for the addition of $J = 1$ and $J = \tfrac{1}{2}$ states.

C.2. ROTATION MATRICES

A state $\phi(j, m)$ is transformed under a rotation through an angle θ about the y-axis into a linear combination of the $(2j + 1)$ states $\phi(j, m')$, where $m' = -j$, $-j + 1, \ldots, j - 1, j$. From the expression (3.14) for the rotation operator, we can write

$$e^{-i\theta J_y} \cdot \phi(j, m) = \sum_{m'} d^j_{m', m}(\theta)\phi(j, m'), \qquad (C.14)$$

where the coefficients $d^j_{m', m}$ are called *rotation matrices*. For fixed m' the expression for $d^j_{m', m}$ is therefore obtained from

$$\phi^*(j, m')e^{-i\theta J_y}\phi(j, m) = d^j_{m', m}(\theta). \qquad (C.15)$$

$j = \tfrac{1}{2}$

For a state of angular momentum $j = \tfrac{1}{2}$, the appropriate operator is the Pauli spin matrix σ_y (see Appendix B):

$$J_y = \tfrac{1}{2}\sigma_y = \frac{1}{2}\begin{vmatrix} 0 & -i \\ i & 0 \end{vmatrix}.$$

Then it is easy to show that

$$e^{-(1/2)i\theta\sigma_y} = \cos\frac{\theta}{2} - i\sigma_y \sin\frac{\theta}{2} = \begin{vmatrix} \cos\dfrac{\theta}{2} & -\sin\dfrac{\theta}{2} \\ \sin\dfrac{\theta}{2} & \cos\dfrac{\theta}{2} \end{vmatrix} \tag{C.16}$$

by expanding the exponential and using the fact that $\sigma_y^2 = 1$. We denote the states ϕ, ϕ^* by column and row matrices; for example, $\phi^*(\frac{1}{2}, \frac{1}{2}) = |1 \quad 0|$, $\phi(\frac{1}{2}, \frac{1}{2}) = \begin{vmatrix} 1 \\ 0 \end{vmatrix}$, etc., so that

$$d_{1/2, 1/2}^{1/2}(\theta) = d_{-1/2, -1/2}^{1/2}(\theta) = |1 \quad 0| \begin{vmatrix} \cos\dfrac{\theta}{2} & -\sin\dfrac{\theta}{2} \\ \sin\dfrac{\theta}{2} & \cos\dfrac{\theta}{2} \end{vmatrix} \begin{vmatrix} 1 \\ 0 \end{vmatrix} = \cos\frac{\theta}{2},$$

$$d_{-1/2, 1/2}^{1/2}(\theta) = -d_{1/2, -1/2}^{1/2}(\theta) = |0 \quad 1| \begin{vmatrix} \cos\dfrac{\theta}{2} & -\sin\dfrac{\theta}{2} \\ \sin\dfrac{\theta}{2} & \cos\dfrac{\theta}{2} \end{vmatrix} \begin{vmatrix} 1 \\ 0 \end{vmatrix} = \sin\frac{\theta}{2}. \tag{C.17}$$

As an example, consider a beam of RH polarized particles described by $\phi(\frac{1}{2}, \frac{1}{2})$ being scattered through angle θ. If the interaction is helicity-conserving (i.e., it is a vector or axial-vector interaction), they will emerge as RH particles, as measured relative to their momentum vector. However, relative to the z-axis (the incident direction) they now represent a superposition of the states $\phi'(\frac{1}{2}, \frac{1}{2})$ and $\phi'(\frac{1}{2}, -\frac{1}{2})$. Angular-momentum conservation in a non-spin-flip interaction, however, allows only the state $\phi'(\frac{1}{2}, \frac{1}{2})$ with the angular distribution

$$|d_{1/2, 1/2}^{1/2}(\theta)|^2 = \cos^2\frac{\theta}{2}. \tag{C.18}$$

For a spin-flip interaction, the final state would be $\phi'(\frac{1}{2}, -\frac{1}{2})$ with a distribution of the form

$$|d_{1/2, -1/2}^{1/2}(\theta)|^2 = \sin^2\frac{\theta}{2}. \tag{C.19}$$

These terms enter the cross-section for electron scattering via the electric and magnetic interactions, respectively (see Section 6.6).

j = 1

In the case $j = 1$, we make use of the expansions

$$e^{-i\theta J_y} = 1 - i\theta J_y - \frac{\theta^2}{2!} J_y^2 + i\frac{\theta^3}{3!} J_y^3 + \cdots \tag{C.20}$$

and

$$J_y = -\frac{i}{2}(J^+ - J^-)$$

from (C.6). It is straightforward to show that

$$J_y\phi(1, 1) = J_y^{(2n+1)}\phi(1, 1) = \frac{i}{\sqrt{2}}\phi(1, 0),$$

(C.21)

$$J_y^2\phi(1, 1) = J_y^{2n}\phi(1, 1) = \tfrac{1}{2}[\phi(1, 1) - \phi(1, -1)].$$

Then

$$\phi^*(1, 1)e^{-i\theta J_y}\phi(1, 1) = 1 - \frac{1}{2}\frac{\theta^2}{2!} + \frac{1}{2}\frac{\theta^4}{4!} + \cdots,$$

$$\phi^*(1, 1)e^{-i\theta J_y}\phi(1, -1) = +\frac{1}{2}\frac{\theta^2}{2!} - \frac{1}{2}\frac{\theta^4}{4!} + \cdots,$$

so that

$$d_{1,1}^1(\theta) = \tfrac{1}{2}(1 + \cos\theta),$$

$$d_{1,-1}^1(\theta) = \tfrac{1}{2}(1 - \cos\theta).$$

(C.22)

Also,

$$\phi^*(1, 0)e^{-i\theta J_y}\phi(1, 1) = \frac{i}{\sqrt{2}}\left[-i\theta + i\frac{\theta^3}{3!} + \cdots\right]$$

or

$$d_{1,0}^1(\theta) = \frac{1}{\sqrt{2}}\sin\theta.$$

(C.23)

These relations have been used in discussing the decay angular distribution of polarized W-bosons in Chapter 7 (see Eq. 7.71)).

A list of the important rotation matrices is given in Table II.

APPENDIX D

The Dirac Equation

An account of the Dirac equation can be found in most standard texts on quantum mechanics. We append a brief account here for convenience only. We start off with the de Broglie prescription of a plane wave representing a free particle. In units $\hbar = c = 1$ this is

$$\psi = e^{i(\mathbf{k} \cdot \mathbf{r} - \omega t)}, \tag{D.1}$$

with $\mathbf{k}$ and ω the propagation vector and angular frequency respectively, which in our units are numerically equal to the momentum $\mathbf{p}$ and energy E. The wavefunction (D.1) is a solution of the *Klein-Gordon equation*:

$$\frac{\partial^2 \psi}{\partial t^2} = \frac{\partial^2 \psi}{\partial x^2} + \frac{\partial^2 \psi}{\partial y^2} + \frac{\partial^2 \psi}{\partial z^2} - m^2 \psi$$

$$= (\nabla^2 - m^2)\psi, \tag{D.2}$$

suitable for describing free relativistic spinless particles. If we insert (D.1) in (D.2) we get the familiar relation

$$E^2 = p^2 + m^2. \tag{D.3}$$

In the nonrelativistic case, expanding (D.3) in powers of p/m gives to first order $E = m + p^2/2m$. In this approximation (D.1) becomes, on dividing by a factor e^{-imt},

$$\phi = e^{i[\mathbf{p} \cdot \mathbf{r} - (p^2/2m)t]},$$

which satisfies the *Schrödinger equation* describing nonrelativistic particles:

$$\frac{\partial \phi}{\partial t} - \frac{i}{2m} \nabla^2 \phi = 0. \tag{D.4}$$

Dirac formulated his relativistic wave equation under the assumption that derivatives of *both* time and space coordinates should occur to *first order*. It turns out that such an equation describes particles of spin-$\frac{1}{2}$. The simplest form that we could write would be the two *Weyl equations*

$$\frac{\partial \psi}{\partial t} = \pm \left(\sigma_x \frac{\partial \psi}{\partial x} + \sigma_y \frac{\partial \psi}{\partial y} + \sigma_z \frac{\partial \psi}{\partial z} \right) = \pm \, \boldsymbol{\sigma} \cdot \frac{\partial}{\partial \mathbf{r}} \, \psi, \tag{D.5}$$

where the σ's are constants. The condition for this to satisfy (D.2), so that (D.3) may hold, is found by squaring (D.5) and equating coefficients, whence we obtain

$$\sigma_x^2 = \sigma_y^2 = \sigma_z^2 = 1,$$

$$\sigma_x \sigma_y + \sigma_y \sigma_x = 0, \quad \text{etc.,} \tag{D.6}$$

$$m = 0.$$

Thus the Weyl equations describe massless particles (in fact, neutrinos). The σ's cannot be numbers, since they do not commute. They can, however, be represented by *matrices*. The 2×2 Pauli spin matrices are a suitable choice (not the only one). They are

$$\sigma_x = \begin{bmatrix} 0 & 1 \\ 1 & 0 \end{bmatrix}, \qquad \sigma_y = \begin{bmatrix} 0 & -i \\ i & 0 \end{bmatrix}, \qquad \sigma_z = \begin{bmatrix} 1 & 0 \\ 0 & -1 \end{bmatrix}, \tag{D.7}$$

where

$$\sigma_x^2 = \sigma_y^2 = \sigma_z^2 = \begin{bmatrix} 1 & 0 \\ 0 & 1 \end{bmatrix},$$

as required. The wavefunction ψ must now have two components, since it is operated on by 2×2 matrices. Thus,

$$\psi = \begin{bmatrix} \psi_1 \\ \psi_2 \end{bmatrix}.$$

Now we wish to include a mass term and, therefore, a further matrix. The simplest set of four anticommuting matrices are 4×4, and ψ is then a four-component spinor. First we write down the space-time coordinates as components of a 4-vector (i.e., in covariant notation),

$$x_1 = x, \qquad x_2 = y, \qquad x_3 = z, \qquad x_4 = it$$

Generally any component is denoted by x_μ, where $\mu = 1, \ldots, 4$, and the relativistically invariant scalar product of two 4-vectors is $x_\mu X_\mu$, where a repeated index implies a summation over μ. Now we write the *Dirac equation* as

$$\left(\gamma_\mu \frac{\partial}{\partial x_\mu} + m\right)\psi = 0,$$ (D.8)

which is just generalizing (D.5) by inserting a mass term. The γ's are 4×4 matrices to be determined. We can do this by requiring (D.8) to satisfy the Klein-Gordon equation (D.2), which we can rewrite in covariant form as

$$\left(\frac{\partial^2}{\partial x_\mu \partial x_\mu} - m^2\right)\psi = 0.$$ (D.9)

Multiplying (D.8) by $[\gamma_\nu(\partial/\partial x_\nu) - m]$, we get

$$\left(\gamma_\nu \frac{\partial}{\partial x_\nu} - m\right)\left(\gamma_\mu \frac{\partial}{\partial x_\mu} + m\right)\psi = \left(\gamma_\nu \gamma_\mu \frac{\partial^2}{\partial x_\nu \partial x_\mu} - m^2\right)\psi = 0,$$ (D.10)

a sum over μ, ν being again understood. Equation (D.10) can be verified by writing out the individual components $\mu, \nu = 1, \ldots, 4$ and proving that the cross terms cancel. Comparing (D.10) with (D.9), we therefore require the following conditions on the γ's:

$$\gamma_\nu \gamma_\mu + \gamma_\mu \gamma_\nu = 2\delta_{\mu\nu}, \qquad \text{where} \quad \delta_{\mu\nu} = \begin{cases} 1, & \nu = \mu, \\ 0, & \nu \neq \mu. \end{cases}$$ (D.11)

The usual representation of the γ-matrices obeying these commutation relations is

$$\gamma_k = \begin{bmatrix} 0 & -i\sigma_k \\ i\sigma_k & 0 \end{bmatrix}, \qquad k = 1,2,3,$$

$$\gamma_4 = \begin{bmatrix} 1 & 0 \\ 0 & -1 \end{bmatrix},$$ (D.12)

where each element stands for a 2×2 matrix, and the σ_k are defined in (D.7). Written out in full

$$\gamma_1 = \begin{bmatrix} 0 & 0 & 0 & -i \\ 0 & 0 & -i & 0 \\ 0 & i & 0 & 0 \\ i & 0 & 0 & 0 \end{bmatrix} \qquad \gamma_2 = \begin{bmatrix} 0 & 0 & 0 & -1 \\ 0 & 0 & 1 & 0 \\ 0 & 1 & 0 & 0 \\ -1 & 0 & 0 & 0 \end{bmatrix}$$

$$\gamma_3 = \begin{bmatrix} 0 & 0 & -i & 0 \\ 0 & 0 & 0 & i \\ i & 0 & 0 & 0 \\ 0 & -i & 0 & 0 \end{bmatrix} \qquad \gamma_4 = \begin{bmatrix} 1 & 0 & 0 & 0 \\ 0 & 1 & 0 & 0 \\ 0 & 0 & -1 & 0 \\ 0 & 0 & 0 & -1 \end{bmatrix}$$ (D.13)

It is also useful to define the product matrix

$$\gamma_5 = \gamma_1\gamma_2\gamma_3\gamma_4 = \begin{bmatrix} 0 & 0 & -1 & 0 \\ 0 & 0 & 0 & -1 \\ -1 & 0 & 0 & 0 \\ 0 & -1 & 0 & 0 \end{bmatrix},$$

with $\gamma_5^2 = 1$, $\gamma_5\gamma_\mu + \gamma_\mu\gamma_5 = 0$, $\mu = 1,\ldots, 4$.

D.1. SOLUTIONS TO THE DIRAC EQUATION

The Dirac equation (D.8) is a set of four simultaneous equations ($\mu = 1$ to 4) and must have four solutions. Separating off the space-time-dependent part, we write the four plane-wave solutions as

$$\psi = u_r e^{ip_\mu x_\mu}, \tag{D.14}$$

$$\psi = v_r e^{-ip_\mu x_\mu},$$

where $r = 1$, 2 and the u_r and v_r are constant four-component spinors. $u_{1,2}$ correspond to a particle of momentum **p** and energy E, and $v_{1,2}$ to a state of momentum $-$**p** and energy $-E$, that is, to a *negative energy state*. This is clearly allowed since (D.3) does not determine the sign of E. Substituting (D.14) in (D.8) we get

$$(i\gamma_\mu p_\mu + m)u = 0, \tag{D.15}$$

$$(-i\gamma_\mu p_\mu + m)v = 0. \tag{D.16}$$

If we now insert the explicit form of the matrices (D.13) into (D.15), we obtain four simultaneous equations, which can be solved up to a constant factor N to give

$$u_1 = \begin{bmatrix} 1 \\ 0 \\ p_3/(E+m) \\ (p_1+ip_2)/(E+m) \end{bmatrix} N \qquad u_2 = \begin{bmatrix} 0 \\ 1 \\ (p_1-ip_2)/(E+m) \\ -p_3/(E+m) \end{bmatrix} N,$$

$$v_1 = \begin{bmatrix} p_3/(|E|+m) \\ (p_1+ip_2)/(|E|+m) \\ 1 \\ 0 \end{bmatrix} N \qquad v_2 = \begin{bmatrix} (p_1-ip_2)/(|E|+m) \\ -p_3/(|E|+m) \\ 0 \\ 1 \end{bmatrix} N. \tag{D.17}$$

Any linear combination of these solutions is also a solution. (D.17) can be written more compactly in two-component form, where each entry is itself a

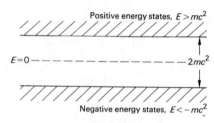

Figure D.1

2×2 matrix. Denoting $\sigma_1 p_1 + \sigma_2 p_2 + \sigma_3 p_3 = \boldsymbol{\sigma} \cdot \mathbf{p}$, where $\sigma_{1,2,3}$ are the Pauli 2×2 matrices (D.7), we obtain

$$u = N \begin{bmatrix} 1 \\ \dfrac{\boldsymbol{\sigma} \cdot \mathbf{p}}{(E + m)} \end{bmatrix} \quad \text{and} \quad v = N \begin{bmatrix} \dfrac{\boldsymbol{\sigma} \cdot \mathbf{p}}{(|E| + m)} \\ 1 \end{bmatrix}. \tag{D.18}$$

The meaning of these solutions becomes more transparent if we go to the particle rest frame where $\mathbf{p} = 0$, when they reduce to the two-component spinors

$$u_1 = \begin{vmatrix} 1 \\ 0 \end{vmatrix} \quad u_2 = \begin{vmatrix} 0 \\ 1 \end{vmatrix} \quad v_1 = \begin{vmatrix} 1 \\ 0 \end{vmatrix} \quad v_2 = \begin{vmatrix} 0 \\ 1 \end{vmatrix},$$

identical with the two-component Pauli spinors of the nonrelativistic theory. u_1 and u_2 are interpreted as positive-energy electron states with two possible *spin* directions, up and down—thus, for electrons with half-integral spin. v_1 and v_2 are the two negative-energy spin states. Dirac supposed that there was a completely filled sea of negative-energy states; a "hole" in this sea of electrons was interpreted as a *positron*, which would be produced whenever a negative energy electron received an energy $\geqslant 2mc^2$ and was kicked into a positive-energy state ($e^{\pm}$ pair creation; see Fig. D.1).

D.2. NORMALIZATION OF DIRAC WAVEFUNCTIONS

For the normalization factor N we take

$$N = \sqrt{(|E| + m)/2m} \tag{D.19}$$

for reasons that will be justified shortly. Particle density is defined by $\rho = \psi^* \psi$, where the asterisk represents the complex conjugate wavefunction. For 4-spinors, ψ is a column matrix and therefore ψ^* has to be a row matrix: We denote it by $\psi^{\dagger}$ to indicate complex conjugation plus transposition (interchange of rows and columns). Thus,

$$\psi = \begin{bmatrix} a_1 \\ a_2 \\ a_3 \\ a_4 \end{bmatrix} \qquad \text{and} \qquad \psi^\dagger = [a_1^* \quad a_2^* \quad a_3^* \quad a_4^*],$$

whence

$$\rho = \psi^\dagger \psi = \sum_i a_i a_i^*. \tag{D.20}$$

With the normalization (D.19) the solutions (D.17) give us*

$$u^\dagger u = v^\dagger v = |E|/m. \tag{D.21}$$

This is a Lorentz-invariant normalization for fermions (see Appendix A) since the probability of finding a fermion of energy E, mass m in a box of volume V is then

$$\int \rho \, dV = u^\dagger u V = \psi^\dagger \psi V = \frac{|E| V_0}{m} \frac{m}{|E|} = V_0 \tag{D.22}$$

independent of E, where V_0 is the box volume in the rest frame of the particle, and in a frame in which the particle has energy E the volume will be Lorentz contracted to $V = V_0 m/E$. Eq. (D.22) therefore shows that the normalization factor (D.19) is Lorentz invariant.

It is also useful to define the spinor

$$\bar{\psi} = \psi^\dagger \gamma_4, \tag{D.23}$$

so that from (D.12)

$$\bar{\psi} \gamma_4 = \psi^\dagger$$

and

$$\rho = \bar{\psi} \gamma_4 \psi.$$

It is straightforward to show that $\bar{\psi}$ satisfies the Dirac equation

$$\frac{\partial}{\partial x_\mu} \bar{\psi} \gamma_\mu - m \bar{\psi} = 0$$

or

$$\bar{\psi}(i\gamma_\mu P_\mu - m) = 0. \tag{D.24}$$

---

* Normalization of wavefunctions is a matter of convention and convenience. If we had used the usual convention for bosons as in (A.15), namely, $\int \rho \, dV = 2EV_0$, then for (D.19) we would have $N = \sqrt{|E| + m}$.

D.3. PARITY OF PARTICLE AND ANTIPARTICLE

Suppose $\psi(\mathbf{r}, t)$ satisfies the Dirac equation (D.8). Spatial inversion of this state, simply by replacing $\mathbf{r}$ by $-\mathbf{r}$, does not satisfy the Dirac equation, since it is first-order in the space coordinates. $\psi(-\mathbf{r}, t) = \psi(-x_k, x_4)$ satisfies

$$\left(\gamma_4 \frac{\partial}{\partial x_4} - \gamma_k \frac{\partial}{\partial x_k} + m\right)\psi(-\mathbf{r}, t) = 0, \qquad k = 1, 2, 3.$$

On multiplying through from the left by γ_4, and using the rule $\gamma_4\gamma_k + \gamma_k\gamma_4 = 0$, we get

$$\left(\gamma_\mu \frac{\partial}{\partial x_\mu} + m\right)\gamma_4\psi(-\mathbf{r}, t) = 0.$$

Thus, $\gamma_4\psi(-\mathbf{r}, t)$ is the inversion of the original state $\psi(\mathbf{r}, t)$, and γ_4 is the parity operator for the Dirac wavefunction. Now consider a spin-up positive-energy state (in the particle rest frame, $p = 0$), Eq. (D.17):

$$u_1 = \begin{bmatrix} 1 \\ 0 \\ 0 \\ 0 \end{bmatrix}, \qquad \gamma_4 u_1 = \begin{bmatrix} 1 \\ 0 \\ 0 \\ 0 \end{bmatrix}.$$

For a spin-up negative-energy state,

$$v_1 = \begin{bmatrix} 0 \\ 0 \\ 1 \\ 0 \end{bmatrix}, \qquad \gamma_4 v_1 = -\begin{bmatrix} 0 \\ 0 \\ 1 \\ 0 \end{bmatrix}.$$

Thus, if the positive-energy state is assigned an even intrinsic parity, then the negative-energy state, or, equivalently, the antiparticle, has odd intrinsic parity.

D.4. COMPLETENESS RELATION AND PROJECTION OPERATORS

In calculating cross-sections and decay rates, it is useful to sum over fermion spins, and this can be achieved with the help of the completeness relation. Let us denote the spin states by r and $s(r, s = 1, 2)$. Then from (D.17) it can be demonstrated that

$$\bar{u}_r u_s = u_r^\dagger \gamma_4 u_s = \delta_{rs}; \qquad \delta_{rs} = \begin{cases} 1 \text{ for } r = s \\ 0 \text{ for } r \neq s \end{cases}$$

and

$$\bar{v}_r v_s = -\delta_{rs}. \tag{D.25}$$

For the spin sum we therefore obtain the relation

$$\sum_{s=1,2} (\bar{u}_s u_s - \bar{v}_s v_s) = \hat{1}, \qquad \text{where} \quad \hat{1} = \begin{bmatrix} 1 & 0 & 0 & 0 \\ 0 & 1 & 0 & 0 \\ 0 & 0 & 1 & 0 \\ 0 & 0 & 0 & 1 \end{bmatrix}, \qquad \text{(D.26)}$$

with $\hat{1}$ denoting the unit 4×4 matrix. This is a spin sum over both positive and negative energy states. For a general matrix Q we get

$$\sum_{r,s=1\rightarrow 4} (\bar{u}_r Q_{rs} u_s - \bar{v}_r Q_{rs} v_s) = \sum Q_{rs}\delta_{rs} = \sum Q_{rr} = \text{Tr } Q,$$

where Tr Q denotes the trace or sum of the diagonal elements of Q. (In the previous case (D.26), Q was a unit matrix with Tr $Q = 4$). In an actual problem, we want a sum over positive-energy states only, so the negative-energy states must be projected out. We introduce the energy projection operators

$$\Lambda^+ = (m - i\gamma_\mu P_\mu)/2m,$$
$$\Lambda^- = (m + i\gamma_\mu P_\mu)/2m, \qquad \text{(D.27)}$$

where

$$\Lambda^+ + \Lambda^- = \hat{1}$$

and

$$\Lambda^+ u = \frac{(m - i\gamma_\mu P_\mu)}{2m} u = \frac{(m + m)}{2m} u = u,$$

$$\Lambda^- v = v; \qquad \Lambda^+ v = \Lambda^- u = 0$$

from (D.15), and (D.16). Thus, for a sum over spins of positive energy states

$$\sum \bar{u}_s u_s = \Lambda^+ \sum \bar{u}_s u_s = \Lambda^+ \sum (\bar{u}_s u_s - \bar{v}_s v_s) = \Lambda^+ \hat{1}$$

and

$$\sum \bar{v}_s v_s = -\Lambda^- \hat{1}.$$

For a 4×4 matrix operator Q acting on a fermion wavefunction, the result of the spin sum over positive-energy states will be

$$\sum_{\text{spin}} \bar{u} Q u = \text{Tr}(\Lambda^+ Q), \qquad \text{(D.28)}$$

with a corresponding expression for the v-functions. These results are applied elsewhere in calculating cross-sections (see Appendices F, G).

D.5. HELICITY STATES

The Dirac wavefunctions u, v are eigenvalues of the operator $\boldsymbol{\sigma} \cdot \mathbf{p}$ in the particular case that $\mathbf{p} = p_3$, i.e., $p_1 = p_2 = 0$. This result holds for the usual representation (D.7) of the Pauli matrices, with σ_3 diagonal and σ_1 and σ_2 nondiagonal.

Define a 4×4 matrix

$$\hat{\sigma}_3 = \begin{bmatrix} \sigma_3 & 0 \\ 0 & \sigma_3 \end{bmatrix}, \quad \text{where} \quad \sigma_3 = \begin{bmatrix} 1 & 0 \\ 0 & -1 \end{bmatrix}. \tag{D.29}$$

With $p = p_3$ one finds

$$\hat{\sigma}_3 p_3 u = \begin{bmatrix} p_3 & & & \\ & -p_3 & & \\ & & p_3 & \\ & & & -p_3 \end{bmatrix} \begin{bmatrix} a_1 \\ a_2 \\ a_3 \\ a_4 \end{bmatrix} = p_3 \begin{bmatrix} a_1 \\ -a_2 \\ a_3 \\ -a_4 \end{bmatrix}.$$

Thus, from (D.17) we see that

$$\hat{\sigma}_3 p_3 u_1 = +p_3 u_1, \qquad \hat{\sigma}_3 p_3 v_1 = +p_3 v_1,$$
$$\hat{\sigma}_3 p_3 u_2 = -p_3 u_2, \qquad \hat{\sigma}_3 p_3 v_2 = -p_3 v_2. \tag{D.30}$$

If we denote $\boldsymbol{\sigma}$ as a vector with components $\hat{\sigma}_1$, $\hat{\sigma}_2$, $\hat{\sigma}_3$—in an obvious extension of (D.29)—then the result (D.30) can be stated in terms of the eigenvalues of the *helicity operator*

$$H = \boldsymbol{\sigma} \cdot \mathbf{p}/|\mathbf{p}|, \tag{D.31}$$

that is,

$$Hu_1 = \left(\frac{\boldsymbol{\sigma} \cdot \mathbf{p}}{|\mathbf{p}|}\right) u_1 = u_1,$$

$$Hu_2 = -u_2,$$

$$Hv_1 = v_1,$$

$$Hv_2 = -v_2.$$

The states u_1 and u_2 (and v_1 and v_2) are said to be states of helicity $H = +1$ and -1, respectively. Thus, the fermion spin is *aligned* with the direction of motion. If the momentum vector defines the z-axis, then the eigenvalues of angular momentum J_z are $\pm\frac{1}{2}\hbar$.

D.6. APPLICATION OF THE DIRAC THEORY TO BETA DECAY

By appealing to Lorentz invariance, it can be shown that there are only five possible forms of operator which can be involved in the four-fermion β-decay coupling. These operators are named after their Lorentz transformation properties:

$$\bar{\psi}\psi = \psi^\dagger \gamma_4 \psi \qquad \text{scalar} \qquad S$$
$$\bar{\psi}\gamma_\mu\psi \qquad \text{4-vector} \qquad V$$
$$i\bar{\psi}\gamma_\mu\gamma_\nu\psi \qquad \text{tensor} \qquad T \qquad (D.32)$$
$$i\bar{\psi}\gamma_5\gamma_\mu\psi \qquad \text{axial 4-vector} \quad A$$
$$\bar{\psi}\gamma_5\psi \qquad \text{pseudoscalar} \quad P$$

By way of example, consider the inversion $\psi(-\mathbf{r}, t)$ of the state $\psi(\mathbf{r}, t)$. As mentioned above, $\psi(-\mathbf{r}, t)$ does not satisfy the Dirac equation, but $\gamma_4\psi(-\mathbf{r}, t)$, or $\bar{\psi}(-\mathbf{r}, t)\gamma_4$, does. Thus, under inversion

$$\bar{\psi}\psi \rightarrow \bar{\psi}\gamma_4^2\psi = \bar{\psi}\psi$$

and is therefore a scalar, whereas

$$\bar{\psi}\gamma_5\psi \rightarrow \bar{\psi}\gamma_4\gamma_5\gamma_4\psi = -\bar{\psi}\gamma_5\psi$$

changes sign and is therefore a pseudoscalar. It can be shown that both $\bar{\psi}\gamma_\mu\psi$ and $i\bar{\psi}\gamma_5\gamma_\mu\psi$ behave under rotations like 4-vectors; however, the first changes sign under inversions, while the second does not, and is therefore an axial vector.

As stated in Section 7.6, the general, parity-nonconserving β-decay matrix element describing neutron decay $n \rightarrow p + e^- + \bar{\nu}_e$, or the equivalent process $\nu_e + n \rightarrow e^- + p$, will have the form

$$M = G(\bar{\psi}_p O_i \psi_n)[\bar{\psi}_e O_i(C_i + \gamma_5 C_i')\psi_\nu], \qquad (D.33)$$

where C and C' are constants, and O_i represents the S, V, T, A, and P interactions. We can rewrite the lepton bracket

$$\bar{\psi}_e O_i(C_i + \gamma_5 C_i')\psi_\nu = (C_i + C_i')\bar{\psi}_e O_i(1 + \gamma_5)\psi_\nu/2$$
$$+ (C_i - C_i')\bar{\psi}_e O_i(1 - \gamma_5)\psi_\nu/2. \qquad (D.34)$$

Let us now find the effect of the operators $\frac{1}{2}(1 \pm \gamma_5)$ on the lepton wavefunctions, for example, on the plane-wave state (D.18):

$$\tfrac{1}{2}(1 + \gamma_5)u = \frac{N}{2}\begin{bmatrix} 1 & -1 \\ -1 & 1 \end{bmatrix}\begin{bmatrix} 1 \\ \boldsymbol{\sigma}\cdot\mathbf{p}/(E + m) \end{bmatrix} = \frac{N}{2}\begin{bmatrix} 1 - \boldsymbol{\sigma}\cdot\mathbf{p}/(E + m) \\ -(1 - \boldsymbol{\sigma}\cdot\mathbf{p}/(E + m)) \end{bmatrix},$$

$$\tfrac{1}{2}(1 - \gamma_5)u = \frac{N}{2}\begin{bmatrix} 1 & 1 \\ 1 & 1 \end{bmatrix}\begin{bmatrix} 1 \\ \boldsymbol{\sigma}\cdot\mathbf{p}/(E + m) \end{bmatrix} = \frac{N}{2}\begin{bmatrix} 1 + \boldsymbol{\sigma}\cdot\mathbf{p}/(E + m) \\ 1 + \boldsymbol{\sigma}\cdot\mathbf{p}/(E + m) \end{bmatrix},$$

where each element stands for a 2×2 matrix.

The amplitude $1 - \boldsymbol{\sigma} \cdot \mathbf{p}/(E + m)$ can be written

$$1 - \frac{\boldsymbol{\sigma} \cdot \mathbf{p}}{(E + m)} = \frac{a}{2}\left(1 + \frac{\boldsymbol{\sigma} \cdot \mathbf{p}}{|\mathbf{p}|}\right) + \frac{b}{2}\left(1 - \frac{\boldsymbol{\sigma} \cdot \mathbf{p}}{|\mathbf{p}|}\right), \qquad \text{(D.35)}$$

$$\uparrow \qquad\qquad \uparrow$$
$$\text{RH} \qquad\qquad \text{LH}$$

where

$$a = 1 - \frac{|\mathbf{p}|}{(E + m)}$$

$$b = 1 + \frac{|\mathbf{p}|}{(E + m)}$$

are the coefficients multiplying the RH and LH helicity states u_R, $u_L = 1 \pm \boldsymbol{\sigma} \cdot \mathbf{p}/|\mathbf{p}|$. The *intensities* of the RH and LH states in the wavefunction are a^2 and b^2, respectively, and the longitudinal lepton polarization is therefore

$$P = \frac{N_R - N_L}{N_R + N_L} = \frac{a^2 - b^2}{a^2 + b^2} = -\frac{p}{E} = -\frac{v}{c}. \qquad \text{(D.36)}$$

When $p \to E$ and $v/c \to 1$, that is, in the extreme relativistic limit,

$$a \to 0,$$

$$b \to 2,$$

and in this limit, applying to massless neutrinos of any energy,

$$\underset{v \to c}{\text{Lt}} \begin{cases} \frac{1}{2}(1 + \gamma_5)u = u_L \\ \frac{1}{2}(1 - \gamma_5)u = u_R \end{cases}. \qquad \text{(D.37)}$$

Thus the operator $(1 + \gamma_5)$ projects out helicity -1 (LH) states, and $(1 - \gamma_5)$ the helicity $+1$ (RH) states.

In (D.34), the operators $(1 \pm \gamma_5)$ operate on the neutrino state. Experimentally, the neutrino is found to have negative helicity. Hence we need the first term of (D.34) but not the second, that is, we require $C_i = C'_i$ so that (D.34) becomes, apart from a numerical factor and using (D.32) and (D.11),

$$\bar{\psi}_e O_i (1 + \gamma_5)\psi_v = \begin{cases} \bar{\psi}_e(1 + \gamma_5)O_i\psi_v & \text{for } i = S, T, P \\ \bar{\psi}_e(1 - \gamma_5)O_i\psi_v & \text{for } i = V, A \end{cases} \qquad \text{(D.38)}$$

Using (D.18) and (D.23) it is straightforward to show that $\bar{\psi}_e(1 + \gamma_5)$ corresponds to generation of RH electrons and $\bar{\psi}_e(1 - \gamma_5)$, to LH electrons. Experimentally, it is found that electrons emitted in β-decay are LH, so the interactions involved are V, A and not $S, T,$ or P. So the matrix element (D.33) has the form (using the fact that $\gamma_5^2 = 1$)

$$M = G\{C_V(\bar{\psi}_p\gamma_\mu\psi_n)[\bar{\psi}_e\gamma_\mu(1 + \gamma_5)\psi_v] + i^2C_A(\bar{\psi}_p\gamma_\mu\gamma_5\psi_n)[\bar{\psi}_e\gamma_\mu\gamma_5(1 + \gamma_5)\psi_v]\}$$
$$= G\{[\bar{\psi}_p\gamma_\mu(C_v - \gamma_5 C_A)\psi_n][\bar{\psi}_e\gamma_\mu(1 + \gamma_5)\psi_v]\},$$

We see that if $C_V = -C_A$, i.e., the axial-vector and vector couplings are equal in magnitude but opposite in sign, we obtain the famous V-A *interaction*, and the matrix element acquires the symmetric form

$$[\bar{\psi}_p\gamma_\mu(1 + \gamma_5)\psi_n][\bar{\psi}_e\gamma_\mu(1 + \gamma_5)\psi_v]. \tag{D.39}$$

While in muon decay or any other purely leptonic weak interaction, $C_A = -C_V$, this is not true for decays involving hadrons. The axial coupling C_A (but not the vector coupling C_V) is affected by strong interaction effects. For nucleons, for example, $C_A = -1.25\, C_V$.

D.7. HELICITY CONSERVATION IN VECTOR INTERACTIONS

Any fermion spinor can be written as a superposition of helicity states

$$u = u_L + u_R.$$

A vector operator γ_μ acting on an extreme relativistic fermion current produces

$$\bar{u}\gamma_\mu u = (\bar{u}_L + \bar{u}_R)\gamma_\mu(u_L + u_R) = \bar{u}_L\gamma_\mu u_L + \bar{u}_R\gamma_\mu u_R, \tag{D.40}$$

where the cross terms cancel, since

$$\bar{u}_L = \tfrac{1}{2}\bar{u}(1 - \gamma_5)$$
$$u_R = \tfrac{1}{2}(1 - \gamma_5)u,$$

so that for a V-interaction

$$\bar{u}_L\gamma_\mu u_R = \frac{\bar{u}}{4}(1 - \gamma_5)\gamma_\mu(1 - \gamma_5)u = \frac{\bar{u}}{4}\gamma_\mu(1 - \gamma_5^2)u = 0,$$

and this is also true for the A-operator. For a scalar operator S,

$$\bar{u}_L u_R = \frac{\bar{u}}{4}(1 - \gamma_5)^2 u = \frac{\bar{u}}{2}(1 + \gamma_5)u \neq 0,$$

whereas

$$\bar{u}_L u_L = \frac{\bar{u}}{4}(1 - \gamma_5)(1 + \gamma_5)u = 0,$$

and similarly for P- and T-interactions. As a result of (D.40), the helicity of a fermion is unchanged by a V- or A-interaction, as in the diagrams (a) and (b) of Fig. D.2. Since an incoming fermion can be replaced by an outgoing

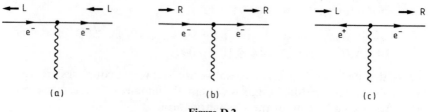

Figure D.2

antifermion without changing the matrix element, we see from Fig. D.2(c) that in the process $e^+e^- \rightarrow \gamma \rightarrow \mu^+\mu^-$, the helicity law tells us that in the relativistic limit $v/c \rightarrow 1$, the fermion and antifermion will have opposite helicity. Therefore, the state must have angular momentum $J = 1$ (it couples to a photon) and $J_z = +1$ or -1, where the z-axis is defined by the incoming e^+ and e^- beams. $J_z = 0$ is excluded, that is, the mediating photon must be transverse.

APPENDIX E

Transition Rates in Perturbation Theory

In time-dependent perturbation theory, we consider transitions—for example, in a collision or decay process—from a well-defined initial state of the system, say ψ, to some set of final states, as a result of the action of a perturbing potential, say V. Suppose at time $t < 0$ the system is in a stationary state of energy E_m, which is one of several possible eigenstates of the unperturbed Hamiltonian H_0. We can write the wavefunction as the product of time-dependent and spatial parts

$$\psi(0) = \phi_m e^{-iE_m t/\hbar}. \tag{E.1}$$

For $t \geq 0$, we suppose that the potential V is "turned on," resulting in transitions to other states of the system, ϕ_n. Then $\psi(t)$ can be expressed as the sum

$$\psi(t) = \sum_{n=0}^{\infty} c_n(t)\phi_n e^{-iE_n t/\hbar}, \tag{E.2}$$

where the functions ϕ_n are a complete set of orthonormal eigenfunctions of the unperturbed Hamiltonian H_0 of the system, that is,

$$H_0 \phi_n = E_n \phi_n. \tag{E.3}$$

The coefficients c_n give the probability amplitude to find the system in the state ϕ_n. Thus, $c_n(0) = 1$ for $n = m$ and zero for all other n-values. The total wavefunction in (E.2) must satisfy the time-dependent Schrödinger equation (in the nonrelativistic case),

$$H\psi = i\hbar\, \partial\psi/\partial t, \tag{E.4}$$

where the total Hamiltonian, for $t > 0$, is given by

$$H = H_0 + V. \tag{E.5}$$

If we substitute in (E.4) the expressions for H and $\psi(t)$, we find, using (E.5), that

$$i\hbar \sum_{n=0}^{\infty} \left(\frac{dc_n}{dt}\right)\phi_n e^{-iE_nt/\hbar} = \sum_{n=0}^{\infty} Vc_n(t)\phi_n e^{-iE_nt/\hbar}.$$

Multiplying through by ψ_k^*, where k is some other eigenstate of H_0, and using the fact that ϕ_n and ϕ_k are orthogonal for all values of $n \neq k$, we obtain

$$i\hbar\left(\frac{dc_k}{dt}\right) = \sum_{n=0}^{\infty} c_n(t)M_{nk}e^{-i(E_n - E_k)t/\hbar},$$

where

$$M_{nk} = \int \phi_k^* V\phi_n\, d\tau \tag{E.6}$$

is defined as the matrix element for the transition from state k to state n, brought about by the perturbation V. M_{nk} has dimensions of energy, since V, like H, is an energy operator, and ϕ_k and ϕ_n are normalized to the total volume.

In first-order perturbation theory, the perturbation is assumed to be so weak and the transition rate so small, that over the time t considered, $c_m(t) \simeq 1$ and $c_n(t) \simeq 0$ for $n \neq m$. Then integrating over time, assuming V is constant for $t > 0$,

$$c_k(t) = \frac{1}{i\hbar} \int_0^t M_{km}e^{-i(E_k - E_m)t/\hbar}\, dt$$

$$= M_{km}\left[\frac{1 - e^{-i(E_k - E_m)t/\hbar}}{E_k - E_m}\right]$$

$$= 2iM_{km}e^{-i(E_k - E_m)t/2\hbar} \cdot \left[\frac{\sin[(E_k - E_m)t/2\hbar]}{E_k - E_m}\right]. \tag{E.7}$$

The total transition rate per unit time to all states E_n is

$$W = \frac{1}{t}\sum_n |c_n(t)|^2 = \frac{1}{t}\int_{-\infty}^{+\infty} |c_n(t)|^2 \cdot \frac{dN}{dE} \cdot dE_n. \tag{E.8}$$

In the last expression, we have replaced the sum over discrete states by an integral over a group of states of closely similar energy, which is the practical situation in collision or decay processes. dN/dE is the density of states per unit energy interval. If we set

$$x = (E_k - E_m)t/2\hbar, \tag{E.9}$$

then

$$W = \frac{2}{\hbar}|M_{km}|^2 \int_{-\infty}^{+\infty} \left(\frac{dN}{dE}\right) \frac{\sin^2 x}{x^2} \, dx.$$

The expression $(\sin^2 x)/x^2$ has a principal maximum of unity at $x = 0$ and a first zero at $x = \pm \pi$. The contribution to the integral beyond $x = \pm \pi$ is small ($\sim 10\%$). The corresponding energy interval is, from (E.9), $dE_k = 2\pi\hbar/t \simeq 4 \times 10^{-15}$ eV for $t = 1$ s. This simply corresponds to the fact that for a finite transition rate, energy must be conserved and $E_k \simeq E_m$ within the limitations of the Uncertainty Principle. In all practical cases, we can assume dN/dE is a constant over the interval dE_n, and using the fact that

$$\int_{-\infty}^{+\infty} \frac{\sin^2 x}{x^2} \, dx = \pi,$$

we obtain

$$\boxed{W = \frac{2\pi}{\hbar} \cdot |M_{if}|^2 \cdot \frac{dN}{dE_f}}, \tag{E.10}$$

where we have used a slightly changed notation,

$$M_{if} = \int \phi_i^* V \phi_f \, d\tau,$$

for the matrix element for the transition from an initial state i to a (set of) final states f, with ϕ_i and ϕ_f as the spatial parts of the wavefunctions. dN/dE_f is the energy density of final states factor and is often denoted ρ_f. Equation (E.10) is often called Fermi's Second Golden Rule of nonrelativistic quantum mechanics.

APPENDIX F

Neutrino-Electron Scattering in the V-A Theory

The cross-section for the reaction

$$v_e + e^- \to \dot{e}^- + v_e \tag{F.1}$$

can be obtained from Eqs. (4.5 and 4.6):

$$\frac{d\sigma}{d\Omega} = \frac{2\pi}{v_0} \frac{p^{*2}}{\hbar} \frac{dp^*}{dE_0} \cdot \frac{|M_{if}|^2}{(2\pi\hbar)^3}, \tag{F.2}$$

which we shall evaluate in the CMS of the collision. Let the 4-momenta of the incoming and outgoing neutrino be p_1 and p_3 and the corresponding quantities for the electron be p_2 and p_4. The total energies of the particles are E_1, E_2, E_3, and E_4. p^* is the value of the CMS 3-momentum, v_0 the relative velocity of the colliding particles. The total CMS energy is E_0, where

$$E_0^2 = s = -(p_1 + p_2)^2 = -(p_3 + p_4)^2.$$

We shall use wavefunctions of leptons normalized in their individual rest frames, so that as explained in Appendix D, we must use a relativistic factor $\sqrt{m/E}$ for normalization in the overall CMS. Then with $c = \hbar = 1$, we obtain

$$\frac{d\sigma}{d\Omega} = \frac{1}{(2\pi)^2} \frac{m_e^2 m_v^2}{E_1 E_2 E_3 E_4} \cdot |M_{if}|^2 \cdot \frac{p^{*2}}{v_0} \frac{dp^*}{dE_0}. \tag{F.3}$$

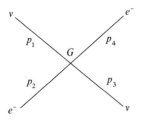

For the charged-current part of the reaction, the V-A theory predicts for the current-current matrix element in the limit of low q^2, so that W-boson propagator effects can be neglected (see Appendix D.6 and Eq. (7.27)):

$$M_{if} = \frac{G}{\sqrt{2}} j_1 j_2 = \frac{G}{\sqrt{2}} \sum_{spins} (\bar{u}_\nu(p_3)\gamma_\mu(1 + \gamma_5)u_e(p_2))(\bar{u}_e(p_4)\gamma_\mu(1 + \gamma_5)u_\nu(p_1)),$$
(F.4)

where u_ν, u_e denote the neutrino and electron spinors and $\gamma_\mu(1 + \gamma_5)$ is the V-A operator. The complex transpose of this matrix element is

$$M^\dagger = \frac{G}{\sqrt{2}} \sum (\bar{u}_\nu(p_1)\gamma_\rho(1 + \gamma_5)u_e(p_4))(\bar{u}_e(p_2)\gamma_\rho(1 + \gamma_5)u_\nu(p_3)),$$
(F.5)

so that

$$|M_{if}|^2 = MM^\dagger = \frac{G^2}{2} \sum A_{\mu\rho} B_{\mu\rho},$$
(F.6)

where

$$A_{\mu\rho} = \bar{u}_\nu(p_3)\gamma_\mu(1 + \gamma_5)u_e(p_2) \cdot \bar{u}_e(p_2)\gamma_\rho(1 + \gamma_5)u_\nu(p_3)$$

and

$$B_{\mu\rho} = \bar{u}_e(p_4)\gamma_\mu(1 + \gamma_5)u_\nu(p_1) \cdot \bar{u}_\nu(p_1)\gamma_\rho(1 + \gamma_5)u_e(p_4).$$
(F.7)

To perform the spin sum, we use the completeness relation (D.28), applying the positive-energy projection operator in (D.27),

$$\Lambda^+ = (m - i\gamma_\sigma P_\sigma)/2m,$$
(F.8)

and so obtain

$$A_{\mu\rho} = Tr[\Lambda_\nu^+(p_3)\gamma_\mu(1 + \gamma_5)\Lambda_e^+(p_2)\gamma_\rho(1 + \gamma_5)]$$

$$= Tr\left[\frac{(m_\nu - i\gamma_\sigma P_{3\sigma})}{2m_\nu} \cdot \gamma_\mu(1 + \gamma_5) \cdot \frac{(m_e - i\gamma_\phi P_{2\phi})}{2m_e} \cdot \gamma_\rho(1 + \gamma_5)\right]$$

$$= -\frac{1}{2m_e m_\nu} P_{3\sigma} P_{2\phi} Tr[\gamma_\sigma \gamma_\mu \gamma_\phi \gamma_\rho(1 + \gamma_5)],$$
(F.9)

where we have made use of the following facts:

$$m_\nu = 0$$

$$\gamma_5\gamma_\mu + \gamma_\mu\gamma_5 = 0$$

$$\gamma_5(1 + \gamma_5) = (1 + \gamma_5)$$

Trace of product of odd number of γ matrices $= 0$

The second term in (F.6) has a similar form,

$$B_{\mu\rho} = -\frac{1}{2m_e m_\nu} P_{4\alpha} P_{1\beta} \, \text{Tr}[\gamma_\alpha \gamma_\mu \gamma_\beta \gamma_\rho (1 + \gamma_5)]. \tag{F.10}$$

The product

$$\text{Tr}[\gamma_\sigma \gamma_\mu \gamma_\phi \gamma_\rho (1 + \gamma_5)] \cdot \text{Tr}[\gamma_\alpha \gamma_\mu \gamma_\beta \gamma_\rho (1 + \gamma_5)] = 64 \delta_{\alpha\sigma} \delta_{\beta\phi}.$$

This equality can be demonstrated through the relations

$$\tfrac{1}{4} \text{Tr}[\gamma_\mu \gamma_\nu \gamma_\rho \gamma_\sigma] = \delta_{\mu\nu} \delta_{\rho\sigma} - \delta_{\mu\rho} \delta_{\nu\sigma} + \delta_{\mu\sigma} \delta_{\nu\rho},$$

$$\tfrac{1}{4} \text{Tr}[\gamma_5 \gamma_\mu \gamma_\nu \gamma_\rho \gamma_\sigma] = \varepsilon_{\mu\nu\rho\sigma}; \qquad \varepsilon_{\rho\sigma\alpha\beta} \varepsilon_{\mu\nu\alpha\beta} = -2\delta_{\mu\rho} \delta_{\nu\sigma} + 2\delta_{\mu\sigma} \delta_{\nu\rho}, \tag{F.11}$$

where $\varepsilon = +1$ or -1 if μ, ν, ρ, σ are an even (odd) permutation of 1, 2, 3, 4 and is zero if any two indices are equal. These relations can be verified by substituting $\mu = \nu$, $\rho = \sigma$, etc., and making use of the fact that $\delta_{\mu\alpha} \delta_{\alpha\rho} = \delta_{\mu\rho}$ and $\delta_{\alpha\beta} \delta_{\alpha\beta} = 4$.

    Dividing through by a factor 2 to average over the spin states of the target electron (since we summed over them in the trace), we get

$$\begin{aligned}
MM^\dagger &= \frac{G^2}{4} \cdot \frac{64}{4m_e m_\nu} \cdot \delta_{\alpha\sigma} \delta_{\beta\phi} \cdot P_{1\beta} P_{2\phi} P_{3\sigma} P_{4\alpha} \\
&= \frac{4G^2}{m_e^2 m_\nu^2} (p_1 \cdot p_2)(p_3 \cdot p_4)
\end{aligned} \tag{F.12}$$

summed over final-state and averaged over initial-state spins. Inserting this into (F.3) we obtain

$$\frac{d\sigma}{d\Omega} = \frac{G^2}{\pi^2} \cdot \frac{(p_1 \cdot p_2)(p_3 \cdot p_4)}{E_1 E_2 E_3 E_4} \cdot \frac{p^{*2}}{v_0} \cdot \frac{dp^*}{dE_0}. \tag{F.13}$$

If we neglect the neutrino mass, we can readily verify the following relations:

$$E_1 E_2 v_0 = \sqrt{(p_1 p_2)^2 - (m_\nu m_e)^2} = p_1 p_2$$

$$p_3 p_4 = (s - m_e^2)/2; \qquad E_3 E_4 = (s^2 - m_e^4)/4s$$

$$p^* = (s - m_e^2)/2\sqrt{s}; \qquad dp^*/dE_0 = (s + m_e^2)/2s$$

whence the CMS differential cross-section becomes

$$\frac{d\sigma}{d\Omega}(\nu e) = \frac{G^2}{\pi^2} p^{*2} \tag{F.14}$$

and the total cross-section

$$\sigma(\nu e) = \frac{G^2}{\pi} 4p^{*2} = \frac{G^2}{\pi} \frac{(s - m_e^2)^2}{s} \simeq \frac{G^2}{\pi} s. \tag{F.15}$$

In terms of the laboratory energy, E, of the incident neutrino on a stationary electron target, $s = 2m_e E + m_e^2$, so that

$$\sigma(ve) = \frac{G^2}{\pi} \frac{2m_e E}{(1 + m/2E)} \simeq \frac{2}{\pi} G^2 m_e E. \tag{F.16}$$

For incident antineutrinos, the reaction

$$\bar{v}_e + e^- \to e^- + \bar{v}_e$$

has the same matrix element as before, if we replace out (in)-going neutrinos by in (out)-going antineutrinos. This amounts to simply interchanging the 4-momenta p_1 and p_3 so that

$$\frac{d\sigma(\bar{v}e)}{d\Omega} = \frac{G^2}{\pi^2} \cdot \frac{(p_1 p_4)(p_2 p_3)}{E_1 E_2 E_3 E_4} \cdot \frac{p^{*2}}{v_0} \cdot \frac{dp^*}{dE_0}. \tag{F.17}$$

If θ^* is the scattering angle of the antineutrino in the CMS, then

$$p_1 p_4 = -(p^{*2} \cos \theta^* + E_1 E_4) \simeq -\frac{s}{4}(1 + \cos \theta^*)$$

in the limit $s \gg m_e^2$. $p_2 p_3$ gives the same result. Thus, one finds that in the CMS of the collision

$$\frac{d\sigma(\bar{v}e)}{d\Omega} = \frac{G^2 p^{*2}}{4\pi^2}(1 + \cos \theta^*)^2 \tag{F.18}$$

and

$$\sigma(\bar{v}e) = \frac{4G^2 p^{*2}}{3\pi} \simeq \frac{G^2 s}{3\pi}. \tag{F.19}$$

Setting the recoil energy of the electron in the laboratory system as $E' = yE$, where E is the incident antineutrino energy, it is easy to show that

$$(1 - y) = \tfrac{1}{2}(1 + \cos \theta^*)$$

for $E \gg m_e$, so the LS differential cross-sections for neutrinos and antineutrinos on electrons can be written in the form

$$\frac{d\sigma(ve)}{dy} = \frac{G^2 s}{\pi}$$

$$\frac{d\sigma(\bar{v}e)}{dy} = \frac{G^2 s}{\pi}(1 - y)^2, \tag{F.20}$$

as quoted in Eq. (7.53).

Finally, it should be stressed that we have calculated cross-sections for *charged-current* scattering and that there are also neutral-current contributions, as described in Chapter 9.

APPENDIX G

Scattering of Pointlike Fermions in QED. The Processes $e^-\mu^+ \to e^-\mu^+$ and $e^+e^- \to \mu^+\mu^-$.

G.1. $e^-\mu^+ \to e^-\mu^+$

We first derive the cross-section (6.24) for the elastic scattering of electrons by pointlike charged fermions (a muon or a pointlike "Dirac" proton). The matrix element will consist of a product of fermion currents and a photon propagator term

$$M_{if} = \frac{e^2}{q^2} j_e j_\mu, \qquad (G.1)$$

where

$$j_e = \sum_{\text{spins}} \bar{u}_e \gamma_\rho u_e,$$

$$j_\mu = \sum_{\text{spins}} \bar{u}_\mu \gamma_\rho u_\mu$$

are the currents describing the electron, of initial (final) 4-momentum $p(p')$, and the muon, initial (final) momentum $k(k')$ (see Fig. G.1). The lepton spinors u_e, u_μ are operated on by the vector operator γ_ρ. Then

$$M = \frac{e^2}{q^2} [\bar{u}_e(p')\gamma_\rho u_e(p)][\bar{u}_\mu(k')\gamma_\rho u_\mu(k)],$$

$$M^\dagger = \frac{e^2}{q^2} [\bar{u}_e(p)\gamma_\sigma u_e(p')][\bar{u}_\mu(k)\gamma_\sigma u_\mu(k')], \qquad (G.2)$$

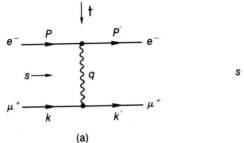
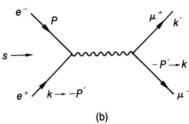

(a)

(b)

Figure G.1

so that

$$MM^\dagger = \frac{e^4}{q^4} \sum A_{\rho\sigma} B_{\rho\sigma},$$

where

$$A_{\rho\sigma} = [\bar{u}_e(p')\gamma_\rho u_e(p)][\bar{u}_e(p)\gamma_\sigma u_e(p')],$$

$$B_{\rho\sigma} = [\bar{u}_\mu(k')\gamma_\rho u_\mu(k)][\bar{u}_\mu(k)\gamma_\sigma u_\mu(k')].$$

(G.3)

The spin sum is made with the help of the projection operator (D.27)

$$\Lambda^+ = (m - i\gamma_\alpha p_\alpha)/2m,$$

so that using the result (D.28) we obtain

$$4m_e^2 A_{\rho\sigma} = \text{Tr}\ [\gamma_\rho(m_e - i\gamma_\alpha p'_\alpha)\gamma_\sigma(m_e - i\gamma_\beta p_\beta)]$$

$$= m_e^2\ \text{Tr}\ [\gamma_\rho\gamma_\sigma] - p'_\alpha p_\beta\ \text{Tr}\ [\gamma_\rho\gamma_\alpha\gamma_\sigma\gamma_\beta],$$

which becomes, with the trace theorem (F.11),

$$m_e^2 A_{\rho\sigma} = m_e^2\delta_{\rho\sigma} - p'_\rho p_\sigma + p_\alpha p'_\alpha\delta_{\rho\sigma} - p_\rho p'_\sigma,$$

and similarly for the muon current,

$$m_\mu^2 B_{\rho\sigma} = m_\mu^2\delta_{\rho\sigma} - k'_\rho k_\sigma + k_\alpha k'_\alpha\delta_{\rho\sigma} - k_\rho k'_\sigma.$$

Thus,

$$\sum_{\rho,\sigma} m_e^2 m_\mu^2 A_{\rho\sigma} B_{\rho\sigma} = 4m_e^2 m_\mu^2 + 2m_e^2(kk') + 2m_\mu^2(pp') + 2(kp)(k'p') + 2(kp')(k'p)$$

$$\simeq 2[m_\mu^2(pp') + (kp)(k'p') + (kp')(k'p)],$$

(G.4)

where we have, in the final line, neglected the electron mass. The 4-momentum transfer in the scattering is q, where

$$q^2 = (k - k')^2 = -2m^2 - 2kk'$$

$$= (p - p')^2 \simeq -2pp',$$

and energy-momentum conservation gives $k' = p + k - p'$. So the sum (G.4) becomes

$$\Sigma = 2\left[-m_\mu^2 \frac{q^2}{2} + kp(pp' + kp' - p'^2) + kp'(p^2 + kp - pp') \right]$$

$$\simeq 2\left[-m_\mu^2 \frac{q^2}{2} - \frac{q^2}{2}(kp - kp') + 2(kp')(kp) \right],$$

where again the electron mass has been neglected in the final line. We evaluate this expression in the lab system, where the muon target is initially at rest. Denoting $m_\mu = M$, E and E' the incident and scattered electron energies, the various 4-momenta and scalar products are

$$p = \mathbf{p}, iE \qquad\quad p' = \mathbf{p}', iE' \qquad k = 0, iM, \qquad k' = \sqrt{W^2 - M^2}, iW$$

$$kp = -ME \qquad kp' = -ME',$$

where W is the total energy of the recoiling muon. Hence

$$\Sigma = 2\left[\frac{-q^2 M^2}{2} + \frac{q^2 M}{2}(E - E') + 2M^2 EE' \right]$$

$$= 4M^2 EE'\left[1 - \frac{q^2}{4EE'} + \frac{q^2 M(E - E')}{4M^2 EE'} \right].$$

In terms of the laboratory scattering angle of the electron, we have

$$q^2 = -2pp' = -2|\mathbf{p}|\,|\mathbf{p}'|\cos\theta + 2EE' \simeq 2EE'(1 - \cos\theta) = 4EE' \sin^2\frac{\theta}{2},$$

and

$$q^2 = 2MT = 2M(E - E'),$$

where $T = W - M$ is the kinetic energy acquired by the muon. So

$$MM^\dagger = |M|^2 = \frac{e^4}{q^4}\sum_{\rho,\sigma} A_{\rho\sigma}B_{\rho\sigma} = \frac{4M^2 EE'}{m_e^2 m_\mu^2}\left[\cos^2\frac{\theta}{2} + \frac{q^2}{2M^2}\sin^2\frac{\theta}{2} \right]\cdot\frac{e^4}{q^4}. \quad \text{(G.5)}$$

Referring to (F.3), we see that the cross-section will then be (in units $\hbar = c = 1$)

$$\frac{d\sigma}{d\Omega} = \frac{1}{(2\pi)^2}\cdot|M|^2\cdot\frac{p'^2}{v_0}\cdot\frac{dp'}{dE_f}\cdot\frac{m_e^2 m_\mu^2}{EE'MW}\cdot\frac{1}{4}, \quad\quad \text{(G.6)}$$

where the factor $\frac{1}{4}$ is for the average over the four spin states of the particles in the initial state. $E_f = E + M = E' + W$ is the total energy of the final state, and p' now stands for the value of 3-momentum of the scattered electron in the lab system. From (6.6) we know that

$$p'^2 \frac{dp'}{dE_f} = E'^2 \frac{W}{M}\frac{E'}{E}.$$

With $v_0 = c = 1$, $e^2 = 4\pi\alpha$, and $q^2 = 4EE' \sin^2 \dfrac{\theta}{2}$, the final result for the Mott scattering cross-section is

$$\frac{d\sigma}{d\Omega} = \left(\frac{\alpha^2}{4E^2 \sin^4 \dfrac{\theta}{2}}\right)\left(\frac{E'}{E}\right)\left[\cos^2 \frac{\theta}{2} + \frac{q^2}{2M^2} \sin^2 \frac{\theta}{2}\right]. \tag{G.7}$$

This is the result used in (6.24), with slightly different notation (replacing E by p_0, E' by p and Eq. (6.6) for the recoil factor E'/E). It holds for scattering of an electron by any singly charged point fermion of mass M, initially stationary in the laboratory system.

G.2. THE PROCESS $e^+e^- \to \mu^+\mu^-$

The expression (G.7) for electron–muon (or electron–pointlike proton) scattering was derived in the lab system. We can expresss it in terms of Lorentz invariant kinematic variables. The quantities used are the so-called Mandelstam variables s, t, and u (see Fig. G.1(a)), which, in terms of the 4-momenta of the particles, are

$$s = (k + p)^2 = (k' + p')^2 \simeq 2kp = 2k'p',$$
$$t = -q^2 = (k - k')^2 = (p - p')^2 \simeq -2pp' = -2kk', \tag{G.8}$$
$$u = (p - k')^2 = (k - p')^2 \simeq -2k'p = -2kp'.$$

In the final expressions on the right-hand side, all particle masses have been neglected. In terms of these variables, it is left as an exercise to show that the cross-section (G.7) can be expressed in the CMS in the form

$$\frac{d\sigma}{d\Omega}(e^-\mu^+ \to e^-\mu^+) = \frac{\alpha^2}{2s} \frac{(s^2 + u^2)}{t^2} \tag{G.9}$$

with the help of (G.4). We now consider the annihilation reaction $e^+e^- \to \mu^+\mu^-$. This may be obtained from the cross-section (G.9) as follows. As shown in Fig. G.1, the momenta of ingoing particles may be replaced by those of outgoing antiparticles, and vice-versa, without changing the form of the matrix element. Figure G.1(b) is the "crossed" diagram to that in Fig. G.1(a), and the cross-section for it is simply obtained by the replacement $k \leftrightarrow -p'$, and hence $\mu \leftrightarrow \mu$, $s \leftrightarrow t$ in the matrix element, that is the second term of (G.9)

So

$$\frac{d\sigma}{d\Omega}(e^+e^- \to \mu^+\mu^-) = \frac{\alpha^2}{2s} \frac{(t^2 + u^2)}{s^2}. \tag{G.10}$$

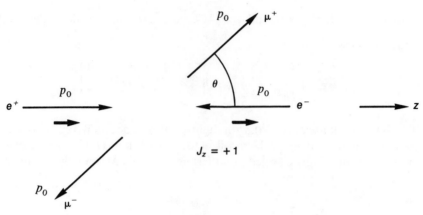

Figure G.2

If θ is the CMS angle of emission of the $\mu^+(\mu^-)$ with respect to the incoming $e^+(e^-)$ and, again, all lepton masses are neglected, we have (see Fig. G.2)

$$s = (p_0 + p_0)^2 = 4p_0^2,$$

$$t = 2p_0^2(1 - \cos\theta),$$

$$u = 2p_0^2(1 + \cos\theta),$$

where p_0 is the numerical value of the CMS 3-momentum of each particle, before and after scattering. So

$$\frac{d\sigma}{d\Omega} = \frac{\alpha^2}{4s}(1 + \cos^2\theta), \tag{G.11}$$

and the total cross-section is

$$\sigma(e^+e^- \to \mu^+\mu^-) = \frac{4\pi\alpha^2}{3s}. \tag{G.12}$$

It is left as an exercise to show that if the muon mass is not neglected, the bracket in (G.11) acquires an additional term, $(4m_\mu^2/s)\sin^2\theta$, and the whole expression is multiplied by a factor $(1 - 4m_\mu^2/s)^{1/2}$. The above expressions are the conventional QED cross-sections to leading order in α. At high energies, an extra contribution from Z^0 (weak neutral-current) exchange becomes important and leads to a backward-forward asymmetry in (G.11); see Section 9.7.2.

The Decays of W- and Z-Bosons

The W-boson has spin 1, and its state of polarization can be described in relativistically invariant form by a 4-vector e_μ with space components e_1, e_2, e_3 and time component e_4. In analogy with the description of photons (3.35), a RH state is, for example, written

$$e_R = \left(\frac{1}{\sqrt{2}}, \ \frac{i}{\sqrt{2}}, 0, 0 \right), \tag{H.1}$$

describing a RH circularly polarized W, with spin component $J_z = +1$ along the z-axis. Similarly,

$$e_L = \left(\frac{1}{\sqrt{2}}, \ \frac{-i}{\sqrt{2}}, 0, 0 \right) \tag{H.2}$$

and

$$e_z = (0, 0, 1, 0) \tag{H.3}$$

correspond to $LH(J_z = -1)$ and unpolarized $(J_z = 0)$ W's, respectively. (The fourth, time component e_4, can always be eliminated by suitable choice of gauge).

The V-A matrix element for the decay of a W of 4-momentum k to an electron and neutrino of 4-momenta p, q, respectively, can be written

$$M = g_w e_\mu(k) \bar{u}_v(q) \gamma_\mu (1 + \gamma_5) u_e(p), \tag{H.4}$$

where g_w denotes the coupling of the W to the lepton pair. Also,

$$M^\dagger = g_w e_v^*(k) u_v(q) \gamma_v (1 + \gamma_5) \bar{u}_e(p).$$

Using the projection operator (D.27) for the sum over spins and neglecting m_e, m_v in comparison with their momentum in evaluating the trace, we find (refer to (F.11)):

$$\sum_{\text{spins}} MM^\dagger = g_w^2 e_\mu e_v^* \text{Tr}[\Lambda_e^+(p)\gamma_\mu(1 + \gamma_5)\Lambda_v^+(q)\gamma_v(1 + \gamma_5)] \tag{H.5}$$

$$= \frac{g_w^2}{4 m_e m_v} e_\mu e_v^* \text{Tr}[\gamma_\rho p_\rho \gamma_\mu (1 + \gamma_5)\gamma_\sigma q_\sigma \gamma_v (1 + \gamma_5)]$$

$$= \frac{g_w^2}{2 m_e m_v} e_\mu e_v^* \text{Tr}[\gamma_\rho p_\rho \gamma_\mu \gamma_\sigma q_\sigma \gamma_v (1 + \gamma_5)]$$

$$= \frac{4 g_w^2}{2 m_e m_v} e_\mu e_v^* [p_\rho q_\sigma \delta_{\mu\rho}\delta_{v\sigma} + p_\rho q_\sigma \delta_{\mu\sigma}\delta_{\rho v} - p_\rho q_\sigma \delta_{\mu v}\delta_{\rho\sigma}$$

$$+ \varepsilon_{\rho\mu\sigma v} p_\rho q_\sigma]$$

$$= \frac{4 g_w^2}{2 m_e m_v} [\delta_{\mu v}(e_\mu p_\mu e_v^* q_v + e_\mu q_\mu e_v^* p_v - pq e_\mu e_v^*) - \varepsilon_{\mu v\rho\sigma} p_\rho q_\sigma e_\mu e_v^*]. \tag{H.6}$$

Let us consider the RH state (H.1) of the W, i.e., with $J_z = +1$, $e_1 = 1/\sqrt{2}$, $e_2 = i/\sqrt{2}$, and $e_3 = e_4 = 0$. Then

$$|MM^\dagger| = \frac{4 g_w^2}{2 m_e m_v} [2(p_1 q_1 |e_1|^2 + p_2 q_2 |e_2|^2) - pq$$

$$- (\varepsilon_{1234} e_1 e_2^* p_3 q_4 + \varepsilon_{1243} e_1 e_2^* p_4 q_3 + \varepsilon_{2134} e_1^* e_2 p_3 q_4$$

$$+ \varepsilon_{2143} e_1^* e_2 p_4 q_3)]$$

$$= \frac{4 g_w^2}{2 m_e m_v} [p_1 q_1 + p_2 q_2 - pq + i(p_3 q_4 - p_4 q_3)]. \tag{H.7}$$

Let p^* (Fig. H.1) denote the numerical value of the (equal and opposite) 3-momenta of the electron and neutrino decay products, θ their angle of emission to the z-axis. Neglecting m_e, m_v terms in the numerator we then find

$$p_3 = p^* \cos \theta = -q_3 \qquad p_4 q_4 = -p^{*2}$$

$$p_3 q_3 = -p^{*2} \cos^2 \theta \qquad p_3 q_4 = -p_4 q_3 = ip^{*2}\cos \theta$$

and

$$|MM^\dagger| = \frac{4 g_w^2 p^{*2}}{2 m_e m_v}(\cos^2 \theta + 1 - 2 \cos \theta) = \frac{2 g_w^2 p^{*2}}{m_e m_v}(1 - \cos \theta)^2 \tag{H.8}$$

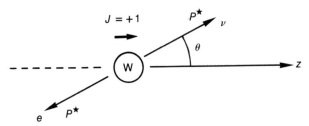

Figure H.1

The decay rate or width, Γ, is given by including the phase-space factor,

$$\frac{d\Gamma}{d\Omega} = \frac{1}{(2\pi)^2} \cdot \frac{1}{(2E_W)} \cdot \frac{p^{*2}\, dp^*}{dE_f} \frac{m_e m_v}{E_e E_v} |MM^\dagger|, \qquad (H.9)$$

using the Lorentz normalization factors $\sqrt{m/E}$ for the fermions, as in (D.22), and $\sqrt{1/2E_W}$ for the boson, as in (A.15). In the W-boson rest frame, $E_W = M_W$ and $p^* = M_W/2$. With

$$g_w^2 = \frac{g^2}{8} = \frac{GM_W^2}{\sqrt{2}},$$

as in (9.28), we obtain for the partial width for the decay $W \to ev$, after integrating (H.9),

$$\Gamma(W \to ev) = \frac{GM_W^3}{16\pi\sqrt{2}} \int_{-1}^{+1} (1 - \cos\theta)^2\, d(\cos\theta) = \frac{GM_W^3}{6\pi\sqrt{2}} = 232 \text{ MeV}. \quad (H.10)$$

H.1. PARTIAL WIDTHS FOR W- AND Z-DECAY

The decays $W \to ev$, μv, τv, $u\bar{d}$, $t\bar{b}$, $c\bar{s}$ into lepton or quark pairs of one color obviously have the same universal coupling and the same partial width (H.10). However, for quarks we must include a color factor 3, so that

$$\Gamma_W(\text{total}) = 3[\Gamma_W(ev) + 3\Gamma_W(ev)] = 12\Gamma_W(ev) \simeq 2.75 \text{ GeV} \quad (H.11)$$

which is a slight overestimate because of the reduced phase-space factor for the heavy quarks, $t\bar{b}$. From the discussions in Section 9.6 (Eq. (9.27)) we know that the matrix element

$$M(W \to ev) = \frac{g}{\sqrt{2}} \cdot J^+ W^- = \frac{g}{2\sqrt{2}} [\bar{v}\gamma_\mu(1 + \gamma_5)e] \cdot W^-,$$

whereas

$$M(Z^0 \to \nu\bar{\nu}) = \frac{g}{\cos\theta_w} \cdot J(\text{NC})Z^0 = \frac{g}{4\cos\theta_w} [\bar{\nu}\gamma_\mu(1 + \gamma_5)\nu] \cdot Z^0.$$

Thus

$$\frac{\Gamma(Z^0 \to \nu\bar{\nu})}{\Gamma(W \to e\nu)} = \left(\frac{M_Z}{M_W}\right) \frac{1}{2\cos^2\theta_w} = \frac{M_Z^3}{2M_W^3},$$

where the first term in brackets on the right-hand side is the phase-space factor ratio, and we have used the relation $M_W^2 = M_Z^2\cos^2\theta_w$ from (9.29). So

$$\Gamma(Z^0 \to \nu\bar{\nu}) = \frac{GM_Z^3}{12\pi\sqrt{2}} = 180 \text{ MeV}, \qquad (\text{H.12})$$

using the values $M_Z = 94$ GeV, $G = 1.166 \times 10^{-5}$ GeV^{-2}. For decay to any fermion pairs, the width relative to that for a neutrino pair will be

$$\Gamma(Z^0 \to f\bar{f}) = \Gamma(Z^0 \to \nu\bar{\nu}) \cdot 2(c_V^2 + c_A^2), \qquad (\text{H.13})$$

where c_V and c_A, the vector and axial-vector coupling coefficients for fermions to the Z^0, are given in (9.33). Allowing for three quark colors, it is left as an exercise to show that for three fermion generations and $\sin^2\theta_w = 0.22$, the total width is $\Gamma_Z = 2.7$ GeV, essentially the same as that of the charged W-boson in (H.11).

APPENDIX J

Color Forces between Quarks, Antiquarks and Gluons

J.1. GLUON COUPLING COEFFICIENTS

The color force between quarks is mediated by a color-anticolor octet of vector gluons as in (8.51), which was written down in analogy with the meson flavor-antiflavor octet states of Table 5.2. Denoting the three colors by r, b, g (for red, blue, and green), we have

$$r\bar{b} \tag{J.1a}$$

$$r\bar{g} \tag{J.1b}$$

$$b\bar{g} \tag{J.1c}$$

$$b\bar{r} \tag{J.1d}$$

$$g\bar{b} \tag{J.1e}$$

$$g\bar{r} \tag{J.1f}$$

$$(r\bar{r} + g\bar{g} - 2b\bar{b})/\sqrt{6} \tag{J.1g}$$

$$(r\bar{r} - g\bar{g})/\sqrt{2}. \tag{J.1h}$$

The coupling between a pair of like quarks, such as $rr \to rr$ in Fig. J.1, is mediated by exchange of states (J.1g) and (J.1h). The first gives a factor $\beta^2(1/\sqrt{6})(1/\sqrt{6})$, and the second, $\beta^2(1/\sqrt{2})(1/\sqrt{2})$, for a total coupling

395

of $+2\beta^2/3$, where β represents the color charge. For $rb \to rb$ as in Fig. J.2, only the gluon state (J.1g) can be exchanged, giving a factor $-\beta^2(2/\sqrt{6})(1/\sqrt{6}) = -\beta^2/3$. For the interaction $rb \to br$ as in Fig. J.3, the gluon state (J.1a) is involved, yielding a factor $+\beta^2$.

For antiquarks, we assume the coupling to gluons is given by $-\beta$, in analogy with the electrical case. So $r\bar{r} \to r\bar{r}$ involves a gluon coupling coefficient of $-2\beta^2/3$, as in Fig. J.4; $r\bar{r} \to b\bar{b}$ as in Fig. J.5 involves the gluon state (J.1a), just as in Fig. J.3, and so has a coupling factor $-\beta^2$. Finally $r\bar{b} \to r\bar{b}$ of Fig. J.6 is the same as for Fig. J.2 with the sign reversed, that is, $+\beta^2/3$.

In summary, the coupling coefficients for the gluon octet to pairs of quarks or antiquarks will be

$$\langle rr|G|rr \rangle = +\tfrac{2}{3}\beta^2, \qquad \langle r\bar{r}|G|r\bar{r} \rangle = -\tfrac{2}{3}\beta^2,$$

$$\langle rb|G|rb \rangle = -\tfrac{1}{3}\beta^2, \qquad \langle r\bar{b}|G|r\bar{b} \rangle = +\tfrac{1}{3}\beta^2, \qquad\qquad \text{(J.2)}$$

$$\langle br|G|rb \rangle = +\beta^2, \qquad \langle r\bar{r}|G|b\bar{b} \rangle = -\beta^2.$$

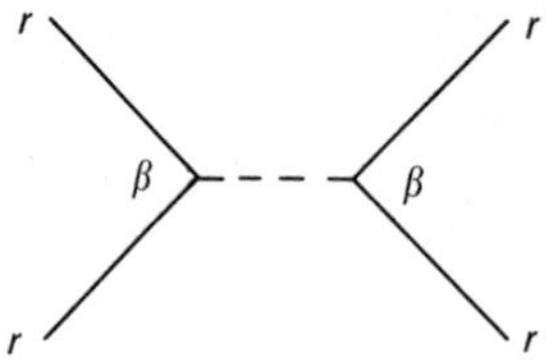

Figure J.1 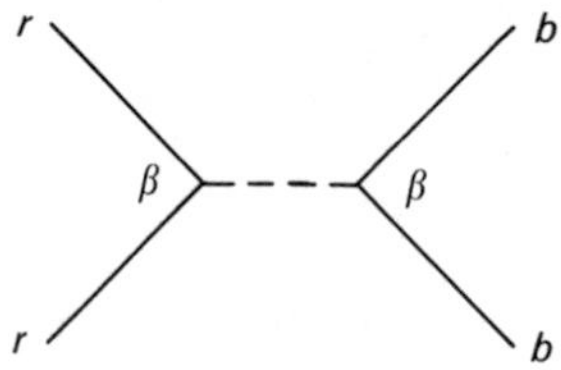   Figure J.2 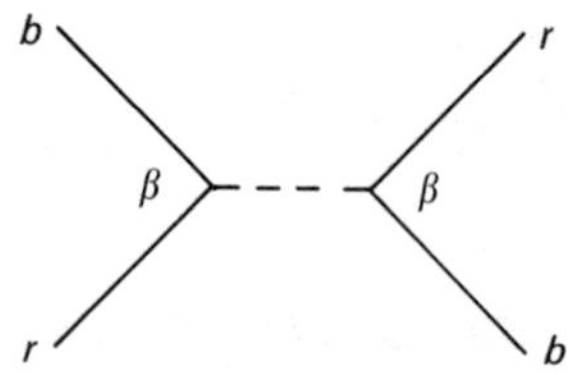   Figure J.3

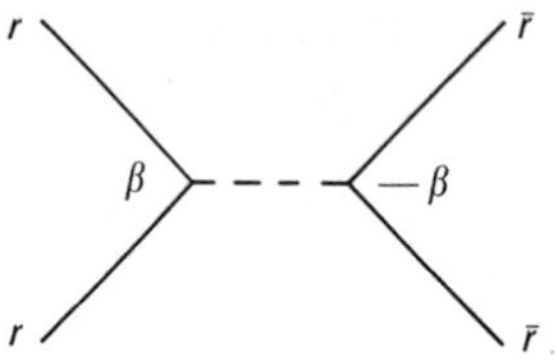

Figure J.4 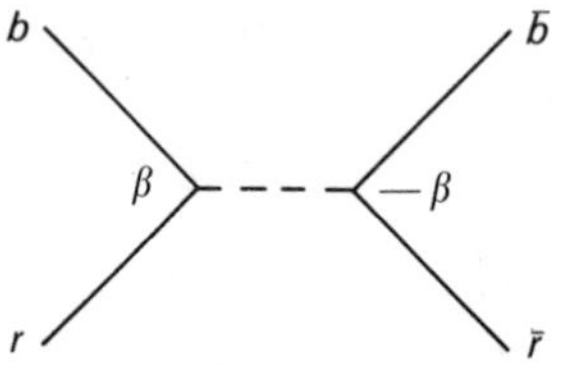   Figure J.5 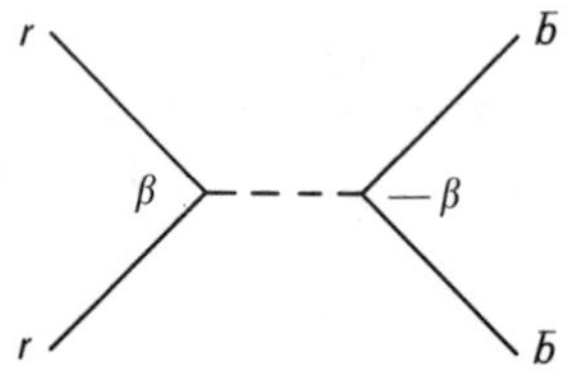   Figure J.6

J.2. COUPLING FOR THE BARYON COLOR SINGLET

We now compute the color coupling between a pair of quarks in a color singlet, that is, a color-antisymmetric baryon state. In analogy with the flavor singlet (5.4), the baryon color wavefunction must be of the form

$$(QQQ)_{\text{singlet}} = \frac{1}{\sqrt{6}}[(rb - br)g + (bg - gb)r + (gr - rg)b]. \qquad \text{(J.3)}$$

Looking at the interaction between two of the three quarks, we see that the first term in round brackets will involve $rb \to rb$ with a coefficient of $\tfrac{1}{6}$, and $rb \to br$ with $-\tfrac{1}{6}$. Including the gluon coupling factors (J.2), these two contributions together give for the rb-term

$$\frac{1}{6}\frac{-\beta^2}{3} + \frac{-1}{6}\beta^2 = -\frac{2\beta^2}{9}. \tag{J.4}$$

Obviously the br-term in (J.3) gives the same number, as do the bg, gb, gr, and rg states. So we need to multiply (J.4) by a factor 6 to obtain the total potential summed over all colors, for a pair of quarks in a three-quark color singlet of

$$V_{(QQ \text{ in } QQQ \text{ singlet})} = -\frac{4\beta^2}{3}\frac{1}{r}. \tag{J.5}$$

The negative sign indicates that the quark-quark interaction is attractive, and the $1/r$ dependence at short range indicates that the color field has the Coulombic form in (5.47).

J.3. COUPLING FOR THE MESON COLOR SINGLET

A quark-antiquark color-singlet wavefunction has the form, in analogy with the flavor singlet in Table 5.2,

$$(Q\bar{Q})_{\text{singlet}} = \frac{1}{\sqrt{3}}(r\bar{r} + b\bar{b} + g\bar{g}). \tag{J.6}$$

The contribution $r\bar{r} \to r\bar{r}$ in (J.6), using (J.2), will be

$$(1/\sqrt{3})(1/\sqrt{3})(-2\beta^2/3) = -2\beta^2/9.$$

There are three terms of this type in (J.6), giving together $-2\beta^2/3$. The interactions $r\bar{r} \to b\bar{b}$ and $r\bar{r} \to g\bar{g}$ each give $-\beta^2/3$. Again, multiplying by 3 for all such terms in (J.6) gives $-2\beta^2$. Hence the total potential for the color-singlet meson state will be

$$V_{(Q\bar{Q} \text{ singlet})} = \frac{-2\beta^2 - 2\beta^2/3}{r} = -\frac{8\beta^2}{3}\frac{1}{r}. \tag{J.7}$$

J.4. COUPLINGS FOR COLOR NONSINGLETS

The couplings for nonsinglet color combinations of QQ and $Q\bar{Q}$ can be obtained in a similar fashion. As an example, consider a "6" (sextet) representation of color SU(3). This will have the form

$$(QQ)_{\text{sextet}} = \frac{1}{\sqrt{6}}\left[rr + bb + gg + \frac{1}{\sqrt{2}}(rb + br) \right.$$
$$\left. + \frac{1}{\sqrt{2}}(rg + gr) + \frac{1}{\sqrt{2}}(gb + bg) \right]. \tag{J.8}$$

It is left as an exercise to show that

$$V_{(QQ\,\text{sextet})} = +\frac{2\beta^2}{3}\frac{1}{r}.\qquad (J.9)$$

As another example we can consider a $Q\bar{Q}$ color octet. The color wavefunction of the $Q\bar{Q}$ pair is the same as that of the gluon octet given in (J.1a)–(J.1h). It is straightforward to show that

$$V_{(Q\bar{Q}\,\text{octet})} = +\frac{\beta^2}{3}\frac{1}{r}.\qquad (J.10)$$

We observe that both the potentials (J.9) and (J.10), which involve quarks and antiquarks with net color, are repulsive. Only the color-singlet combinations (J.5) and (J.7) have positive binding.

The foregoing results can be obtained from more general formulae, or with the help of Young diagrams; the interested reader is referred to the review lectures by Feynman (1973) and by Rosner (1980).

The above expressions for the QQ and $Q\bar{Q}$ color-singlet potentials (J.5) and (J.7) lead to the formulae for the hyperfine splitting in hadron multiplets in Section 5.10. The strong coupling constant is defined as $\alpha_s = 2\beta^2$. The values of ΔE in the quark case are obtained from the expression (5.30) in the electrical case by the substitutions $e_i e_j = 4\pi\alpha_s/3$ and $2\pi\alpha_s/3$ in (5.31) and (5.32), respectively.

J.5. QQG AND GGG VERTEX COUPLINGS

The coupling strengths for the quark-quark, quark-gluon, and gluon-gluon scattering via single gluon exchange are required in discussing the $p\bar{p} \to 2$ jet events (Chapter 8, Fig. 8.20).

QQG Coupling

The relevant diagrams are shown in Fig. J.7. Starting off with a red quark, this can transform to a g-, b-, or r-quark with emission of an $r\bar{g}$, $r\bar{b}$, or $r\bar{r}$ gluon state. From (J.1(a), (b)) we get β^2, β^2 for the first two and from

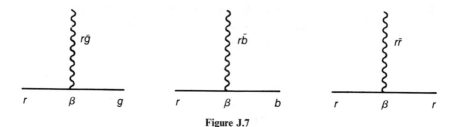

Figure J.7

(J.1(g), (h)) we obtain factors $\beta^2/6$ and $\beta^2/2$ for the third state, adding to $2\beta^2/3$. The total coupling factor is then $8\beta^2/3$. Obviously, the same result would obtain if we started with a g- or b-quark color. With $\alpha_s = 2\beta^2$, we therefore find, for the square of the coupling amplitude of the QQG vertex,

$$A^2(QQG) = \frac{4\alpha_s}{3} \tag{J.11}$$

GGG Coupling

How do we treat the gluon-gluon coupling? Let us take two of the gluons as $Q\bar{Q}$ virtual states, as in Fig. J.8. First, assume the gluon G1 is composed of an $r\bar{X}$ pair, where $\bar{X} = \bar{b}$, $\bar{g}$, or $\bar{r}$, and let us consider one particular color state, $r\bar{b}$, for the second gluon G2. The first diagram in Fig. J.8 gives us for the probability that G3 is in the state $b\bar{b}$, the value $(-2\beta/\sqrt{6})^2 = 2\beta^2/3$. The second diagram gives simply β^2. The third diagram gives a factor, from (J.1g, h), of $(1/6 + 1/2)\beta^2 = \beta^2/3$. Adding these diagrams together we find, for $r\bar{X} \rightarrow r\bar{b}$ + gluon, a total factor of $2\beta^2$.

Now, G2 is only one of eight possible color gluon states. Another factor 3 arises because for G1 we could start off with $b\bar{X}$ or $g\bar{X}$ as well as $r\bar{X}$. This covers the eight possible G1 states and the eight possible G2 states. By symmetry, all possible color states of the gluon G3 will have automatically been included. Dividing by the eight states of G1, therefore, we get—for any particular gluon state G1—an overall coupling coefficient, squared, of

$$A^2(GGG) = 2\beta^2 \times 8 \times \tfrac{3}{8} = 6\beta^2 = 3\alpha_s \tag{J.12}$$

The results (J.11) and (J.12) were used in Fig. 8.20. They are more properly derived using group theory. It should be noted that it does not appear possible, using these simple arguments, to derive the coupling for the $GGGG$ vertex.

In passing, we might ask why a triple photon vertex could not also be represented like Fig. J.8, replacing gluons by photons and quark pairs by electron-positron pairs. Such a process would have zero amplitude, however, because $C = -1$ for each photon and therefore C could not be conserved.

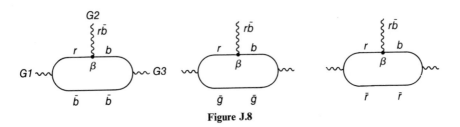

Figure J.8

TABLE I Atomic constants

Avogadro's number	N_0	$6.02205(\pm 3) \times 10^{23}$ mole^{-1}
Velocity of light *in vacuo*	c	$2.99792458 \times 10^{10}$ cm s^{-1} $= 2.99792458 \times 10^{8}$ m s^{-1}
Charge on electron	e	$4.80325(\pm 2) \times 10^{-10}$ esu $= 1.602189(\pm 5) \times 10^{-19}$ coulomb
Planck's constant reduced	$h/2\pi = \hbar$	$6.58217(\pm 2) \times 10^{-22}$ MeV s $= 1.054589(\pm 6) \times 10^{-27}$ erg s
Fine-structure constant	$\alpha = e^2/4\pi\hbar c$	$1/137.0360(\pm 1)$
Mass of electron	$m_e c^2$	$9.10953(\pm 5) \times 10^{-28}$ g $= 0.511003(\pm 1)$ MeV
Mass of proton	$m_p c^2$	$938.280(\pm 3)$ MeV $= 1836.152 m_e$
Classical radius of electron	$r_e = e^2/4\pi m_e c^2$	$2.81794(\pm 1) \times 10^{-13}$ cm
Compton wavelength of electron	$\hbar/m_e c = r_e/\alpha$	$3.86159(\pm 1) \times 10^{-11}$ cm
First Bohr radius for infinitely heavy nucleus	$a_\infty = 4\pi\hbar^2/e^2 m_e = r_e/\alpha^2$	$0.5291771(\pm 4) \times 10^{-8}$ cm
Thomson cross-section	$\frac{8}{3}\pi r_e^2$	$665.245(\pm 3)$ mb (or $\times 10^{-31}$ m^2)
Bohr magneton	$\mu_B = e\hbar/2m_e c$	$0.578838(\pm 1) \times 10^{-14}$ MeV G^{-1}
Nuclear magneton	$\mu_n = e\hbar/2m_p c$	$3.15245(\pm 1) \times 10^{-18}$ MeV G^{-1}
Unit of precession frequency	$\omega = \dfrac{\mu_B}{\hbar} = \dfrac{e}{2m_e c}$	$8.79401(\pm 3)$ $\times 10^6$ rad s^{-1} G^{-1}
	$\omega = \dfrac{\mu_n}{\hbar} = \dfrac{e}{2m_p c}$	$4.78948(\pm 3)$ $\times 10^3$ rad s^{-1} G^{-1}

Units

1 MeV $= 1.602189(\pm 5) \times 10^{-6}$ erg
1 fermi $= 1$ fm $= 10^{-13}$ cm
1 barn $= 1$ b $= 10^3$ mb $= 10^6$ μb $= 10^{-24}$ cm$^2 = 10^{-28}$ m^2.

Useful approximations

$$\hbar c \simeq 200 \text{ MeV fm}$$

Classical radius of electron:

$$e^2/4\pi m_e c^2 \simeq 2.8 \text{ fm} \simeq 2 \times \hbar/m_\pi c$$
$$\simeq 2 \times \text{(range of nuclear forces)}$$

Radius of curvature for momentum p:

$$pc \simeq 0.03 H\rho \ (pc \text{ in GeV, } H \text{ in kG, } \rho \text{ in m})$$

TABLE II Spherical harmonics and D-functions

$$Y_l^m(\theta, \phi) = \sqrt{\frac{(2l+1)(l-m)!}{4\pi(l+m)!}} \, P_l^m(\cos\theta) e^{im\phi}$$

$$P_l^m(\cos\theta) = (-1)^m \sin^m\theta \left[\left(\frac{d}{d(\cos\theta)} \right)^m P_l(\cos\theta) \right] \qquad (m \le l)$$

$$P_l(\cos\theta) = \frac{1}{2^l l!} \left[\left(\frac{d}{d(\cos\theta)} \right)^l (-\sin^2\theta)^l \right]$$

$$Y_l^{-m}(\theta, \phi) = (-1)^m [Y_l^m(\theta, \phi)]^*$$

$$P_l^{-m}(\cos\theta) = (-1)^m \frac{(l-m)!}{(l+m)!} P_l^m(\cos\theta)$$

$l = 0$	$Y_0^0 = \dfrac{1}{\sqrt{4\pi}}$	
$l = 1$	$Y_1^0 = \sqrt{\dfrac{3}{4\pi}} \cos\theta$	$Y_1^1 = -\sqrt{\dfrac{3}{8\pi}} \sin\theta e^{i\phi}$
$l = 2$	$Y_2^0 = \sqrt{\dfrac{5}{16\pi}} (3\cos^2\theta - 1)$	$Y_2^1 = -\sqrt{\dfrac{15}{8\pi}} \sin\theta \cos\theta e^{i\phi}$:
	$Y_2^2 = \sqrt{\dfrac{15}{32\pi}} \sin^2\theta e^{2i\phi}$	
$l = 3$	$Y_3^0 = \sqrt{\dfrac{7}{16\pi}} (5\cos^3\theta - 3\cos\theta)$	$Y_3^1 = -\sqrt{\dfrac{21}{64\pi}} \sin\theta (5\cos^2\theta - 1)e^{i\phi}$
	$Y_3^2 = \sqrt{\dfrac{105}{32\pi}} \sin^2\theta \cos\theta e^{2i\phi}$	$Y_3^3 = -\sqrt{\dfrac{35}{64\pi}} \sin^3\theta e^{3i\phi}$

D-functions

$$d_{m',m}^j = (-1)^{m-m'} d_{m,m'}^j = d_{-m,-m'}^j$$

$$d_{1,1}^1 = \frac{(1+\cos\theta)}{2} \qquad d_{1,0}^1 = -\frac{1}{\sqrt{2}} \sin\theta \qquad d_{1,-1}^1 = \frac{(1-\cos\theta)}{2}$$

$$d_{0,0}^1 = \cos\theta$$

$$d_{1/2,\,1/2}^{1/2} = \cos\frac{\theta}{2} \qquad d_{1/2,\,-1/2}^{1/2} = -\sin\frac{\theta}{2}$$

TABLE III   Clebsch-Gordan coefficients

As an example of the use of this table, take the case of combining two angular momenta $j_1 = 1$, $m_1 = 1$ and $j_2 = 1$, $m_2 = -1$. We look up the entry for combining 1×1, and the fourth line gives for the coefficients in Eq. (C1) of Appendix C

$$\phi_1(1, 1)\phi_2(1, -1) = \sqrt{\tfrac{1}{6}}\psi(2, 0) + \sqrt{\tfrac{1}{2}}\psi(1, 0) + \sqrt{\tfrac{1}{3}}\psi(0, 0).$$

This tells us how two particles of angular momentum (or isospin) unity combine to form states of angular momentum $j = 0$, 1, or 2. Alternatively, a state of particular (j, m) can be decomposed into constituents. Thus $j = 2$, $m = 0$ can be decomposed into products of states of $j = 1$, with of course $m_1 + m_2 = m = 0$. The fourth column of the 1×1 table gives

$$\psi(2, 0) = \sqrt{\tfrac{1}{6}}\phi_1(1, 1)\phi_2(1, -1) + \sqrt{\tfrac{2}{3}}\phi_1(1, 0)\phi_2(1, 0) + \sqrt{\tfrac{1}{6}}\phi_1(1, -1)\phi_2(1, 1).$$

The sign convention in the table follows that of Condon and Shortley (1951).

$\frac{1}{2} \times \frac{1}{2}$

		J: 1	1	0	1		
		M: +1	0	0	-1		
m_1	m_2						
$+\frac{1}{2}$	$+\frac{1}{2}$	1					
$+\frac{1}{2}$	$-\frac{1}{2}$		$\sqrt{\frac{1}{2}}$	$\sqrt{\frac{1}{2}}$			
$-\frac{1}{2}$	$+\frac{1}{2}$		$\sqrt{\frac{1}{2}}$	$-\sqrt{\frac{1}{2}}$			
$-\frac{1}{2}$	$-\frac{1}{2}$				1		

$1 \times \frac{1}{2}$

		J: $\frac{3}{2}$	$\frac{3}{2}$	$\frac{1}{2}$	$\frac{3}{2}$	$\frac{1}{2}$	$\frac{3}{2}$
		M: $+\frac{3}{2}$	$+\frac{1}{2}$	$+\frac{1}{2}$	$-\frac{1}{2}$	$-\frac{1}{2}$	$-\frac{3}{2}$
m_1	m_2						
$+1$	$+\frac{1}{2}$	1					
$+1$	$-\frac{1}{2}$		$\sqrt{\frac{1}{3}}$	$\sqrt{\frac{2}{3}}$			
0	$+\frac{1}{2}$		$\sqrt{\frac{2}{3}}$	$-\sqrt{\frac{1}{3}}$			
0	$-\frac{1}{2}$				$\sqrt{\frac{2}{3}}$	$\sqrt{\frac{1}{3}}$	
-1	$+\frac{1}{2}$				$\sqrt{\frac{1}{3}}$	$-\sqrt{\frac{2}{3}}$	
-1	$-\frac{1}{2}$						1

TABLE III Continued

$\frac{3}{2} \times \frac{1}{2}$

		J:	2	2	1	2	1	2	1	2
		M:	$+2$	$+1$	$+1$	0	0	-1	-1	-2
m_1	m_2									
$+\frac{3}{2}$	$+\frac{1}{2}$	1								
$+\frac{3}{2}$	$-\frac{1}{2}$			$\sqrt{\frac{1}{4}}$	$\sqrt{\frac{3}{4}}$					
$+\frac{1}{2}$	$+\frac{1}{2}$			$\sqrt{\frac{3}{4}}$	$-\sqrt{\frac{1}{4}}$					
$+\frac{1}{2}$	$-\frac{1}{2}$					$\sqrt{\frac{1}{2}}$	$\sqrt{\frac{1}{2}}$			
$-\frac{1}{2}$	$+\frac{1}{2}$					$\sqrt{\frac{1}{2}}$	$-\sqrt{\frac{1}{2}}$			
$-\frac{1}{2}$	$-\frac{1}{2}$							$\sqrt{\frac{3}{4}}$	$\sqrt{\frac{1}{4}}$	
$-\frac{3}{2}$	$+\frac{1}{2}$							$\sqrt{\frac{1}{4}}$	$-\sqrt{\frac{3}{4}}$	
$-\frac{3}{2}$	$-\frac{1}{2}$									1

$2 \times \frac{1}{2}$

		J:	$\frac{5}{2}$	$\frac{5}{2}$	$\frac{3}{2}$	$\frac{5}{2}$	$\frac{3}{2}$	$\frac{5}{2}$	$\frac{3}{2}$	$\frac{5}{2}$	$\frac{3}{2}$	$\frac{5}{2}$
		M:	$+\frac{5}{2}$	$+\frac{3}{2}$	$+\frac{3}{2}$	$+\frac{1}{2}$	$+\frac{1}{2}$	$-\frac{1}{2}$	$-\frac{1}{2}$	$-\frac{3}{2}$	$-\frac{3}{2}$	$-\frac{5}{2}$
m_1	m_2											
$+2$	$\frac{1}{2}$	1										
$+2$	$-\frac{1}{2}$		$\sqrt{\frac{1}{5}}$	$\sqrt{\frac{4}{5}}$								
$+1$	$+\frac{1}{2}$		$\sqrt{\frac{4}{5}}$	$-\sqrt{\frac{1}{5}}$								
$+1$	$-\frac{1}{2}$				$\sqrt{\frac{2}{5}}$	$\sqrt{\frac{3}{5}}$						
0	$+\frac{1}{2}$				$\sqrt{\frac{3}{5}}$	$-\sqrt{\frac{2}{5}}$						
0	$-\frac{1}{2}$						$\sqrt{\frac{3}{5}}$	$\sqrt{\frac{2}{5}}$				
-1	$+\frac{1}{2}$						$\sqrt{\frac{2}{5}}$	$-\sqrt{\frac{3}{5}}$				
-1	$-\frac{1}{2}$								$\sqrt{\frac{4}{5}}$	$\sqrt{\frac{1}{5}}$		
-2	$+\frac{1}{2}$								$\sqrt{\frac{1}{5}}$	$-\sqrt{\frac{4}{5}}$		
-2	$-\frac{1}{2}$											1

TABLE III Continued

1×1

	J:	2	2	1	2	1	0	2	1	2
	M:	+2	+1	+1	0	0	0	−1	−1	−2
m_1	m_2									
+1	+1	1								
+1	0		$\sqrt{\frac{1}{2}}$	$\sqrt{\frac{1}{2}}$						
0	+1		$\sqrt{\frac{1}{2}}$	$-\sqrt{\frac{1}{2}}$						
+1	−1				$\sqrt{\frac{1}{6}}$	$\sqrt{\frac{1}{2}}$	$\sqrt{\frac{1}{3}}$			
0	0				$\sqrt{\frac{2}{3}}$	0	$-\sqrt{\frac{1}{3}}$			
−1	+1				$\sqrt{\frac{1}{6}}$	$-\sqrt{\frac{1}{2}}$	$\sqrt{\frac{1}{3}}$			
0	−1							$\sqrt{\frac{1}{2}}$	$\sqrt{\frac{1}{2}}$	
−1	0							$\sqrt{\frac{1}{2}}$	$-\sqrt{\frac{1}{2}}$	
−1	−1									1

$\frac{3}{2} \times 1$

	J:	$\frac{5}{2}$	$\frac{5}{2}$	$\frac{3}{2}$	$\frac{5}{2}$	$\frac{3}{2}$	$\frac{1}{2}$	$\frac{5}{2}$	$\frac{3}{2}$	$\frac{1}{2}$	$\frac{5}{2}$	$\frac{3}{2}$	$\frac{5}{2}$
	M:	$+\frac{5}{2}$	$+\frac{3}{2}$	$+\frac{3}{2}$	$+\frac{1}{2}$	$+\frac{1}{2}$	$+\frac{1}{2}$	$-\frac{1}{2}$	$-\frac{1}{2}$	$-\frac{1}{2}$	$-\frac{3}{2}$	$-\frac{3}{2}$	$-\frac{5}{2}$
m_1	m_2												
$+\frac{3}{2}$	+1	1											
$+\frac{3}{2}$	0		$\sqrt{\frac{2}{5}}$	$\sqrt{\frac{3}{5}}$									
$+\frac{1}{2}$	+1		$\sqrt{\frac{3}{5}}$	$-\sqrt{\frac{2}{5}}$									
$+\frac{3}{2}$	−1				$\sqrt{\frac{1}{10}}$	$\sqrt{\frac{2}{5}}$	$\sqrt{\frac{1}{2}}$						
$+\frac{1}{2}$	0				$\sqrt{\frac{3}{5}}$	$\sqrt{\frac{1}{15}}$	$-\sqrt{\frac{1}{3}}$						
$-\frac{1}{2}$	+1				$\sqrt{\frac{3}{10}}$	$-\sqrt{\frac{8}{15}}$	$\sqrt{\frac{1}{6}}$						
$+\frac{1}{2}$	−1							$\sqrt{\frac{3}{10}}$	$\sqrt{\frac{8}{15}}$	$\sqrt{\frac{1}{6}}$			
$-\frac{1}{2}$	0							$\sqrt{\frac{3}{5}}$	$-\sqrt{\frac{1}{15}}$	$-\sqrt{\frac{1}{3}}$			
$-\frac{3}{2}$	+1							$\sqrt{\frac{1}{10}}$	$-\sqrt{\frac{2}{3}}$	$\sqrt{\frac{1}{2}}$			
$-\frac{1}{2}$	−1										$\sqrt{\frac{3}{5}}$	$\sqrt{\frac{2}{5}}$	
$-\frac{3}{2}$	+0										$\sqrt{\frac{2}{5}}$	$-\sqrt{\frac{3}{5}}$	
$-\frac{3}{2}$	−1												1

TABLE IV Elementary particles

Long-lived particles (stable or decaying by weak or electromagnetic transitions)[a]

Particle	J^{Pb}	I^{Gc}	Mass,[d] MeV	Mean life,[d] s	Decay		
					Mode	Fraction	p_{max}, MeV/c^e
γ	1^-	—	0	stable	—	—	—
Leptons							
ν_e	$\frac{1}{2}$	—	0(<30 eV)	stable	—	—	—
ν_μ	$\frac{1}{2}$	—	0(<0.25)		—	—	—
ν_τ	$\frac{1}{2}$	—	0(<70)		—	—	—
$e^\pm$	$\frac{1}{2}$	—	0.511003(±1)	stable	—	—	—
$\mu^\pm$	$\frac{1}{2}$	—	105.6595(±2)	$2.1971(\pm1) \times 10^{-6}$	$e\nu\bar{\nu}$	100%	53
$\tau^\pm$	$\frac{1}{2}$	—	1784(±4)	$2.8(\pm2) \times 10^{-13}$	$\mu\nu\bar{\nu}$	18%	889
					$e\nu\bar{\nu}$	18%	892
					$\pi\nu$	8%	887
					$\rho\nu$	22%	723
					hadrons $+\nu$	34%	—

TABLE IV Continued

Particle	J^P	I^G	Mass, MeV	Mean life, s	Decay Mode	Fraction	p_{max}, MeV/c
Nonstrange mesons							
$\pi^\pm$	0^-	1^-	139.567($\pm$1)	2.603($\pm$2) $\times 10^{-8}$	$\mu\nu$	$\simeq 100\%$	30
					$e\nu$	1.27×10^{-4}	70
					$\mu\nu\gamma$	1.24×10^{-4}	30
					$\pi^0 e\nu$	1.02×10^{-8}	5
					$e\nu\gamma$	6×10^{-8}	70
π^0	0^-	1^-	134.963($\pm$4)	0.83($\pm$6) $\times 10^{-16}$	$\gamma\gamma$	98.8%	67
					$\gamma e^+ e^-$	1.17%	67
					$e^+ e^-$	2.10^{-7}	
η	0^-	0^+	548.8($\pm$6)	(Width = 0.9 keV)	$\gamma\gamma$	39%	274
					$\pi^0\gamma\gamma$	0.1%	258
					$3\pi^0$	32%	179
	d				$\pi^+\pi^-\pi^0$	23%	174
					$\pi^+\pi^-\gamma$	5%	236
η'	0^-	0^+	957.6($\pm$3)	(Width = 0.3 MeV)	$\eta\pi\pi$	66%	—
					$\rho^0\gamma$	30%	—
					$\gamma\gamma$	$\simeq 2\%$	—
					$\omega\gamma$	2.8%	—

TABLE IV Continued

Particle	J^P	I^G	Mass, MeV	Mean life, s	Decay Mode	Fraction	p_{max}, MeV/c
Strange mesons							
$K^\pm$	0^-	$\frac{1}{2}$	493.67($\pm$2)	1.237($\pm$3) $\times 10^{-8}$	$\mu^\pm \nu$	63.5% ($K_{\mu 2}$)	236
					$\pi^\pm \pi^0$	21.2% ($K_{\pi 2}$)	205
					$\pi^\pm \pi^+ \pi^-$	5.6% ($K_{\pi 3}$)	126
					$\pi^\pm \pi^0 \pi^0$	1.7%	133
					$\mu^\pm \pi^0 \nu$	3.2% ($K_{\mu 3}$)	215
					$e^\pm \pi^0 \nu$	4.8% (K_{e3})	228
					$e^\pm \nu$	1.5×10^{-5} (K_{e2})	247
					and others		
$K^0, \bar{K}^0$	0^-	$\frac{1}{2}$	497.7($\pm$1)	50% K_S, 50% K_L			
K_S				$0.892(\pm 2) \times 10^{-10}$	$\pi^+ \pi^-$	68.6%	206
					$\pi^0 \pi^0$	31.3%	209
					$\pi^+ \pi^- \gamma$	1.9×10^{-3}	206
K_L				$5.18(\pm 4) \times 10^{-8}$	$\pi^0 \pi^0 \pi^0$	21.5%	139
					$\pi^+ \pi^- \pi^0$	12.6%	133
					$\pi \mu \nu$	26.8%	216
					$\pi e \nu$	38.8%	229
					$\pi^+ \pi^-$	2.0×10^{-3}	206
					and others		

$$m_{K_L} - m_{K_S} = \frac{0.477}{\tau_S}$$

TABLE IV Continued

Particle	J^P	I^G	Mass, MeV	Mean life, s	Decay Mode	Fraction	p_{max}, MeV/c
Charmed mesons							
$D^\pm$	0^-	$\tfrac{1}{2}$	$1869.4(\pm 6)$	$(9 \pm 3) \times 10^{-13}$	D^+ $\begin{cases} K^-\pi^+\pi^+ \\ K^- + \text{any} \\ K^+ + \text{any} \\ e^\pm + \text{any} \\ \bar{K}^0 + \text{any} \end{cases}$	4 % 16 % 6 % 19 % 48 %	845 — — — —
$D^0, \bar{D}^0$	0^-	$\tfrac{1}{2}$	$1864.7(\pm 6)$	$(5 \pm 1) \times 10^{-13}$	D^0 $\begin{cases} K^-\pi^+ \\ K^-\pi^+\pi^0 \\ K^-\pi^+\pi^+\pi^- \\ K^- + \text{any} \\ \bar{K}^0 + \text{and} \\ e^\pm + \text{any} \\ \phi\pi^+ \text{ and others} \end{cases}$	2.4 % 9.3 % 4.6 % 44 % 33 % 5 %	860 843 812 — — —
$F^\pm$	0^-	0	$1971(\pm 6)$	$(1.9 \pm 1) \times 10^{-13}$			
Bottom mesons							
$B^\pm$	0^-	$\tfrac{1}{2}$	$5271(\pm 3)$	$\sim 10^{-13}$	$\begin{cases} \bar{D}^0\pi^+ \\ D^{x-}\pi^-\pi^+ \\ e^\pm \nu + \text{hadrons} \\ \mu^\pm \nu + \text{hadrons} \\ D^0 + \text{any} \end{cases}$	4 % 5 % 13 % 13 % 80 %	2303 2243 — — —
B^0	0^-	$\tfrac{1}{2}$	$5274(\pm 3)$				

TABLE IV Continued

Particle	S	J^P	I	Mass, MeV	Mean life, s	Decay Mode	Fraction	p_{max}, MeV/c
Baryons								
p	0	$\frac{1}{2}^+$	$\frac{1}{2}$	938.280($\pm$3)	stable ($>10^{32}$ yr)			
n	0	$\frac{1}{2}^+$	$\frac{1}{2}$	939.573($\pm$3)	898 $\pm$ 16	$pe^-\bar{\nu}$	100%	1
Λ	-1	$\frac{1}{2}^+$	0	1115.60($\pm$5)	2.63($\pm$2) $\times$ 10^{-10}	$p\pi^-$	65%	100
						$n\pi^0$	35%	104
						$pe\nu$	0.84×10^{-3}	163
						$p\mu\nu$	1.6×10^{-4}	131
Σ^+	-1	$\frac{1}{2}^+$	1	1189.4($\pm$1)	0.800($\pm$4) $\times$ 10^{-16}	$p\pi^0$	52%	189
						$n\pi^+$	48%	185
						$\Lambda e^+\nu$	2×10^{-5}	72
						$p\gamma$	1.2×10^{-3}	225
						$n\pi^+\gamma$	9×10^{-4}	185
Σ^0	-1	$\frac{1}{2}^+$	1	1192($\pm$2)	6×10^{-20}	$\Lambda\gamma$	100%	75
Σ^-	-1	$\frac{1}{2}^+$	1	1197.3($\pm$7)	1.48($\pm$1) $\times$ 10^{-10}	$n\pi^-$	100%	193
						$\Lambda e^-\nu$	6×10^{-5}	79
						$ne^-\nu$	1.1×10^{-3}	230
						$n\mu^-\nu$	0.5×10^{-3}	210
						$n\pi^-\gamma$	10^{-4}	193

TABLE IV Continued

Particle	S	J^P	I	Mass, MeV	Mean life, s	Decay Mode	Decay Fraction	p_{max} MeV/c
Ξ^0	-2	$\frac{1}{2}^+$	$\frac{1}{2}$	$1314.7(\pm 6)$	$2.9(\pm 1) \times 10^{-10}$	$\Lambda\pi^0$	100%	135
Ξ^-	-2	$\frac{1}{2}^+$	$\frac{1}{2}$	$1321.3(\pm 1)$	$1.64(\pm 2) \times 10^{-10}$	$\Lambda\pi^-$ and others	100%	139
Ω^-	-3	$\frac{3}{2}^+$	0	$1672.5(\pm 3)$	$0.82(\pm 3) \times 10^{-10}$	$\Xi^0\pi^-$ $\Xi^-\pi^0$ ΛK^-	23% 8% 69%	293 289 210
Λ_c $(C = +1)$	0	$\frac{1}{2}^+$	0	$2282(\pm 3)$	$2.3(\pm 8) \times 10^{-13}$	$pK^-\pi^+$	$\simeq 3\%$ and others	814

TABLE IV Continued

Particle	J^P	I^G	Mass, MeV	Width Γ (MeV)	Decay Mode	Fraction
Meson resonances (nonstrange)						
$\rho(770)$	1^-	1^+	769 ± 3	154 ± 5	$\pi\pi$	100%
					e^+e^-	6×10^{-5}
					$\mu^+\mu^-$	6×10^{-5}
$\omega(783)$	1^-	0^-	$782.6(\pm 2)$	9.9 ± 0.3	$\pi^+\pi^-\pi^0$	90%
					$\pi^0\gamma$	9%
					$\pi^+\pi^-$	1.4%
					e^+e^-	7×10^{-5}
$\delta(980)$	0^+	1^-	983 ± 2	54 ± 7	$\eta\pi, K\bar{K}$	
$S^*(980)$	0^+	0^+	975 ± 4	33 ± 6	$\pi\pi, K\bar{K}$	
$\phi(1020)$	1^-	0^-	$1019.5(\pm 1)$	4.2 ± 0.1	K^+K^-	49%
					$K_L K_S$	36%
					$\pi^+\pi^-\pi^0$	15%
					e^+e^-	3×10^{-4}
					$\mu^+\mu^-$	3×10^{-4}
$A1(1270)$	1^+	1^-	1275 ± 30	$\simeq 300$	$\rho\pi$	100%
$B(1235)$	1^+	1^+	1234 ± 10	130 ± 10	$\omega\pi$	100%
$f(1270)$	2^+	0^+	1273 ± 5	178 ± 20	$\pi\pi$	$\simeq 80\%$ and others
$D(1285)$	1^+	0^+	1283 ± 5	26 ± 5	$K\bar{K}\pi, \eta\pi\pi$	—
$A2(1320)$	2^+	1^-	1317 ± 5	110 ± 5	$\rho\pi$	40%
					$K\bar{K}$	$\simeq 5\%$
					$\eta\pi$	15%
					$\omega\pi\pi$	11%
$E(1420)$	1^+	0^+	1418 ± 10	50 ± 10	$K\bar{K}\pi$	
$f'(1525)$	2^+	0^+	1525 ± 5	67 ± 10	$K\bar{K}, \pi\pi$	
$\rho'(1600)$	1^-	1^+	1590 ± 20	260 ± 100	4π	60%
					2π and others	23%
$A3(1680)$	2^-	1^-	1680 ± 30	250 ± 50	$f\pi$	60%
					$\rho\pi$	30%

TABLE IV Continued

Particle	J^P	I^G	Mass, MeV	Width Γ (MeV)	Decay Mode	Fraction
$\omega(1670)$	3^-	0^-	1666 ± 5	166 ± 15	$\rho\pi$	
$J/\psi(3100)$	1^-	0^-	3097 ± 0.1	0.063 ± 0.009	e^+e^- $\mu^+\mu^-$ hadrons $\pi^0\pi^+\pi^+\pi^-\pi^-$ $3(\pi^+\pi^-)\pi^0$ and others	7% 7% 86% 3.7% 3%
$\chi(3415)$	(0^+)	0^+	3415 ± 1	—	$2(\pi^+\pi^-)$ $\pi^+\pi^-K^+K^-$ $\gamma\psi(3100)$ $3(\pi^+\pi^-)$ and others	5% 4% 3% 2%
$\chi(3510)$	—	0^+	3510 ± 0.6	—	$\gamma\psi(3100)$ $2(\pi^+\pi^-)$ $3(\pi^+\pi^-)$ and others	30% 2% 3%
$\chi(3555)$	—	0^+	3556 ± 0.6	—	$\gamma\psi(3100)$ $2(\pi^+\pi^-)$ $\pi^+\pi^-K^+K^-$ and others	15% 3% 2%
$\psi(3685)$	1^-	0^-	3686 ± 0.1	0.22	e^+e^- $\mu^+\mu^-$ $\gamma\chi$ $\pi^+\pi^-\psi(3100)$ $2\pi^0\psi(3100)$ and others	1% 1% 20% 33% 17%
$\psi(3770)$	1^-	—	3770 ± 3	25 ± 3	$D\bar{D}$	$\sim 100\%$
$\Upsilon(9460)$	1^-	—	9460 ± 0.3	0.04	l^+l^-	5%
$\Upsilon(10025)$	1^-	—	10023 ± 0.3	0.03	l^+l^-	4%
$\Upsilon(10355)$	1^-	—	10355 ± 0.5	0.02	l^+l^- and others	5%
$\Upsilon(10575)$	1^-	—	10573 ± 4	14 ± 5	e^+e^-	0.2%

TABLE IV Continued

Particle	J^P	I^G	Mass, MeV	Width Γ (MeV)	Decay Mode	Decay Fraction
$\chi_b(9875)$			9873 ± 6		$\gamma\Upsilon(9460)$	—
$\chi_b(9895)$			9895 ± 4		$\gamma\Upsilon(9460)$	43%
$\chi_b(9915)$			9915 ± 2		$\gamma\Upsilon(9460)$	20%
$\chi_b(10255)$			10255 ± 3		$\gamma\Upsilon(9460)$ $\gamma\Upsilon(10025)$	
$\chi_b(10270)$			10271 ± 2		$\gamma\Upsilon(9460)$ $\gamma\Upsilon(10025)$	
Meson resonances (strange)						
$K^*(890)$	1^-	$\frac{1}{2}$	$892.1(\pm 4)$	51 ± 1	$K\pi$ $K\gamma$	$\simeq 100\%$ 0.1%
$K^*(1420)$	2^+	$\frac{1}{2}$	1425 ± 5	100 ± 10	$K\pi$ $K^*(890)\pi$ $K\rho$ $K\omega$ $K\eta$	45% 25% 9% 4% 2%
$K(1780)$	3^-	$\frac{1}{2}$	1780 ± 10	160 ± 20	$K\pi\pi$ $[K^*(890)\pi$ or $K\rho]$ $K\pi$	large large 17%
Meson resonances (charmed)						
$D^{*+}(2010)$	1^-	$\frac{1}{2}$	2010 ± 1	<2	$D^0\pi^+$ $D^+\pi^0$ $D^+\gamma$	65% 29% 8%
$D^{*0}(2010)$	1^-	$\frac{1}{2}$	2007 ± 2	<5	$D^0\pi^0$ $D^0\gamma$	60% 40%

TABLE IV Continued

Resonance	J^P	I	Mass, MeV	Width Γ (MeV)	Decay Mode	Decay Fraction, %
Baryon resonances (nonstrange)						
$N(1470)$	$\frac{1}{2}^+$	$\frac{1}{2}$	1435–1505	200–400	$N\pi$	60
					$N\pi\pi$	40
$N(1520)$	$\frac{3}{2}^-$	$\frac{1}{2}$	1510–1540	100–150	$N\pi$	50
					$N\pi\pi$	50
$N(1535)$	$\frac{1}{2}^-$	$\frac{1}{2}$	1500–1600	100–250	$N\pi$	35
					$N\eta$	35
$N(1650)$	$\frac{1}{2}^-$	$\frac{1}{2}$	1655–1680	105–175	$N\pi$	60
					$N\pi\pi$	30
$N(1670)$	$\frac{5}{2}^-$	$\frac{1}{2}$	1660–1690	120–180	$N\pi$	40
					$N\pi\pi$	60
$N(1688)$	$\frac{5}{2}^+$	$\frac{1}{2}$	1680–1692	105–180	$N\pi$	60
					$N\pi\pi$	40
$N(1700)$	$\frac{3}{2}^-$	$\frac{1}{2}$	1665–1730	70–120	$N\pi$	10
					$N\pi\pi$	90
$N(1710)$	$\frac{1}{2}^+$	$\frac{1}{2}$	1650–1750	100–150	$N\pi$	20
					$N\pi\pi$	>50
$N(1720)$	$\frac{3}{2}^+$	$\frac{1}{2}$	1700–1810	150–250	$N\pi$	17
					$N\pi\pi$	70
$N(2190)$	$\frac{7}{2}^-$	$\frac{1}{2}$	2120–2230	100–400	$N\pi, N\eta, \Lambda K$	
$N(2220)$	$\frac{9}{2}^+$	$\frac{1}{2}$	2150–2300	300–500	$N\pi, N\eta, \Lambda K$	
$N(2250)$	$\frac{9}{2}^-$	$\frac{1}{2}$	2130–2270	200–500	$N\pi, N\eta, \Lambda K$	
$N(2600)$	$\frac{11}{2}^-$	$\frac{1}{2}$	2580–2700	>300	$N\pi$	
$\Delta(1232)$	$\frac{3}{2}^+$	$\frac{3}{2}$	1230–1234	115 ± 5	$N\pi$	99.4
					$N\gamma$	0.6
$\Delta(1620)$	$\frac{1}{2}^-$	$\frac{3}{2}$	1600–1650	120–160	$N\pi$	27
					$N\pi\pi$	73
$\Delta(1700)$	$\frac{3}{2}^-$	$\frac{3}{2}$	1630–1740	175–300	$N\pi, N\pi\pi$	
$\Delta(1900)$	$\frac{1}{2}^-$	$\frac{3}{2}$	1850–2000	150–350	$N\pi$	10
					ΣK	10
$\Delta(1905)$	$\frac{5}{2}^+$	$\frac{3}{2}$	1890–1930	250–400	$N\pi, N\pi\pi$	
$\Delta(1910)$	$\frac{1}{2}^+$	$\frac{3}{2}$	1850–1950	200–330	$N\pi, N\pi\pi$	
$\Delta(1920)$	$\frac{3}{2}^+$	$\frac{3}{2}$	1860–2160	190–300	$N\pi, \Sigma K$	
$\Delta(1930)$	$\frac{5}{2}^-$	$\frac{3}{2}$	1890–1960	150–350	$N\pi, \Sigma K$	
$\Delta(1950)$	$\frac{7}{2}^+$	$\frac{3}{2}$	1910–1950	200–340	$N\pi, \Sigma K, N\pi\pi$	

TABLE IV Continued

Resonance	J^P	I	Mass, MeV	Width Γ (MeV)	Decay Mode	Fraction, %
Baryon resonances ($S = -1$)						
$\Lambda(1405)$	$\frac{1}{2}^-$	0	1405 ± 5	40 ± 10	$\Sigma\pi$	100
$\Lambda(1520)$	$\frac{3}{2}^-$	0	1520 ± 1	16 ± 1	$N\bar{K}$	46
					$\Sigma\pi$	41
					$\Lambda\pi\pi$	10
					and others	
$\Lambda(1600)$	$\frac{1}{2}^+$	0	1560–1700	50–250	$N\bar{K}, \Sigma\pi$	
$\Lambda(1670)$	$\frac{1}{2}^-$	0	1670	$\simeq 30$	$N\bar{K}$	15
					$\Lambda\eta$	35
					$\Sigma\pi$	50
$\Lambda(1690)$	$\frac{3}{2}^-$	0	1690	~ 60	$N\bar{K}$	25
					$\Sigma\pi$	30
					$\Lambda\pi\pi$	25
					$\Sigma\pi\pi$	20
$\Lambda(1800)$	$\frac{1}{2}^-$	0	1700–1850	200–400	$N\bar{K}$	40
					$\Sigma\pi$ etc.	
$\Lambda(1800)$	$\frac{1}{2}^+$	0	1750–1850	50–250	$N\bar{K}, \Sigma\pi$	
					and others	
$\Lambda(1820)$	$\frac{5}{2}^+$	0	1815 ± 5	75 ± 10	$N\bar{K}$	65
					$\Sigma\pi$	11
					$\Sigma(1385)\pi$	17
$\Lambda(1830)$	$\frac{5}{2}^-$	0	$\simeq 1820$	60–110	$N\bar{K}$	10
					$\Sigma\pi$	30
$\Lambda(1890)$	$\frac{3}{2}^+$	0	$\simeq 1890$	60–200	$N\bar{K}$	30
					$\Sigma\pi$ etc.	
$\Lambda(2100)$	$\frac{7}{2}^-$	0	2100	100–250	$N\bar{K}$	25
					and others	
$\Lambda(2110)$	$\frac{5}{2}^+$	0	$\simeq 2100$	150–250	$N\bar{K}, \Sigma\pi$	—
$\Lambda(2350)$	$\frac{9}{2}^+$	0	$\simeq 2350$	100–250	$N\bar{K}, \Sigma\pi$	
$\Sigma(1385)$	$\frac{3}{2}^-$	1	1382 ± 0.4	35 ± 1	$\Lambda\pi$	90
					$\Sigma\pi$	10
$\Sigma(1660)$	$\frac{1}{2}^+$	1	1580–1690	30–200	$N\bar{K}, \Sigma\pi, \Lambda\pi$	
$\Sigma(1670)$	$\frac{3}{2}^-$	1	1670	50	$\Sigma\pi, N\bar{K}, \Lambda\pi$	
$\Sigma(1750)$	$\frac{1}{2}^-$	1	1750	80	$N\bar{K}, \Lambda\pi, \Sigma\pi$	
$\Sigma(1775)$	$\frac{5}{2}^-$	1	1775	100–150	$N\bar{K}, \Lambda\pi, \Sigma\pi$	
$\Sigma(1915)$	$\frac{5}{2}^+$	1	1910	80–160	$N\bar{K}, \Lambda\pi, \Sigma\pi$	
$\Sigma(1940)$	$\frac{3}{2}^-$	1	1900–1950	150–300	$N\bar{K}, \Lambda\pi, \Sigma\pi,$	
					and others	
$\Sigma(2030)$	$\frac{7}{2}^+$	1	2030	150–200	$N\bar{K}$	30
					$\Lambda\pi$	20
					$\Sigma\pi$	7

TABLE IV Continued

Resonance	J^P	I	Mass, MeV	Width Γ (MeV)	Decay Mode	Fraction, %
Baryon resonances ($S = -2$)						
$\Xi(1530)$	$\frac{3}{2}^+$	$\frac{1}{2}$	1532 ± 0.3	9 ± 0.5	$\Xi\pi$	100
$\Xi(1820)$	$\frac{3}{2}$	$\frac{1}{2}$	1823 ± 6	$\simeq 30$	$\Lambda\bar{K}$	45
					$\Xi\pi$	small
					$\Xi(1530)\pi$	45
					$\Sigma\bar{K}$	10
$\Xi(2030)$	(?)	$\frac{1}{2}$	2024 ± 6	$\simeq 20$	$\Lambda\bar{K}$	$\simeq 20$
					$\Sigma\bar{K}$	$\simeq 80$
					and others	

[a] Adapted from "Review of particle properties" of Particle Data Group, C. G. Wohl *et al.*, *Reviews of Modern Physics* **56**, 2 (Apr. 1984).

[b] J^P, spin and parity of particle. ( ), spin-parity assignment still in some doubt.

[c] I^G, isospin and G-parity.

[d] Errors given in a bracket [thus (± 2)] refer to the last decimal place.

[e] p_{max}, momentum of each secondary in two-body decay, or maximum momentum of any secondary in three-body decay.

Solutions to Problems

CHAPTER 1

1.1 (a) Bohr radius for muon $r_0 = 137\hbar/Zm_\mu c$.
Nuclear radius $R = R_0 A^{1/3}$, where $R_0 = 1.2 \ fm$.
So $f = (R/r_0)^3 \simeq 0.27 \ A(Z/137)^3 = \underline{6.2 \times 10^{-3}}$.
Velocity of muon in S-state $v = Zc/137$
Distance traveled in time t is $l = fct(Z/137)$

(b) Capture probability is $1 - e^{-\lambda_c t} = 1 - e^{-l/\lambda}$, where $\lambda_c = \lambda - \lambda_d$ and interaction mean free path $\Lambda = l/\lambda_c t = fcZ/(137\lambda_c)$.
Inserting numerical values, $\underline{\Lambda = 26.5 \ cm.}$

(c) Ratio of mean free paths for interaction in nuclear matter is $\Lambda_{weak}/\Lambda_{strong} = (g_{strong}/g_{weak})^2$, where $g_{strong} \simeq 1$ and $\Lambda_{strong} \simeq 1.5 \ fm$.
Hence, $\underline{g_{weak} = \sqrt{\Lambda_{strong}/\Lambda_{weak}} \simeq 10^{-7}}$

1.2 (a) $E_\gamma = \frac{1}{2}E_\pi(1 + \beta \cos \theta)$ from Lorentz transformation (see Appendix A).

(b) If $J_\pi = 0$, no direction is preferred and CMS angular distribution is isotropic. So $dN/dE_\gamma = 2/\beta E_\pi \cdot dN/d(\cos \theta) = $ constant, with limits given in (a) for $\cos \theta = \pm 1$.

(c) $D = E_1/E_2 \simeq (1 + \cos \theta)/(1 - \cos \theta)$ for $\beta \to 1$. So $dN/dD = 2/(D + 1)^2$. Integration gives the results stated.

1.3 Binding of $(H_2\mu)^+$ larger than $(H_2 e)^+$. Reduced mass $\mu_H = m_\mu/(1 + m_\mu/M_H) < \mu_D = m_\mu/(1 + m_\mu/M_D)$. Internuclear distance $\simeq 3 \times 10^{-11}$ cm. $HD \to He^3 + \mu + 5.4$ MeV. [References: L. Alvarez *et al.*, *Phys. Rev.* **105**, 1127 (1957); G. Feinberg and L. Lederman, *Ann. Rev. Nucl. Science* **13**, 431 (1963).]

1.4   The difference in muon energies for neutrino masses m_1, m_2 is

$$\delta E_\mu = \frac{m_1^2 - m_2^2}{2m_\pi}.$$

The irreducible error on E_μ due to the finite pion lifetime is

$$\Delta E_\mu = \frac{\hbar}{2\tau_\pi}\left(1 - \frac{m_\mu^2}{m_\pi^2}\right).$$

Requiring $\delta E_\mu > \Delta E_\mu$, the result follows.

1.5   The range of strong interactions is $R \simeq 1$ fm, with a cross-section of order $\pi R^2 \simeq 30$ mb, and a characteristic time $R/c \simeq 3 \times 10^{-24}$ s. For a cross-section of 1 mb for the associated production reaction, the characteristic time will be 2×10^{-23} s. Hence,

$$\left(\frac{g_{\text{production}}}{g_{\text{decay}}}\right)^2 = \frac{t_{\text{weak}}}{t_{\text{strong}}} = \frac{10^{-10}}{2 \times 10^{-23}}, \quad \text{or} \quad \left(\frac{g_{\text{production}}}{g_{\text{decay}}}\right) \simeq 10^6.$$

1.6   (i): Allowed. (ii): Forbidden for free protons, allowed for nuclei when p-n binding-energy difference is sufficient. (iii): Forbidden by conservation of muon number. (iv): Forbidden by strangeness conservation.

1.7 $|\Delta e/e| > (KM^2/e^2)^{1/2} = 10^{-18}$.

CHAPTER 2

2.1   (a) 0.38. (b) 0.12. (c) 0.038.

2.2 $y_{\text{rms}} = (21/p\beta\sqrt{2})(s^{3/2}/\sqrt{3})$ radiation lengths, $p\beta$ in MeV. (a) 5.1 cm. (b) 1.2 cm.

2.3   Mean range $s = E_0/(dE/dx) \simeq 1.6$ km rock, where $E_0 = 10^3$ GeV is initial muon energy. For a constant rate of energy loss, so that $E(x) = E_0 - x(dE/dx)$, radial spread at the point where muons stop is $r_{\text{rms}} = 21s^{3/2}/E_0$ radiation lengths, where E_0 is in MeV and s in radiation lengths. With $X_0 = 25$ g cm^{-2}, one obtains $r_{\text{rms}} = 4.9$ m.

2.5   The threshold energy is obtained by requiring that the available kinetic energy in the CMS be $2m$, in order to create a pair. This means the total CMS energy must be $4m$, corresponding to three electrons and one positron mutually at rest in the CMS frame. If E, p are the energy and momentum of the incident electron in the laboratory frame, the total (4-momentum)^2 of the colliding particles is

$$(E + m)^2 - (p + 0)^2 = 2m^2 + 2mE.$$

This must be equal to the (total energy)^2 in the CMS frame, where the total momentum is zero. Therefore

$$(4m)^2 = 2m^2 + 2mE,$$

or

$$\underline{E^{\text{threshold}} = 7m.}$$

2.6 4-momentum squared is $s = -(p + p_f)^2$, where p, p_f are the 4-momenta of the incident and target protons. So if $p_f \ll M$, the nucleon mass

$$s = E_{CMS}^2 = -(p^2 + p_f^2 + 2pp_f) = -2M^2 + 2(EM - \mathbf{p} \cdot \mathbf{p}_f)$$
$$= -2M^2 + 2ME(1 - (\mathbf{p} \cdot \mathbf{p}_f)/ME),$$

where E is the total energy of the incident proton. The available kinetic energy in the CMS is $E_{CMS} - 2M$. If $\mathbf{p}_f$ and $\mathbf{p}$ are parallel (antiparallel), $E_{CMS}^2 = -2M^2 + 2ME(1 \mp |\mathbf{P}/E\|\mathbf{P}_f/M|)$.
If $\mathbf{p}_f$ and $\mathbf{p}$ are orthogonal, $E_{CMS}^2 \simeq -2M^2 + 2ME$.

2.7 The equation for the maximum transferable energy is

$$E'_{max} = 2m\beta^2\gamma^2c^2,$$

where mc^2 is the electron rest energy, and γ is the Lorentz factor of the primary particle. To identify a particle as a pion, it is necessary that $E' > (E'_{max})_K$. For kaons of 5 GeV/c, $\gamma = 10.2$, $\beta = 1$, so that $(E'_{max})_K = 104 \times 1.02 = 106$ MeV. For a pion, $(E'_{max})_\pi = 1320$ MeV.

The probability of observing such a δ-ray in liquid hydrogen is, with $\beta \simeq 1$,

$$P(> E') = 2\pi \left(\frac{e^2}{mc^2}\right)^2 \times mc^2 \times N_0 \frac{1}{E'} \left[1 - \frac{E'}{E'_{max}}\left(1 + \beta^2 \log_e \frac{E'_{max}}{E'}\right)\right]$$

per g/cm^2 traversed. Inserting the numerical values, and with $E' = 106$ MeV, $E'_{max} = 1320$ MeV, one finds

$$P(> 106 \text{ MeV}) = 1.04 \times 10^{-3} \text{ g}^{-1} \text{ cm}^2.$$

Therefore in 1 m path, the probability is $\underline{6.2 \times 10^{-3}}$.

2.8 Size of detector required to contain all energy from proton decay must be large compared with $t_{max} = 1.2$ m—see Eq. (2.22). Track length integral is $M_p c^2/E_{critical}$ radiation lengths, or 340 g cm^{-2}, taking $E_{critical} = 100$ MeV and $X_0 = 36$ g cm^{-2} in water. Using the value of 200 photons cm^{-1} in Section 2.4.6, gives a total of $N = 68,000$ photons of $\lambda = 400$–700 nm. The average number of photons recorded will be $N \times 0.15 \times 0.20 \, f = 2040 \, f$, where f is the fractional surface area of photocathode. A 10% resolution requires of order 100 photons recorded, or $f = 0.05$. (For details, see reference in Section 9.10.2).

CHAPTER 3

3.1 (a) z-component of orbital angular momentum of Λ, K^0 produced along z-axis is $xp_y - yp_x = 0$. Thus, $J_z(\Lambda) = \pm\frac{1}{2}$, the same as that of the target proton, irrespective of the value of J_Λ.

(b) One can use the same argument as for the decay angular distribution of $\Delta(1232)$ in Section 4.9. For example, for $J_\Lambda = \frac{5}{2}$, the amplitude will be (see Table III)

$$\psi(\tfrac{5}{2}, \tfrac{1}{2}) = \sqrt{\tfrac{2}{5}} \, \phi(2, 1)\alpha(\tfrac{1}{2}, -\tfrac{1}{2}) + \sqrt{\tfrac{3}{5}} \, \phi(2, 0)\alpha(\tfrac{1}{2}, \tfrac{1}{2}),$$

where $\alpha(s, s_z)$ is the spin factor for the decay proton, and $\phi(l, m) = Y_l^m(\theta)$ is the angular momentum wavefunction of the pion. From Table II the angular distribution is, therefore,

$$|\psi|^2 = \tfrac{2}{5}|Y_2^1(\theta)|^2 + \tfrac{3}{5}|Y_2^0(\theta)|^2 = 5\cos^4\theta - 2\cos^2\theta + 1.$$

(c) Same as (a) or (b). S-state capture ensures $J_z(\Sigma) = \pm\tfrac{1}{2}$.

3.2 Let $\mathbf{k}$, ε be the propagation and polarization vectors of the photon, $\boldsymbol{\sigma}$ the spin vector of the Λ, in the decay $\Sigma^0 \to \Lambda + \gamma$. If parity

$$P_\Sigma = P_\Lambda, \ P_\gamma = +1; \text{ one gets an } M1 \text{ (or } E2) \text{ photon.}$$

$$P_\Sigma = -P_\Lambda, \ P_\gamma = -1; \text{ one gets an } E1 \text{ (or } M2) \text{ photon.}$$

Simplest linear matrix elements from combining $\mathbf{k}$, ε, $\boldsymbol{\sigma}$ are

$$M_+ = \boldsymbol{\sigma}\cdot(\mathbf{k}_\Lambda\varepsilon) \qquad P_\gamma = +1 \text{ (scalar)}$$

$$M_- = \boldsymbol{\sigma}\cdot\varepsilon \qquad P_\gamma = -1 \text{ (pseudoscalar)}$$

Invariant mass of e^+ and e^- from Dalitz decay is

$$M_{e^+e^-} = \sqrt{(E_{e^+} + E_{e^-})^2 - \mathbf{k}^2} \simeq \sqrt{(M_\Sigma - M_\Lambda)^2 - \mathbf{k}^2},$$

where the Λ recoil is neglected. $|M_-|^2$ is independent of $\mathbf{k}$, whereas $|M_+|^2 \propto \mathbf{k}^2$ and therefore small (large) for large (small) $M_{e^+e^-}$.

3.3 Let $\mathbf{k}_1, \mathbf{k}_2, \mathbf{k}_3$ be the CMS momenta of the three secondary particles. Since the parent meson M has $J_M = 0$, the matrix element can only be a function of $\mathbf{k}_1, \mathbf{k}_2$, and $\mathbf{k}_3$, and appropriate linear combinations of these vectors are

(a) $\mathbf{k}_1 \cdot (\mathbf{k}_{2\Lambda} \mathbf{k}_3)$ pseudoscalar

(b) $\mathbf{k}_1 \cdot (\mathbf{k}_2 - \mathbf{k}_3)$ scalar.

Since $P_M = +1$, and the three secondary mesons have intrinsic parity $(-1)^3 = -1$, we need the pseudoscalar combination (a). But since $\mathbf{k}_1 + \mathbf{k}_2 + \mathbf{k}_3 = 0$, then $\mathbf{k}_1 \cdot \mathbf{k}_{2\Lambda}\mathbf{k}_3 = -\mathbf{k}_1 \cdot \mathbf{k}_{2\Lambda}(\mathbf{k}_1 + \mathbf{k}_2) = -\mathbf{k}_1 \cdot \mathbf{k}_{2\Lambda}\mathbf{k}_1 = 0$, so the matrix element vanishes.

3.4 Two Λ's are in ^1S state if $P(\Xi) = +1$, ^2P state if $P(\Xi) = -1$.

3.5 If l is the orbital angular momentum and the spins of all the particles involved are zero, it follows that, equating parities in initial and final states, $(-1)^l P_K = (-1)^l P_\pi$ and thus $P_K = P_\pi = -1$.

3.6 Since $J_D = 1$ and S-state capture is involved, the reaction would involve $J = 1$ in initial and final states. Since the Q-value is only 0.5 MeV, the final state $nn\pi^0$ must be an S-state. Then the two neutrons must be in a triplet spin state, forbidden by the Pauli Principle.

3.7 Refer to Section 7.14.

3.8 (a) J_K is even by Bose symmetry. (b) None, parity is not conserved in weak interactions.

CHAPTER 4

4.1 As for $\pi^- p \to n\pi^0$, $\pi^- p \to p\pi^-$, $\pi^+ p \to \pi^+ p$.

4.2 1:2.

4.3 2:1.

4.4 (a) No: $I = 0$ or 2 only, by Bose symmetry.
(b) Yes.
(c) No: $I = 2$ since $I_3 = 2$.
(d) No: $I = 0$ or 2 only.
(e) Yes.

4.5 (a) $I = 0, 1, 2$, or 3. (b) $I = 1$ or 3.

4.6 $I = 0$ or 1; $\sigma_a/\sigma_b = 1$ if $I = 0$ only, and zero if $I = 1$ only.

4.7 Ratio $= 1$.

4.8 ρ-ω interference in $\pi^+\pi^-$ mode, with amplitude α, typical of G-violating electromagnetic interactions. A narrow dip (or peak) will occur in $\pi^+\pi^-$ mass spectrum in ω-region.

4.9 The $p\bar{p}$ system has C-parity $(-1)^{l+S}$, where l is the relative orbital angular momentum, and S is the total spin (0 or 1).

The space parity of the $p\bar{p}$ system is $P = (-1)^{l+1}$, since particle and antiparticle have opposite intrinsic parity. Thus, the initial state has

$$(CP)_{p\bar{p}} = (-1)^{2l+S+1} = (-1)^{S+1} \qquad \text{for all } l.$$

Now let the total angular momentum of the two K^0's be J, where

$$|l + S| \geq J \geq |l - S|.$$

Measured in the K^0 rest frame, the K_1^0 and K_2^0 mesons have CP-eigenvalues $+1$ and -1 respectively. Thus, the final state has

$$CP = \begin{cases} (\pm 1)(\pm 1)(-1)^J = (-1)^J & \text{for } 2K_1^0 \text{ or } 2K_2^0, \\ (+1)(-1)(-1)^J = (-1)^{J+1} & \text{for } K_1^0 K_2^0, \end{cases}$$

$l = 0$: For annihilation from an S-state, $J = S$, so that the initial state has

$$(CP)_{p\bar{p}} = (-1)^{J+1}, \qquad J = 0, 1.$$

Thus, the $K_1^0 K_2^0$ final state is allowed, and $2K_1^0$ or $2K_2^0$ is forbidden.

$l = 1$: For the triplet state $(S = 1)$, $J = 0, 1$, or 2. $(CP)_{p\bar{p}}$ is even, so that $J = 0, 2$ allows $2K_1^0$ or $2K_2^0$ in the final state, while $J = 1$ gives $K_1^0 K_2^0$ only. For the singlet state $(S = 0)$, $J = 1$, $(CP)_{p\bar{p}}$ is odd, and only the states $2K_1^0$ or K_2^0 are allowed.

Experimentally, it is found that for annihilation at rest, only $K_1^0 K_2^0$ is observed, proving that annihilation takes place from an $l = 0$ state. On the contrary, annihilation of antiprotons in flight takes place from p-states also, and the modes $K_1^0 K_2^0$, $2K_1^0$, and $2K_2^0$ all appear.

4.11 If we let E, T represent total kinetic energies of the pions respectively, and Q the total kinetic energy available in the decay, it is readily verified that the conditions are

$$(E_1 E_2 E_3) = \text{maximum} \qquad \text{when} \quad T_1 = T_2 = T_3 = Q/3,$$

$$(E_1 E_2 E_3) = \text{minimum} \qquad \text{when} \quad T_1 = 0, \quad T_2 = T_3 = Q/2.$$

With $x = Q/m$, where m is the pion mass, then

$$(E_1 E_2 E_3)_{\text{max}} = m^3 \left[1 + x + \frac{x^2}{3} + \frac{x^3}{27} \right],$$

$$(E_1 E_2 E_3)_{\text{min}} = m^3 \left[1 + x + \frac{x^2}{4} \right],$$

so that

$$\varepsilon = \frac{(E_1 E_2 E_3)_{\text{max}} - (E_1 E_2 E_3)_{\text{min}}}{(E_1 E_2 E_3)_{\text{max}}} \simeq \frac{x^2}{12(1 + x)}, \qquad \text{where} \quad x \simeq \frac{75}{140} \simeq \frac{1}{2}.$$

Thus, $\varepsilon \simeq 0.014$.

4.12 $2\sqrt{E_1^2 - m^2} \sqrt{E_2^2 - m^2} = \pm [M^2 + 2E_1 E_2 - 2M(E_1 + E_2) + m^2]$.

4.14 Referring to Appendix A and writing the transverse momentum component $p_T = \sqrt{p_y^2 + p_z^2} = p_T'$, we see that the lab system angle of emission, θ, is given by

$$\tan \theta = \frac{p_T'}{p_x} = \frac{\sin \theta'}{\gamma^*(\cos \theta' + \beta^*/\beta')},$$

where θ' and β' are the angle and velocity of a particle in the CMS, and β^* is the CMS velocity in the LS, with $\gamma^* = 1/\sqrt{1 - \beta^{*2}}$. Differentiating and using the condition for a maximum, $\partial(\tan \theta)/\partial \theta' = 0$ gives

$$\gamma^* \tan \theta_{\text{max}} = \beta'/(\beta^{*2} - \beta'^2), \qquad \text{(for } \beta' < \beta^* \text{ only!)}.$$

The total CMS energy, squared, is

$$s = (E_\pi + M)^2 - \mathbf{p}_\pi^2 = m_\pi^2 + M^2 + 2ME_\pi,$$

where E_π, $\mathbf{p}_\pi$ are the energy and momentum of the pion, M the nucleon mass. By energy conservation,

$$\gamma^* \sqrt{s} = E_\pi + M \quad \text{or} \quad \gamma^* = (E_\pi + M)/\sqrt{s}.$$

If E' is the proton CMS energy, $\mathbf{p}'$ its momentum, it is easy to show that

$$E' = (s + M^2 - M_X^2)/2\sqrt{s}.$$

Inserting the values: $p_\pi \simeq E_\pi = 12 \text{ GeV}$, $M_X = 2.4 \text{ GeV}$, and with $M = 0.938 \text{ GeV}$, $m_\pi = 0.14 \text{ GeV}$, we find

$$s = 23.411 \text{ GeV}^2, \qquad \gamma^* = 2.674, \qquad \beta^* = 0.9274, \qquad E' = 1.9147,$$
$$p' = 1.6693, \qquad \beta' = 0.8719,$$

and hence

$$\tan \theta_{max} = 1.0317, \underline{\theta_{max} = 0.801 \text{ rdn.}}$$

For the LS momentum of the proton, the Lorentz transformations give $p_x = 7.8549$ GeV, $p_T = 1.30861$ GeV, and hence $\underline{p = 7.962 \text{ GeV.}}$

The 4-momentum transfer to the proton, if it is initially at rest, is readily shown to be $q^2 = 2MT$, where T is its final kinetic energy (see Eq. 6.17(b)). q^2 is maximum when T is maximum, that is, when the proton is projected forward ($\theta' = 0$). The total energy is given by

$$E = \gamma^*(E' + \beta^* p') = 9.2595 \text{ GeV,}$$

giving

$$\underline{q^2_{max} = 15.61 \text{ GeV}^2 \text{ and } \theta = 0.}$$

4.15 The Lorentz-invariant phase-space factor (Appendix A) for a two-body final state is proportional to

$$p^2 \frac{dp}{dE} \times \frac{1}{E_1 E_2},$$

where E_1 and E_2 are the total energies of the particles and p their momentum, in the CMS. The total energy $E = E_1 + E_2$. Writing

$$E = \sqrt{p^2 + m_1^2} + \sqrt{p^2 + m_2^2},$$

we find that

$$\frac{dp}{dE} = \frac{E_1 E_2}{p(E_1 + E_2)}.$$

Thus, the factor

$$\frac{p^2 \, dp}{E_1 E_2 \, dE} = \frac{p}{E}$$

enters into the width Γ.

The treatment here refers to an S-wave resonance. For one decaying to two (spinless) particles of orbital angular momentum l, it is necessary also to include a centrifugal barrier factor, familiar from reaction theory in nuclear physics (see, for example, J. M. Blatt and V. F. Weisskopf, *Theoretical Nuclear Physics*, John Wiley, 1952, pp. 320 *et seq.* and Appendix A). This factor is associated with the behavior of partial-wave amplitudes near the origin (in the interaction region), which have a radial dependence of approximately $(kr)^l$ when $kr < l$. The width Γ then includes a factor $(pR_0)^{2l}$, where R_0 is some (unknown) range parameter (of order 1 fm). Then it is plausible to write

$$\frac{\Gamma(E)}{\Gamma(E_0)} = \left(\frac{p}{p_0}\right)^{2l+1} \times \left(\frac{E_0}{E}\right),$$

where E_0 and p_0 refer to values at the resonance peak. For decay into particles with spin, the formula is more complicated [see, for example, J. D. Jackson, *Neuvo Cimento* **34**, 1644 (1964)].

4.16 $y = \frac{1}{2}\ln\left(\frac{E + p_L}{E - p_L}\right) = \ln\frac{(E + p_L)}{\sqrt{p_T^2 + m^2}} \simeq \ln((E + p_L)/p_T) \simeq \ln(2p_L/p_T) = \ln(2 \cot \theta)$,

where m has been neglected compared with p_T and the secondary is assumed to be highly relativistic, so that $p_L \gg p_T$ and $p_L \simeq E$. Also, the angular transformations give (see (4.14))

$$\gamma^* \tan \theta = \frac{\sin \theta^*}{(\cos \theta^* + \beta/\beta^*)} \simeq \frac{\sin \theta^*}{(\cos \theta^* + 1)} = \tan(\theta^*/2).$$

Hence

$$\ln(2 \cot \theta) = \ln\left(2\gamma \cot \frac{\theta^*}{2}\right) = \ln \cot \frac{\theta^*}{2} + \ln 2\gamma.$$

4.17 (i) Yes, for strong interaction.
(ii) No, because of Bose symmetry.
(iii) No, because of C symmetry. $C_\rho = -1$ since $\rho^0 \to e^+e^-$. $C_\eta = C_\pi = +1$, since both decay to two gamma rays.
(iv) Yes, for electromagnetic interaction.

CHAPTER 5

5.1 $\mu_p = 3, \mu_n = -2, \mu_{\Xi^0} - \mu_{\Xi^-} = -1, \mu_{\Sigma^+} - \mu_{\Sigma^-} = 4$

5.2 Ω^- has $S = -3, I = 0, J^P = \frac{3}{2}^+, B = 1, M_\Omega = 1672$ MeV

$M_\Omega - M_p = 734$ MeV $< 3M_K = 1485$ MeV; $\qquad M_\Omega - M_\Xi = 357$ MeV $< M_K$;

$M_\Omega - M_\Lambda = 554$ MeV $< 2M_K = 990$ MeV.

Therefore the strong, strangeness-conserving decays

$$\Omega^- \to p + 2K^- + \bar{K}^0, \qquad \Omega^- \to \Lambda + K^- + \bar{K}^0, \qquad \Omega^- \to \Xi^0 + K^-$$

are forbidden by energy conservation. Thus, weak decay ($\Delta S = 1$) is the only possibility ($\Omega \to \Xi^0 \pi^-, \Xi^- \pi^0, \Lambda K^-$—see Table IV in the appendix).

5.3 $\Gamma(V \to e^+e^-) = 16\pi\alpha^2 |\Sigma Q|^2 \ |\psi(0)|^2/M_V^2$—see Eq. (5.27). $|\psi(0)|^2 = 1/(\text{volume of hadron}) = 3/4\pi R_0^3$, where $R_0 \simeq 1$ fm. So

$$\Gamma = 12\alpha^2 M_V \left(\frac{\hbar c}{M_V R_0}\right)^3 |\Sigma Q|^2,$$

where $\hbar c = 0.2$ GeV fm. With $R_0 = 1$ fm we obtain for the vector mesons

ω: $M_V = 0.78$ GeV,	$\|\Sigma Q\|^2 = \frac{1}{18}$,	$\Gamma(e^+e^-) = 0.46$ keV
ρ: $M_V = 0.77$ GeV,	$\|\Sigma Q\|^2 = \frac{1}{2}$,	$\Gamma(e^+e^-) = 4.3$ keV
ϕ: $M_V = 1.02$ GeV,	$\|\Sigma Q\|^2 = \frac{1}{9}$.	$\Gamma(e^+e^-) = 0.55$ keV

5.4 Denote the *uu*, *dd*, and *ud* cross-sections for *u*- and *d*-quark or antiquark scattering by $\sigma(nn)$. Denote *su* or *sd* cross-sections by $\sigma(ns)$ and $\bar{s}u$ or $\bar{s}d$ cross-sections by $\sigma(n\bar{s})$. Then for all three equations, both LHS and RHS give $6\sigma(nn) + 3\sigma(ns)$.

CHAPTER 6

6.1 $\rho(R) = \rho_0 \exp(-MR)$, where $M = M_V$. From (6.19),

$$F(q^2) = \int_0^\infty \rho(R)\left(\frac{\sin qR}{qR}\right)4\pi R^2\, dR$$

$$= \left(\frac{2\pi\rho_0}{iq}\right)\int\{\exp[-(M-iq)R] - \exp[-(M+iq)R]\}R\, dR$$

$$= \left(\frac{2\pi\rho_0}{iq}\right)\left[\frac{1}{(M-iq)^2} - \frac{1}{(M+iq)^2}\right] = \frac{8\pi M\rho_0}{(q^2+M^2)^2},$$

corresponding to the dipole formula. The rms radius of the charge distribution is given by

$$R^2 = \frac{\int\rho_0 \exp(-MR)\cdot R^2 4\pi R^2\, dR}{\int\rho_0 \exp(-MR)\cdot 4\pi R^2\, dR} = \frac{12}{M^2},$$

so that

$$\underline{R_{rms}} = \frac{\sqrt{12}\hbar c}{M_V} = \frac{\sqrt{12}\times 0.2}{0.84} = \underline{0.8\text{ fm.}}$$

6.2 The process $e^+e^- \to$ virtual $\gamma \to \pi^+\pi^-$ is equivalent to scattering of a pion by a point charge via photon exchange, except that q^2 is negative, i.e., the scattering measures the form factor (charge distribution) of the pion for timelike q^2. Now, for a Yukawa-type charge distribution of the form

$$\rho(R) = \rho_0 \exp(-MR)/R,$$

it is a matter of straightforward integration of (6.19) to show that the form factor has the dependence

$$F(q^2) = \frac{M^2}{(q^2+M^2)}.$$

The process $e^+e^- \to \pi^+\pi^-$ is dominated by the ρ resonance, i.e., by $e^+e^- \to$ virtual photon $\to \rho \to \pi^+\pi^-$. The q^2 dependence is determined by the ρ-propagator, of the form $1/(M_\rho^2 + q^2)$, as above. So, the charge distribution of the pion will be of the Yukawa form, with rms radius given by

$$\overline{R^2} = \frac{\int\rho(R)R^2\cdot 4\pi R^2\, dR}{\int\rho(R)\cdot 4\pi R^2\, dR} = \frac{6}{M_\rho^2}$$

or

$$\underline{R_{rms}} = \frac{\sqrt{6}}{M_\rho}\left(=\frac{\sqrt{6}\hbar c}{M_\rho}\right) = \underline{0.64\text{ fm.}}$$

6.3   A current of 10 mA of relativistic particles in a ring of radius 10 m corresponds to a circulating charge $q = (2\pi r/c)i$, or, inserting appropriate numbers, $N = 1.3 \times 10^{10}$ circulating electrons or positrons. If the cross-sectional area of the beam is A, the particle density transverse to the beam will be N/A. The reaction rate will therefore be

$$R = \sigma \left(\frac{N}{A}\right)^2 \times Afn,$$

where f is the revolution frequency, and the bunches meet n times per revolution. With $n = 2$, $f = c/2\pi r$, $A = 0.1$ cm^2, and $\sigma = 1.5 \times 10^{-30}$ cm^2, this formula gives $\underline{L = R/\sigma = 1.6 \times 10^{28} \text{ cm}^{-2} \text{ s}^{-1}}$.

Referring to the Breit-Wigner formula, we obtain for the peak cross-section $\sigma = 3\pi \lambdabar^2 \alpha^2$, assuming that the branching ratio to $e^+ e^-$ is α^2 and the ω has $J = 1$. $\lambdabar = 2/M_\omega$ is the de Broglie wavelength of the e^+ and e^- at resonance. So

$$\sigma = 12\pi\alpha^2(\hbar c/M_\omega)^2 = 1.32\mu b.$$

From the value of L the reaction rate is found to be $\underline{77/h}$.

6.4   The 4-momentum transfer squared is

$$q^2 = 2E_0 E(1 - \cos\theta),$$

where E_0 and E are the incident and scattered electron energies, θ the angular deflection, and the electron mass is neglected. Let W, E' and p' be the mass, energy, and 4-momentum of the recoiling hadronic state, M the nucleon mass. Then also

$$q^2 = (p - p')^2 = 2ME' - M^2 - W^2$$

and

$$E' = E_0 + M - E.$$

Hence

$$W^2 = 2M(E_0 - E) + M^2 - q^2.$$

Substituting the numbers given, we find

$$q^2 = 2.127 \text{ GeV}^2 \qquad W = 2.09 \text{ GeV}.$$

6.5   (a) For pointlike spinless protons, the cross-section is, from (6.24)

$$\frac{d\sigma}{d\Omega} = \left(\frac{\hbar c}{p_0}\right)^2 \frac{\alpha^2[\cos^2(\theta/2) + (q^2/2M^2)\sin^2(\theta/2)]}{4\sin^4(\theta/2)[1 + (2p_0/M)\sin^2(\theta/2)]},$$

where $p_0 = 15$ GeV/c is incident momentum, $\theta = 0.1$ rdn is electron deflection, $d\Omega = 10^{-4}$ sr, $M =$ proton mass, $\hbar c = 0.2$ GeV fm. Then

$$q^2 = 4p_0^2 \sin^2(\theta/2) = 2.248 \text{ GeV}^2, \qquad \cos^2(\theta/2) = 0.9975$$

and

$$d\sigma = 3.4 \times 10^{-34} \text{ cm}^2.$$

For a target of density 0.06 and length 1 m, the number of target protons per square centimetre of cross-section is $N = 3.6 \times 10^{24}$ and the number of electrons scattered per second is

$$\mathbf{R}_1 = d\sigma\phi N = \underline{1.23 \times 10^5 \text{ s}^{-1}}.$$

(b) For the form factors of Eq. (6.25), the above number must be multiplied by the factor

$$\frac{1}{(1 + q^2/M_V^2)^4} \cdot \left[\frac{1}{\cos^2(\theta/2)} \right] \cdot \left[\frac{(1 + \mu^2 q^2/4M^2)}{(1 + q^2/4M^2)} + \left(\frac{q^2}{4M^2} \right) 2\mu^2 \tan^2\left(\frac{\theta}{2} \right) \right],$$

where $\mu = 2.79$ is the proton magnetic moment in nuclear magnetons, $M_V = 0.84$ GeV. The first term has magnitude 3.26×10^{-3}, the last has value 3.669, giving

$$\mathbf{R}_2 = \underline{1.47 \times 10^3 \text{ s}^{-1}}.$$

6.6 (a) For muonium, the transition frequency expected is

$$f = \frac{1420(1 + m_e/M_p)M_p}{(1 + m_e/m_\mu)\mu_p M_\mu},$$

taking into account reduced-mass effects (see (6.1)) and the fact that the magnetic moment of the muon is $M_p/(\mu_p M_\mu)$ times that of the proton, where $\mu_p = 2.7925$ is the proton moment. Inserting the numbers we find

$$f(\mu^+ e^-) = \underline{4497.2 \text{ MHz}} \text{ (cf. 4463.3 MHz observed)}.$$

(b) For positronium, including the $\frac{7}{16}$ annihilation factor, we get

$$f = \frac{1420(1 + m_e/M_p)}{(1 + m_e/m_e)} \cdot \frac{7}{16} \cdot \frac{1}{\mu_p} \cdot \frac{M_p}{m_e}$$

$$= \underline{204394 \text{ MHz}} \text{ (cf. 203387 MHz observed)}$$

The reason for the discrepancy is neglect of higher-order contributions $(\simeq \alpha)$.

CHAPTER 7

7.1
$$\text{Cl}^{34} \rightarrow \text{S}^{34} + e^+ + v_e \qquad \text{is a } 0^+ \rightarrow 0^+ \text{ transition}$$

$$\pi^+ \rightarrow \pi^0 + e^+ + v_e \qquad \text{is a } 0^- \rightarrow 0^- \text{ transition}$$

so both have the same (Fermi) matrix element. Using Sargent's Rule, the decay rate α (endpoint energy)^5. The endpoint energies are 4.6 MeV and 4.5 MeV, respectively, i.e., the ratio of widths is near unity. Hence pion branching ratio is $\tau_\pi/\tau_{Cl} = 10^{-8}$.

7.2 (a) Let $E_v =$ lab energy of neutrino, $\bar{E}$ that in rest frame of decaying meson, θ the lab angle of neutrino relative to beam direction. Let M, E, β be the mass, energy, and velocity of mesons, $\gamma = 1/\sqrt{(1 - \beta^2)}$. Then

$$\bar{E} = \gamma E_v(1 - \beta \cos \theta) = \frac{E_v}{2\gamma}(1 + \gamma^2\theta^2) \qquad \text{for } \gamma \gg 1, \theta \ll 1.$$

So

$$E_v = 2\gamma \bar{E}/(1 + \gamma^2\theta^2) = E_v(\text{max})/(1 + \gamma^2\theta^2),$$

where a simple calculation gives

$$\bar{E} = (M^2 - m_\mu^2)/2M$$

and therefore

(b) $$E_v(\text{max}) = E(1 - m_\mu^2/M^2) = 0.42E_\pi = 84 \text{ GeV}$$
$$= 0.96E_K = 192 \text{ GeV}$$

and

$$E(\text{min}) \simeq 0.$$

(c) Inserting $\theta = 2m/400m = 0.005$, $\gamma_K = 405$ gives $E_v = 34$ GeV.
For pion parents, the corresponding number is 1.6 GeV only, that is, nearly all neutrinos from pion decay traverse the detector.

(d) The proportion of pions decaying in $l = 300$m is $l/\gamma c\tau_\pi = 0.0242$. Since nearly all neutrinos traverse the detector in this case, the number is

$$10^{10} \times 0.024 = 2.4 \times 10^8.$$

(e) The v_π spectrum is flat from $\simeq 0$ to 84 GeV, with mean energy 42 GeV and mean cross-section $\bar{\sigma} = 42 \times 0.6 \times 10^{-38} = 2.52 \times 10^{-37}$ cm^2 nucleon^{-1}. Interaction mean free path is $1/N\sigma = 6.6 \times 10^{12}$ g cm^{-2}, where $N =$ Avogadro's number. Depth of neutrino detector along axis = 800 g cm^{-2}. These numbers give a mean number of interactions of pion neutrinos of 0.029 per burst.

(f) To range out the muons.

7.3 The endpoint electron energies are $E_n = 1.29$ MeV and $E_\mu = 53$ MeV. So on the basis of phase space the Sargent Rule gives

$$\frac{G_n^2}{G_\mu^2} = \left(\frac{\tau_\mu}{\tau_n}\right)\left(\frac{E_\mu}{E_n}\right)^5 = 0.27,$$

so in order of magnitude, $G_n \sim G_\mu$.

7.4 The endpoint energies are $E_\Sigma = 79$ MeV, $E_n = 1.29$ MeV. The neutron and Σ^- lifetimes are 900 s and 1.5×10^{-10} s, respectively, giving for the branching ratio

$$\left(\frac{\tau_\Sigma}{\tau_n}\right)\left(\frac{81}{1.29}\right)^5 = \underline{1.4 \times 10^{-4}}.$$

7.5 (a) 2:1. (b) 2:1.

7.6 (a) 1:2. (b) 2:1.

7.7   Adding a "spurion" of I, $I_3 = \frac{1}{2}, -\frac{1}{2}$ we get for the amplitudes

$$\Sigma^+ + S \to \sqrt{\tfrac{1}{3}}A_3 + \sqrt{\tfrac{2}{3}}A_1$$

$$\Sigma^- + S \to A_3,$$

where A_1, A_3 are the $I = \frac{1}{2}$ and $I = \frac{3}{2}$ contributions. Also the nucleon-pion amplitudes are

$$n\pi^+ = \sqrt{\tfrac{1}{3}}A_3 + \sqrt{\tfrac{2}{3}}A_1,$$

$$n\pi^- = A_3,$$

$$p\pi^0 = \sqrt{\tfrac{2}{3}}A_3 - \sqrt{\tfrac{1}{3}}A_1.$$

Hence,

$$a_+ = \langle \Sigma^+ | n\pi^+ \rangle = \tfrac{1}{3}A_3^2 + \tfrac{2}{3}A_1^2,$$

$$a_- = \langle \Sigma^- | n\pi^- \rangle = A_3^2,$$

$$a_0 = \langle \Sigma^+ | p\pi^0 \rangle = \frac{\sqrt{2}}{3}A_3^2 - \frac{\sqrt{2}}{3}A_1^2,$$

yielding

$$a_+ + \sqrt{2}a_0 = a_-.$$

7.8   The Sargent rule gives for the partial width for $\Delta \to pev$,

$$\Gamma_{(\Delta \to pev)} = \Gamma_n (E_\Delta / E_n)^5 = 4.5 \times 10^{-13} \text{ MeV},$$

where $E_\Delta = 300$ MeV, $E_n = 1.3$ MeV, and $\Gamma_n = \hbar/\tau_n = 7 \times 10^{-25}$ MeV. These numbers give a branching ratio of 3×10^{-15}.

7.9   See Section 7.14 and Eqs. (7.83) and (7.84).

7.10 The reaction rate is given by

$$R = \sigma\phi N,$$

where σ is the cross-section per nucleus for neutrino absorption, N is the total number of nuclei in the detector, and ϕ is the neutrino flux in s^{-1} cm^{-2}.

   We take 164 as the molecular weight of C_2Cl_4, and the total mass of liquid as 6×10^8 g; the number of ^{37}CL nuclei is $N = 2.2 \times 10^{30}$. The solar heat flux is 2 cal cm^{-2} min^{-1}, or 8.8×10^{11} MeV cm^{-2} sec^{-1}. Of this, 10% appears as neutrinos, of mean energy 1 MeV, and 1% of the neutrinos are supposed sufficiently energetic to produce a reaction, so that $\phi = 8.8 \times 10^8$ cm^{-2} s^{-1}. Thus,

$$R = \sigma\phi N = (10^{-45})(8.8 \times 10^8)(2.2 \times 10^{30}) = 1.9 \times 10^{-6} \text{ s}^{-1}$$
$$= \underline{0.17 \text{ day}^{-1}}.$$

7.11 We apply the $\Delta I = 1$ rule by combining the baryon ($I = \frac{1}{2}$) with a "spurion" of $I = 1$, to give a final hadronic state of $I = \frac{3}{2}$ and $I_3 = \frac{3}{2}$ or $\frac{1}{2}$.

   Referring to Table III of Clebsch-Gordan coefficients, and using the $I = 1$ and $I = \frac{1}{2}$ combination, we may write for reaction (i)

$$\phi(1, 1)\phi(\tfrac{1}{2}, \tfrac{1}{2}) = \psi(\tfrac{3}{2}, \tfrac{3}{2}),$$
$$\underset{\text{spurion}}{\uparrow} \quad \underset{\text{nucleon}}{\uparrow}$$

and for (ii)

$$\phi(1, 1)\phi(\tfrac{1}{2}, -\tfrac{1}{2}) = \sqrt{\tfrac{1}{3}}\psi(\tfrac{3}{2}, \tfrac{1}{2}) + \sqrt{\tfrac{2}{3}}\psi(\tfrac{1}{2}, \tfrac{1}{2}).$$

If the pion-nucleon system is in a pure $I = \tfrac{3}{2}$ state, the cross-section ratio obtained by squaring the above amplitudes is $\sigma_{(i)}/\sigma_{(ii)} = 3/1$. For a $\Delta I = 2$ transition, we use $I = 2$ and $I = \tfrac{1}{2}$ entry of the table, and find for reaction (i)

$$\phi(2, 1)\phi(\tfrac{1}{2}, \tfrac{1}{2}) = -\sqrt{\tfrac{1}{5}}\psi(\tfrac{3}{2}, \tfrac{3}{2}) + \sqrt{\tfrac{4}{5}}\psi(\tfrac{5}{2}, \tfrac{3}{2}),$$

and for (ii)

$$\phi(2, 1)\phi(\tfrac{1}{2}, -\tfrac{1}{2}) = \sqrt{\tfrac{3}{5}}\psi(\tfrac{3}{2}, \tfrac{1}{2}) + \sqrt{\tfrac{2}{5}}\psi(\tfrac{5}{2}, \tfrac{1}{2}).$$

For a final state of $I = \tfrac{3}{2}$ only, the ratio is then $\sigma_{(i)}/\sigma_{(ii)} = \tfrac{1}{3}$.

7.12 We assume that the three pions are in a relative S-state. Then, by Bose symmetry, any pair must be in a symmetric isospin state, i.e., $I = 0$ or $I = 2$. Call the amplitudes for an $I = 2$ and an $I = 0$ dipion state A and B, respectively. From the $\Delta I = \tfrac{1}{2}$ rule, we add a "spurion" of $\Delta I = \tfrac{1}{2}$ to the kaon, of $\Delta I = \tfrac{1}{2}$, to form states of $I = 0$ or 1. $I = 0$ is forbidden for the three-pion state, which is formed from a dipion of $I = 0$ or 2 and a third pion of $I = 1$. So we consider a three-pion state of $I = 1$, obtained by adding together $I = 1$ with $I = 0$ or 2. Referring to Table III of coefficients we find, in self-evident notation,

Charged kaon:

$$\psi(1, 1) = A[\sqrt{\tfrac{3}{5}}\phi(2, 2)\phi(1, -1) - \sqrt{\tfrac{3}{10}}\phi(2, 1)\phi(1, 0) + \sqrt{\tfrac{1}{10}}\phi(2, 0)\phi(1, 1)]$$
$$+ B[\phi(0, 0)\phi(1, 1)], \tag{a}$$

Neutral kaon:

$$\psi(1, 0) = A[\sqrt{\tfrac{3}{10}}\phi(2, 1)\phi(1, -1) - \sqrt{\tfrac{2}{5}}\phi(2, 0)\phi(1, 0) + \sqrt{\tfrac{3}{10}}\phi(2, -1)\phi(1, 1)]$$
$$+ B[\phi(0, 0)\phi(1, 0)]. \tag{b}$$

The next step is to express the various pion combinations in terms of the isospin functions ϕ. The three-pion wave function must be completely symmetric under pion label interchange, as required for identical bosons. So we write the $\pi^+\pi^+\pi^-$ combination as

$$(+ + -) = \sqrt{\tfrac{1}{6}}(\pi_1^+\pi_2^+\pi_3^- + \pi_2^+\pi_1^+\pi_3^- + \pi_3^-\pi_2^+\pi_1^+$$
$$+ \pi_3^-\pi_1^+\pi_2^+ + \pi_2^+\pi_3^-\pi_1^+ + \pi_1^+\pi_3^-\pi_2^+), \tag{c}$$

the factor $\sqrt{\tfrac{1}{6}}$ being to normalize the amplitude to unity. Referring to 1×1 entry in Table III, treating the first two pions as the "pair" gives

$$(\pi^+\pi^+\pi^-) = \sqrt{\tfrac{3}{5}}A; \qquad (\pi^+\pi^-\pi^+) = (\pi^-\pi^+\pi^+) = \sqrt{\tfrac{1}{60}}A + \sqrt{\tfrac{1}{3}}B.$$

The second result, for example, follows from the fact that the coefficient for combining $I = 1, I_3 = +1$ and $I = 1, I_3 = -1$ to give $I = 2, I_3 = 0$ is $\sqrt{\tfrac{1}{6}}$; and to give $I = 0$, $I_3 = 0$ the coefficient is $\sqrt{\frac{1}{3}}$. These factors are then multiplied into the appropriate terms in (a), in order to find $\langle \psi(1, 1) | \pi^+ \pi^+ \pi^- \rangle$, etc. Adding together all terms in (c) gives us

$$\langle \psi(1, 1) | + + - \rangle = 2\sqrt{\tfrac{2}{3}}C, \qquad \text{where} \quad C = \sqrt{\tfrac{4}{15}}A + \sqrt{\tfrac{1}{3}}B.$$

Similarly, we find for the other charge combinations

$$\langle \psi(1, 1) | + 00 \rangle = -\sqrt{\tfrac{2}{3}}C,$$

$$\langle \psi(1, 0) | 000 \rangle = -C,$$

$$\langle \psi(1, 0) | + - 0 \rangle = \sqrt{\tfrac{2}{3}}C.$$

Squaring these amplitudes, we obtain the ratios of the decay rates

$$\Gamma(K_L \to 3\pi^0) = C^2 = \tfrac{3}{2}\Gamma(K_L \to \pi^+ \pi^- \pi^0),$$
$$\Gamma(K^+ \to \pi^+ \pi^+ \pi^-) = \tfrac{8}{3}C^2 = 4\Gamma(K^+ \to \pi^+ \pi^0 \pi^0).$$

We have actually calculated the transition for $K^0 \to \pi^+ \pi^- \pi^0$. The weak conservation laws only allow half the K^0's to decay in this mode, called K_2^0 or K_L, which has CP-eigenvalue -1. The other half is called K_1^0 or K_s, has $CP = +1$, and does not decay to three pions. Thus,

$$\langle K^0 | T | \pi^+ \pi^- \pi^0 \rangle = \frac{1}{\sqrt{2}} \langle K_L | T | \pi^+ \pi^- \pi^0 \rangle.$$

Using this result, we obtain

$$\Gamma(K_L \to \pi^+ \pi^- \pi^0) = 2\Gamma(K_0 \to \pi^+ \pi^- \pi^0) = 2\Gamma(K^+ \to \pi^+ \pi^0 \pi^0).$$

Experimentally, there is a small deviation from this prediction, indicating that $\Delta I = \tfrac{3}{2}$ as well as $\Delta I = \tfrac{1}{2}$ transitions are involved.

For a more complete discussion of the $K \to 3\pi$ decay modes, the reader is referred to G. Källén, *Elementary Particle Physics*, Addison-Wesley, 1964, Ch. 16; and for a more general treatment of three-pion decays, to the classic paper by C. Zemach, *Phys. Rev.* **133**, B1201 (1964).

7.13 The first process involves Z^0-exchange only; the second, both Z^0- and $W^\pm$-exchange.

In the small angle approximation, with E_ν the neutrino energy is large compared with the electron mass m, the application of the Lorentz transformations gives

$$\theta = \sqrt{2m\left(\frac{1}{E_e} - \frac{1}{E_\nu}\right)},$$

where E_e is the electron recoil energy.

7.14 From the Sargent rule for three-body decay we get for the τ lifetime

$$\tau_\tau = \tau_\mu (E_\mu/E_\tau)^5 \cdot B = \tau_\mu (m_\mu/m_\tau)^5 B,$$

where B is the leptonic branching ratio (0.18) and the endpoint electron energies are $m_\mu/2$ and $m_\tau/2$. Inserting the values given yields the result $\tau_\tau = \underline{2.9 \times 10^{-13}}$ s. The assumption made here is universality of μ and τ weak couplings.

CHAPTER 8

8.1 From (8.12)
$$(\xi P + q)^2 = -m^2.$$

Defining
$$x = -q^2/2Pq$$

and solving the above quadratic equation gives
$$\xi = \left[\frac{-q^2}{x} \pm \sqrt{\frac{q^4}{x^2} + 4M^2(q^2 + m^2)}\right]/2M^2,$$

where M, m are the nucleon and parton masses. Expanding we obtain
$$\xi = x\left[1 - \frac{(M^2x^2 - m^2)}{q^2} + \cdots\right].$$

8.2 The expression for the cross-section from (8.46) and (8.47) with $N_c = 3$ and m the invariant mass of the muon pair is

$$\begin{aligned}
\frac{d\sigma}{dm^2} &= \frac{4\pi\alpha^2}{9m^4} \int\int_0^1 x_i \bar{x}_i dx_i d\bar{x}_i \delta(x_i \bar{x}_i - \tau) \sum e_i^2 Q(x_i)\bar{Q}(\bar{x}_i) \\
&= \frac{4\pi\alpha^2}{9m^4} \sum e_i^2 \int_\tau^1 [x_i Q(x_i)][\bar{x}_i \bar{Q}(\bar{x}_i)]x_i\, dx_i \qquad \text{(with } \bar{x}_i = \tau/x_i) \\
&= \frac{4\pi\alpha^2}{9m^4} \sum e_i^2 \int_\tau^1 AB(1 - x)^3(1 - \tau/x)x\, dx \\
&= \frac{4\pi\alpha^2}{9m^4} AB \sum e_i^2 \frac{(1 - \tau)^5}{20}.
\end{aligned}$$

For $\tau \ll 1$ this becomes
$$2m^4 \frac{d\sigma}{dm^2} = m^3 \frac{d\sigma}{dm} = \frac{8\pi\alpha^2}{180} AB \sum e_i^2(1 - 5\tau).$$

8.3 From (8.21) we can write for the differential cross-sections
$$\frac{d^2\sigma^\nu}{dy} = (2G^2ME/\pi)[xQ(x) + (1 - y^2)x\bar{Q}(x)]\, dx,$$

$$\frac{d^2\sigma^{\bar\nu}}{dy} = (2G^2ME/\pi)[x\bar{Q}(x) + (1 - y^2)xQ(x)]\, dx.$$

Set $P = \int xQ(x)\, dx$, $\bar{P} = \int x\bar{Q}(x)\, dx$, then integrating over y we find
$$\sigma^\nu = (2G^2ME/\pi)(P + \bar{P}/3); \qquad \sigma^{\bar\nu} = (2G^2ME/\pi)(\bar{P} + P/3)$$

and
$$R = \sigma^{\bar\nu}/\sigma^\nu = (\bar{P} + P/3)/(P + \bar{P}/3)$$

or
$$\bar{P}/P = (3R - 1)/(3 - R).$$

With $R = 0.5$ we obtain $\overline{P}/P = 0.20$. The mean values of y are obtained through the relation

$$\langle y \rangle = \int y \, d\sigma/dy \Big/ \int d\sigma/dy.$$

For neutrinos,

$$\langle y \rangle = \frac{6 + \overline{P}/P}{12 + 4\overline{P}/P} = \underline{0.484}.$$

For antineutrinos

$$\langle y \rangle = \frac{1 + 6\overline{P}/P}{4 + 12\overline{P}/P} = \underline{0.344}.$$

8.4 If the "struck" quark receives momentum transfer q, this has to be transferred to the two spectator quarks. Single gluon exchange introduces a propagator factor $1/q^2$ in the amplitude in each case, or an overall factor $1/q^4$. This q dependence of the amplitude (or $1/q^8$ in the intensity) would be expected to hold for very large q^2, so that α_s is small and single gluon exchange dominates.

8.5 Integration of the momentum distributions gives $a_1 = 2a_2 = \frac{4}{3}$. The double differential cross-section will have the Drell-Yan form

$$d^2\sigma = \sigma_{u\bar{d}}(s) \cdot u(x_1) \cdot \bar{d}(x_2)\delta(1 - x_1 x_2 s/M_W^2) \cdot dx_1 \cdot dx_2,$$

where x_1, x_2 are the fractional momenta of the u- and $\bar{d}$-quarks, and where

$$x_1 x_2 = p = M_W^2/s.$$

Integration over the width of the Breit-Wigner resonance of the W-particle gives

$$\sigma(p\bar{p} \to W) = \frac{8\pi\Gamma_W\sigma_0}{9M_W} \int_p^1 \frac{(1 - x_1)^3}{x_1} \cdot (1 - p/x_1)^3 \cdot dx_1,$$

where σ_0 is the peak $u\bar{d}$ cross-section, Γ_W is the total W width. The integral has the value

$$(1 + p^3 + 9p + 9p^2) \ln(1/p) - (11/3 + 9p - 9p^2 - 11p^3/6),$$

and

$$\sigma_0 = 5.2 \times 12 = 62 \text{ nb (for all decay modes)}.$$

Substituting the numerical values given, we obtain

$\sqrt{s}$	0.3	1.0	10	TeV
σ	0.99	8.78	33.3	nb

CHAPTER 9

9.1　With $A = G^2 mE/2\pi$, $x = \sin^2 \theta_w$

$$\sigma_{\nu_\mu} = A[(2x - 1)^2 + (2x)^2/3]$$
$$\sigma_{\bar{\nu}_\mu} = A[(2x)^2 + (2x - 1)^2/3]$$
$$\sigma_{\nu_e} = A[(2x + 1)^2 + (2x)^2/3]$$
$$\sigma_{\bar{\nu}_e} = A[(2x)^2 + (2x + 1)^2/3]$$

σ_{ν_μ} has a minimum at $x = \frac{3}{8}$, σ_{ν_μ} at $x = \frac{1}{8}$. The interaction is pure axial-vector when $x = \frac{1}{4}$.

9.2　The cross-sections have the form

$$\sigma^\nu = K(c_A^2 + c_V^2 + c_A c_V) \quad \text{with } K = G^2 mE/3\pi,$$
$$\sigma^{\bar{\nu}} = K(c_A^2 + c_V^2 - c_A c_V).$$

There are four solutions for c_A, c_V. The above equations define ellipses in the c_A, c_V plane with principal axes at $45°$ and $135°$ to the x-axis.

9.3　1 rad corresponds to an energy release in ionization of 100 erg g^{-1}, or 6.25×10^7 MeV g^{-1}. Let τ be the proton lifetime in years, E the energy released in ionization by decay. In one year, the radiation dosage will be

$$\frac{N}{2} \frac{E}{\tau} (6.25 \times 10^7) \text{rads} < 500 \text{ rads},$$

where $N = 6 \times 10^{23}$ is Avogadro's number, $E \simeq 1000$ MeV. This inequality yields

$$\underline{\tau > 10^{16} \text{ yr.}}$$

9.5　The atomic mass of an atom with nuclear mass number A and atomic number Z can be written

$$M = AM_n \left[1 - \frac{Z(M_n - M_h)}{AM_n} - \frac{W}{AM_n} \right],$$

where W is the nuclear binding energy and M_n and M_h are the masses of the neutron and the hydrogen atom. Inserting the appropriate values of the constants, the fractional difference in baryon number B for the same mass of two different elements is

$$\Delta B/B = \left[8.3\Delta\left(\frac{Z}{A}\right) + 10.6\Delta\left(\frac{W}{A}\right) \right] \times 10^{-4},$$

where $\Delta(Z/A)$ and $\Delta(W/A)$ are the differences in Z/A and W/A between the two elements, with W in MeV. Nuclear tables show that for Al and Pt, the first term is about 10% of the second, and that $\underline{\Delta B/B = 4 \times 10^{-4}}$.

9.6 The partial widths for decay are $\Gamma(\text{partial}) = \text{constant} \ [(2c_V)^2 + (2c_A)^2]$. From
(9.33) with $\sin^2 \theta_w = 0.22$, the widths in terms of $\Gamma(Z \to \nu\bar{\nu})$ are then

$$\Gamma(\nu\bar{\nu}) = 1,$$

$$\Gamma(e\bar{e}) = \Gamma(\mu\bar{\mu}) = \Gamma(\tau\bar{\tau}) = 0.507,$$

$$\Gamma(u\bar{u}) = \Gamma(c\bar{c}) = \Gamma(t\bar{t}) = 0.585,$$

$$\Gamma(d\bar{d}) = \Gamma(s\bar{s}) = \Gamma(b\bar{b}) = 0.750.$$

Allowing for the color factor 3 for the quarks we get

$\Gamma_{\text{total}} = \Gamma(\nu\bar{\nu})[(3 \times 1) + (3 \times 0.507) + (9 \times 0.585) + (9 \times 0.750)] = 16.54 \ \Gamma(\nu\bar{\nu})$,
and $\Gamma(\nu\bar{\nu}) = GM_Z^3/12\pi\sqrt{2} = 0.175$ GeV. Hence, $\Gamma_{\text{total}} = 2.89$ GeV. This is an
overestimate since the masses of t- and b-quarks have been neglected in
calculating the phase space.

9.7 From the Breit-Wigner formula (4.55) we obtain

$$\sigma(e^+e^- \to Z^0 \to \text{anything}) = \frac{4\pi\lambdabar^2(2J + 1)\Gamma\Gamma_e/4}{(2s + 1)^2[(M_Z - E)^2 + \Gamma^2/4]},$$

where $s = \frac{1}{2}$ is the electron spin and $J = 1$ is that of the Z^0, while $\lambdabar = 2/M_Z$ is the
CMS wavelength, on the resonance peak. Γ_e is the partial width for $Z \to e^+e^-$.
The peak cross-section is then

$$\sigma_{\text{max}}(e^+e^- \to Z^0) = \frac{12\pi\Gamma_e}{M_Z^2\Gamma}.$$

From (6.31) the pointlike cross-section (for $e^+e^- \to \mu^+\mu^-$) in the absence of any
resonance would be

$$\sigma_{\text{point}} = \frac{4\pi\alpha^2}{3M_Z^2},$$

so that

$$\sigma_{\text{max}}/\sigma_{\text{point}} = \frac{9}{\alpha^2} \frac{\Gamma_e}{\Gamma} = 5180,$$

where $\Gamma_e/\Gamma = 0.507/16.54 = 0.0307$.

References

Abrams, G. S., *et al., Phys. Rev. Lett.* **33**, 1452 (1974).

Adair, R. K., *Phys. Rev.* **100**, 1540 (1955).

Adelberger, E. G., *et al., Phys. Rev. Lett.* **34**, 402 (1975).

Alper, B., *et al., Phys. Lett.* **47B**, 75 (1973).

Altarelli, G., and G. Parisi, *Nucl. Phys.* **B126**, 298 (1977).

Altarev, I. S., *et al., Phys. Lett.* **102B**, 13 (1981).

Anderson, C. D., *Phys. Rev.* **43**, 491 (1933).

Anderson, H. L., Fermi, E., Martin, R., and Nagle, D. E., *Phys. Rev.* **91**, 155 (1953).

Anderson, K. J., *et al.,* Proc. 19th Int. Conf. on HEP, Tokyo, 1978.

Antinucci, M., *et al., Lett. Nuov. Cim.* **6**, 121 (1973).

Argento, A., *et al., Phys. Lett.* **120B**, 245 (1982).

Arnison, G., *et al., Phys. Lett.* **122B**, 103 (1983); **126B**, 398 (1983).

Arnison, G., *et al., Phys. Lett.* **136B**, 294 (1984).

Arnison, G., *et al., Phys. Lett.* **147B**, 493 (1984a).

Aubert, J. J., *et al., Phys. Rev. Lett.* **33**, 1404 (1974).

Augustin, J. E., *et al., Phys. Rev. Lett.* **33**, 1406 (1974).

Bacino, W., *et al., Phys. Rev. Lett.* **41**, 13 (1978).

Bagnaia, P., *et al., Phys. Lett.* **129B**, 130 (1983).

Bagnaia, P., *et al., Phys. Lett.* **138B**, 430 (1984).

Banner, M., *et al., Phys. Lett.* **122B**, 476 (1983).

Barnes, V., *et al., Phys. Rev. Lett.* **12**, 204 (1964).

Bartel, W., B. Dudelzak, H. Krehbiel, J. McElroy, U. Meyer-Berkheut, W. Schmidt, V. Walther, and G. Weber, *Phys. Lett.* **28B**, 148 (1968).

Bathow, G., *et al., Nucl. Phys.* **B20**, 592 (1970).

Berko, S., and Pendleton, H. N., *Ann. Rev. Prog. Nucl. Part. Science* **30**, 543 (1980).

Bergkvist, K. E., *Nucl. Phys.* **B39**, 319 (1972).

Bethe, H. A., and J. Ashkin, "Passage of radiations through matter," *Exptl. Nucl. Phys.* **1**, 166 (1953).

Bionta, R. M., *et al.*, *Phys. Rev. Lett.* **54**, 22 (1985).

Bjorken, J. D., *Phys. Rev.* **163**, 1767 (1967).

Bodwin, G. T., and D. T. Yennie, *Phys. Rep.* **43**, 268 (1978).

Boehm, F., Proc. 5th Workshop on Grand Unification, Brown University, Providence, R.I. (World Scientific, 1984).

Cabibbo, N., *Phys. Rev. Lett.* **10**, 531 (1963).

Callan, C. G., and D. G. Gross, *Phys. Rev. Lett.* **21**, 311 (1968); **22**, 156 (1969).

Cartwright, W. F., C. Richman, M. Whitehead, and H. Wilcox, *Phys. Rev.* **91**, 677 (1953).

Charpak, G., Bouclier, T. Bressani, J. Favier, and C. Zupancic, *Nucl. Instr. Methods* **62**, 262 (1968).

Charpak, G., *et al.*, *Nucl. Inst. Meth.* **80**, 13 (1970).

Chew, G., S. Frautschi, and S. Mandelstam, *Phys. Rev.* **126**, 1202 (1962).

Christenson, J. H., J. Cronin, V. Fitch, and R. Turlay, *Phys. Rev. Lett.* **13**, 138 (1964).

Clark, D. L., A. Roberts, and R. Wilson, *Phys. Rev.* **83**, 649 (1951); **85**, 523 (1952).

Combridge, B. L., *et al.*, *Phys. Lett.* **70B**, 234 (1977).

Condon, E. U., and G. H. Shortley, *The Theory of Atomic Spectra*, Cambridge University Press, 1951.

Dalitz, R. H., *Phil. Mag.* **44**, 1068 (1953).

Davis, R., *et al.*, *Phys. Rev. Lett.* **20**, 1205 (1968).

Deutsch, M., *Prog. Nucl. Phys.* **3**, 131 (1953).

Dirac, P. A. M., *Proc. Roy. Soc.* **A117**, 610 (1928); also *The Principles of Quantum Mechanics*, Oxford University Press, 1947.

Dirac, P. A. M., *Proc. Roy. Soc.* **133**, 60 (1931).

Drell, S. D., and T. M. Yan, *Phys. Rev. Lett.* **24**, 181 (1970); *Ann. Phys.* (*N.Y.*) **66**, 595 (1971).

Dress, W. B., J. K. Baird, P. D. Miller, and N. F. Ramsey, *Phys. Rev.* **170**, 1200 (1968).

Dress, W. B., *et al.*, *Phys. Rep.* **43**, 410 (1978).

Durbin, R., H. Loar, and J. Steinberger, *Phys. Rev.* **84**, 581 (1951).

Fabri, E., *Nuovo Cim.* **11**, 479 (1954).

Fermi, E., *Z. Physik* **88**, 161 (1934).

Feynman, R. P., *Phys. Rev. Lett.* **23**, 1415 (1969).

Feynman, R. P., and M. Gell-Mann, *Phys. Rev.* **109**, 193 (1958).

Feynman, R. P., Proc. 5th Hawaii Topical Conf. in Particle Phys. Hawaii University Press, 1973.

Fortson, E. N., and L. Wilets, *Adv. in Atomic and Mol. Phys.* July (1981).

Fortson, E. N., and L. L. Lewis, *Phys. Rep.* **113**, 289 (1984).

Fox, D. J., *et al.*, *Phys. Rev. Lett.* **33**, 1504 (1974).

Friedman, J. T., and H. W. Kendall, *Ann. Rev. Nucl. Science* **22**, 203 (1972).

Gaillard, M. K., *et al.*, *Rev. Mod. Phys.* **47**, 227 (1975).

Gallinaro, G., *et al.*, *Phys. Rev. Lett.* **38**, 1255 (1977).

Gell-Mann, M., *Phys. Rev.* **92**, 833 (1953).

Gell-Mann, M., *Phys. Lett.* **8**, 214 (1964).

Gell-Mann, M., and F. E. Low, *Phys. Rev.* **95**, 1300 (1954).

Gell-Mann, M., and A. Pais, *Phys. Rev.* **97**, 1387 (1955).

Georgi, H., and S. L. Glashow, *Phys. Rev. Lett.*, **32**, 438 (1974).

Geweniger, C., *et al.*, *Phys. Lett.* **B48**, 487 (1974).

Glaser, D., "The Bubble Chamber," Encycl. Phys. *45* (Springer, Berlin, 1955).

Glashow, S. L., *Nucl. Phys.* **22**, 579 (1961).

Glashow, S. L., J. Iliopoulos, and L. Maiani, *Phys. Rev. D* **2**, 1285 (1970).

Goldhaber, A. S., and M. M. Nieto, *Rev. Mod. Phys.* **43**, 277 (1971).

Goldhaber, M., L. Grodzins, and A. Sunyar, *Phys. Rev.* **109**, 1015 (1958).

de Groot, J. G. H., *et al.*, *Z. Physik* **C1**, 143 (1979).

Gross, D. J., and F. Wilczek, *Phys. Rev. D* **8**, 3633 (1973); **9**, 980 (1974).

Gross, D. J., and C. H. Llewellyn-Smith, *Nucl. Phys.* **B14**, 337 (1969).

Hasert, F. J., *et al.*, *Phys. Lett.* **46B**, 138 (1973), *Nucl. Phys.* **B73**, 1 (1974).

Heisenberg, W., *Z. Physik* **77**, 1 (1932).

Herb, S. W., *et al.*, *Phys. Rev. Lett.* **39**, 252 (1977).

Higgs, P. W., *Phys. Lett.* **12**, 132 (1964); *Phys. Rev.* **145**, 1156 (1966).

Hofstadter, R., *Rev. Mod. Phys.* **28**, 214 (1956).

Holder, M., *et al.*, *Nucl. Inst. Meth.* **151**, 69 (1978).

't Hooft, G., *Phys. Lett.* **37B**, 195 (1971).

Innes, W. R., *et al.*, *Phys. Rev. Lett.* **39**, 1240 (1977).

Jackson, J. D., *Rev. Mod. Phys.* **37**, 484 (1965).

Kobayashi, M., and K. Maskawa, *Prog. Theor. Phys.* **49**, 282 (1972).

Langacker, P., *Phys. Rep.* **72**, 185 (1981).

Langer, L., and R. Moffat, *Phys. Rev.* **88**, 689 (1952).

LaRue, G. S., *et al.*, *Phys. Rev. Lett.* **38**, 1011 (1977); **46**, 967 (1981).

Lattes, C. M. G., H. Muirhead, C. F. Powell, and G. P. Occhialini, *Nature* **159**, 694 (1947).

Lee, T. D., and C. N. Yang, *Phys. Rev.* **98**, 1501 (1955).

Lee, T. D., and C. N. Yang, *Phys. Rev.* **104**, 254 (1956).

Lehraus, I., *et al.*, *Nucl. Inst. Meth.* **153**, 347 (1978).

Llewellyn-Smith, C. H., and J. F. Wheater, *Phys. Lett.* **105B**, 486 (1981).

Lyubimov, V. A., E. G. Novikov, V. Z. Nozik, E. F. Tretyakov, and V. S. Kosik, *Phys. Lett.* **94B**, 266 (1980).

Maki, Z., *et al.*, *Prog. Theor. Phys.* **28**, 870 (1962).

Marciano, W. J., and A. Sirlin, *Nucl. Phys.* **189**, 442 (1981).

Marinelli, M., and G. Morpurgo, *Phys. Rep.* **85**, 161 (1982).

Marshak, R., and H. Bethe, *Phys. Rev.* **72**, 506 (1947).

Marshak, R., and E. Sudarshan, *Phys. Rev.* **109**, 1860 (1958).

Ne'eman, Y., *Nucl. Phys.* **26**, 222 (1961).

Neubeck, N., *et al.*, *Phys. Rev. C* **10**, 320 (1974).

Nishijima, K., *Prog. Theor. Phys.* **13**, 285 (1955).

Orear, J., G. Harris, and S. Taylor, *Phys. Rev.* **102**, 1676 (1956).

Pais, A., *Phys. Rev.* **86**, 663 (1952).

Pais, A., and O. Piccioni, *Phys. Rev.* **100**, 1487 (1955).

Panofsky, W. (data of E. Bloom *et al.*), Int. Conf. High Energy Physics, Vienna, 1968.

Pauli, W., *Handbuch der Physik* **24**, 1, 233 (1933).

Pauli, W., *Phys. Rev.* **58**, 716 (1940).

Pendlebury, J. M., *et al.*, *Phys. Lett.* **136B**, 327 (1984).

Perkins, D. H., *Ann. Rev. Nucl. Part. Science* **34**, 1 (1984).

Perl, M. L., *et al.*, *Phys. Rev. Lett.* **35**, 1489 (1975); *Phys. Lett.* **63B**, 366 (1976).

Plano, R., A. Prodell, N. Samios, M. Schwartz, and J. Steinberger, *Phys. Rev. Lett.* **3**, 525 (1959).

Politzer, H. D., *Phys. Rep.* **14C**, 129 (1974).

Pontecorvo, B., *Sov. Phys. JETP* **26**, 984 (1968).

Prescott, C. Y., *et al.*, *Phys. Lett.* **77B**, 347 (1978); **84B**, 524 (1979).

Pryce, M. H. L., and J. C. Ward, *Nature* **160**, 435 (1947).

Ramsey, N. F., *Ann. Rev. Nucl. Part. Science* **32**, 211 (1982).

Reines, F., and C. Cowan, *Phys. Rev.* **113**, 273 (1959).

Rosner, J., Proc. Advanced Study Inst. on Techniques and Concepts in HEP, St. Croix, USVI (ed. T. Ferbel), 1980.

Rossi, B., *High Energy Particles*, Prentice-Hall, New York, 1952.

Sakharov, A., *JETP Lett.* **5**, 24 (1967).

Sakurai, J. J., *Invariance Principles and Elementary Particles*, Princeton University Press, Princeton, New Jersey, 1964.

Salam, A., *Elementary Particle Theory* (ed. N. Svartholm), Almquist and Wiksells, Stockholm, 1968.

Shafer, J., J. Murray, and D. Huwe, *Phys. Rev. Lett.* **10**, 176 (1963).

Snyder, L., *et al.*, *Phys. Rev.* **63**, 440 (1948).

Tadic, C., *Rep. Prog. Phys.* **43**, 67 (1980).

Tretyakov, E. F., *et al.*, Proc. Neutrino Conf. Aachen, 1976.

Tripp, R. D., "Spin and parity determination of elementary particles," *Ann. Rev. Nucl. Science* **15**, 325 (1965).

Van Royen, R., and V. F. Weisskopf, *Nuovo Cim.* **50**, 617 (1967); **51**, 583 (1967).

Von Witsch, W., A. Richter, and P. von Brentano, *Phys. Rev.* **169**, 923 (1968).

Weber, G., Proc. 1967 Int. Sym. on Electron and Photon Interactions at High Energies, Stanford, California, 1967, p. 59.

Weinberg, S., *Phys. Rev. Lett.* **19**, 1264 (1967).

Weizsacker, K. F., *Z. Phys.* **88**, 612 (1934); Williams, E. J., *Kgl. Danske Vid. Selsk. Mat. Fys. Medd.* **13**, 14 (1935).

Weyl, H., *Z. Physik* **56**, 330 (1929).

Wolfenstein, L., *Phys. Lett.* **13**, 562 (1964).

Wu, C. S., and I. Shaknov, *Phys. Rev.* **77**, 136 (1950).

Wu, C. S., E. Ambler, R. Hayward, D. Hoppes, and R. Hudson, *Phys. Rev.* **105**, 1413 (1957).

Wu, S. L., *Phys. Rep.* **107**, 59 (1984).

Yang, C. N., and R. L. Mills, *Phys. Rev.* **96**, 191 (1954).

Yoh, J. K., *et al.*, *Phys. Rev. Lett.* **41**, 684 (1978).

Yukawa, H., *Proc. Phys. Math. Soc. Japan* **17**, 48 (1935).

Zweig, G., CERN Report 8419/Th 412, 1964.

Index